Thomas Böllinghaus

Horst Herold

Hot Cracking Phenomena in Welds

Thomas Böllinghaus

Horst Herold (Eds.)

Hot Cracking Phenomena in Welds

With 322 figures and 46 tables

Springer

Dr.-Ing. Thomas Böllinghaus
Vizepräsident und Prof. der BAM
Bundesanstalt für Materialforschung und -prüfung
Unter den Eichen 87
12205 Berlin
Germany
Thomas.Boellinghaus@bam.de

Prof. Dr.-Ing. habil. Dr. E.h. Horst Herold
Institut für Füge- und Strahltechnik
Fakultät für Maschinenbau
Otto-von-Guericke-Universität
Universitätsplatz 2
39106 Magdeburg
Germany
Horst.Herold@mb.uni-magdeburg.de

Library of Congress Control Number: 2005921916

ISBN 3-540-22332-0 **Springer Berlin Heidelberg New York**

Springer is a part of Springer Science+Business Media
springeronline.com

Printed in Germany

Typesetting: Data conversion by the author.
Final processing by PTP-Berlin Protago-TEX-Production GmbH, Germany
Cover-Design: deblik, Berlin
Printed on acid-free paper 62/3141/Yu – 5 4 3 2 1 0

Preface

During modern fabrication welding of welded components the avoidance of hot cracking still represents a major topic, sometimes also under new aspects. Austenitic stainless steels, for instance, widely used in industry and known to be crack-free joinable by arc welding might turn their primary solidification mode from ferrite to austenite and thus, might become increasingly susceptible to hot cracking during increasingly applied modern laser and hybrid processes.

Additionally, the phenomena of hot cracking in welds have not completely been understood up to the present. Hydrogen added to the shielding gas in arc welding processes, for instance, might enhance solidification cracking by an increasing the heat input, but has been also tentatively considered to contribute to ductility dip cracking by embrittlement. Quite numerous technological hot cracking test procedures have been developed all over the world to rank the hot cracking resistance of base and filler materials ahead of fabrication welding. Standardization of such tests appears as very challenging, because the different results are difficult to compare and to transfer to real component welds.

In order to provide a forum to define the present state of knowledge, to exchange recent research results, to discuss different viewpoints and to contribute to the ongoing standardization work on hot cracking phenomena in welds an international workshop has been organized in March 2004.

The present book contains the 20 individual contributions form experts all over the world covering four major subjects.

By seven contributions the first chapter provides a complete overview of the different hot cracking phenomena. Different mechanisms of solidification cracking proposed in the past decades are summarized and new insight is particularly given into the mechanism of ductility dip cracking.

In the second chapter, metallurgy and materials, the effect of different alloying elements on the hot cracking resistance of various materials are shown. The initiation of stress corrosion cracking at hot cracks has additionally been included in one contribution as a special metallurgical effect.

Since numerical analyses and other simulation techniques represent very helpful tools to explain cracking phenomena, three individual contributions show in the third chapter how modelling of hot cracking can be performed and how such results might support the explanation of mechanisms.

In the final chapter, the various hot cracking tests are presented in seven individual contributions with a special emphasis on the ongoing process of standardization. As a final contribution the necessary linking between testing and practise is outlined on the basis of actual extraordinary cases.

In total, the extensive contributions from eight different countries do not only provide the latest insight and define the international state of knowledge on hot cracking phenomena in welds. As a particular item, the authors included numerous future research perspectives, fairly enough to excite also the next generation of scientists. By touching all three types of hot cracking, namely solidification cracking, liquation cracking and ductility dip cracking and also by explanations of their differences various articles represent also a very helpful tool for metallurgical and mechanical engineering students of the higher semesters. Furthermore, the text contains helpful individual advices, particularly for international welding engineers confronted with hot cracking in practise.

The editors convey their sincere gratitude to the authors and to all the participants of the workshop for their individual contributions and their eager discussions and, in particular, for pushing the scientific knowledge about hot cracking phenomena in welds a huge step forward.

We especially thank Karen Stelling for her tremendous work in formatting the individual articles and to prepare this book for printing, Margit Bauer for her very helpful translation assistance and, in particular, Thomas Kannengiesser as well as BAM Division V.5 Joining Technology for the organization and their support during the workshop.

Berlin and Magdeburg, January 2005

Thomas Böllinghaus
Horst Herold

Contents

I Phenomena and Mechanisms

II Metallurgy and Materials

I Phenomena and Mechanisms

On the Origin of Weld Solidification Cracking

C.E. Cross

Department of Metallurgical & Materials Engineering,
Montana Tech of the University of Montana, USA

Abstract

A review is made of theories regarding weld solidification cracking, identifying the numerous factors controlling crack initiation. A new approach to understanding this phenomenon is discussed, considering the conditions necessary to achieve rupture of liquid films.

Introduction

Weld solidification cracking consists of the rupture of liquid films present at grain boundaries in the mushy zone trailing the weld pool. The initiation of such cracks involves a complex interaction between metallurgical and mechanical factors, driven by the temperature gradients generated during welding, as suggested in Fig. 1. Thermal-metallurgical interactions control the solidification microstructure. Thermal-mechanical interactions control the local and global stresses and strains.

Numerous theories have been proposed to describe the conditions required for crack initiation, normally oriented toward either metallurgical or mechanical aspects of the problem. However, seldom is the actual failure mechanism confronted (i.e. the rupture of a liquid film). In this paper, a review of solidification theories is made, categorizing each theory in terms of the factors controlling crack initiation. A new approach is then introduced which addresses a liquid rupture mechanism, which enables many of the controlling factors to be combined into a unified theory.

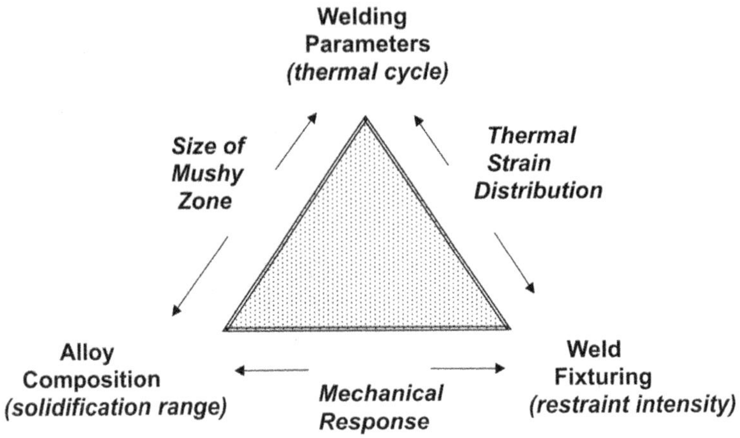

Fig. 1. Diagram indicating complex interaction between process parameters affecting weld solidification cracking

Identifying Controlling Factors

The path to understanding solidification cracking requires an appreciation of how various factors interact to affect cracking susceptibility. There is usually no simple interaction between factors, but instead many complex interrelationships, making direct correlations difficult between observation and theory. These factors are identified below, grouped for comparison as either metallurgical or mechanical in nature. Metallurgical factors are particular to phase relationships, whereas mechanical factors involve stress and strain behavior.

Metallurgical

Solidification Range. It has long been appreciated that solidification temperature range, often referred to as a brittle temperature range (BTR), plays an important role in solidification cracking susceptibility. Its value can be defined as the difference between liquidus and solidus temperatures. For non-equilibrium solidification, the solidus temperature is usually defined by the lowest melting eutectic in the system. A classical example of this is the deleterious effect of sulfur and phosphorous impurities on the cracking susceptibility of ferrous and nickel based alloys. Sulfur can extend the solidus temperature in steel from around 1400°C to 988°C (Fe-S eutectic temperature).

In aluminum alloys, combinations of copper and magnesium are known to result in unweldable alloys [1], attributed to the formation of a low melting Al-Cu-Mg eutectic. However, Al-Mg alloys with an equally large solidification range, exhibit exceptionally good weldability. Thus, there is clearly more to this problem than just solidification range.

The solidification range, divided by the temperature gradient (G), gives the extent of the two-phase mushy zone (i.e. dendrite length). In several models [2, 3] it is argued that an extended mushy zone will experience more shrinkage strain and will thus be more susceptible to cracking. It follows that, when welding a susceptible alloy with high heat input (i.e. low G), there may be a higher likelihood for cracking to occur. However, other effects of high heat input (e.g. thermal-mechanical response) may negate this effect.

In certain weldability tests where an augmented strain is applied (e.g. varestraint test), the maximum crack length generated at large values of imposed strain corresponds to a characteristic temperature range equal to or less than the solidification range [4]. Referred to by Lippold et al. [5] as a solidification cracking temperature range (SCTR), this parameter has been effectively used to rank the relative weldability of various different alloys (also see Recent Developments in Weldability Testing in this text).

Back-filling. Back-filling refers to the drawing of liquid back through the dendritic network to feed solidification shrinkage. It can also be driven by capillary flow to fill or "heal" cracks, sometimes observed in metallography as pools of high solute material (e.g. eutectic) deposited in crack-shaped defects. This interdendritic flow of liquid has been modeled extensively by numerous researchers [6], with the rate of flow controlled by dendrite tortuosity, liquid fraction, fluidity, and surface tension. The concept of back-filling comes from foundry practice, where risers are strategically placed to feed hot spots to avoid hot tearing. In the case of welding, the molten weld pool serves as the reservoir of liquid, taking the place of a riser in a casting.

Dendrite Coherency. Coherency refers to the degree of solid-solid bonding between secondary dendrite arms occurring at the latter stage of solidification within the mushy zone (Fig. 2). Theory of Pumphrey and Jennings [2] proposed that cracking is associated with the thermal contraction experienced in the coherent region (also defined as a brittle temperature range), where an alloy with a large coherent region is expected to have a higher susceptibility to cracking. The problem with this reasoning, as will be pointed out later, is that alloys with extensive coherent bonding (e.g. Al-Mg alloys) may actually be more resistant to strain and resultant cracking.

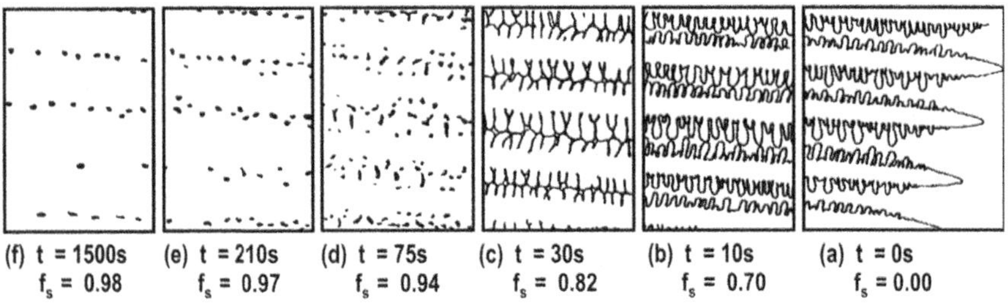

Fig. 2. Schematic drawing showing progressive stages of dendritic solidification demonstrating regions of liquid back-filling **a–b**, thin liquid film **c**, and dendrite coherency **d–f** [7]

The maximum coherent region under equilibrium conditions is seen to occur at the point of maximum solid solution (point B in Fig. 3). This phase equilibria has been used [2] to explain the peak cracking susceptibility often observed for aluminum alloys, noting that point B will be shifted to lower solute values for non-equilibrium solidification.

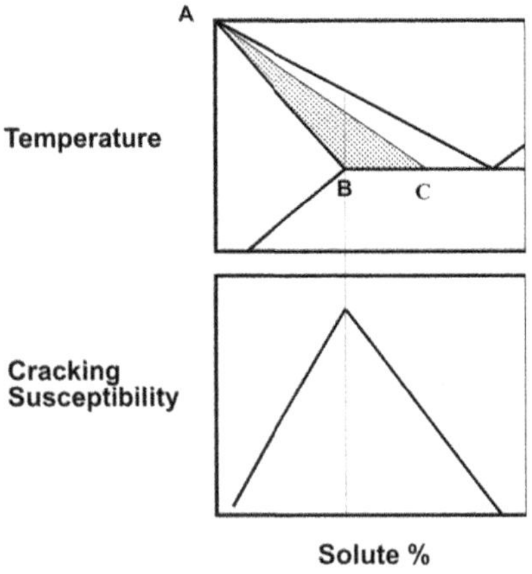

Fig. 3. Schematic diagram suggesting connection between phase equilibrium and peak susceptibility in solidification cracking per theory of Pumphrey et al. [2]. Shaded region represents coherent zone

Eutectic Fraction. The weight fraction of interdendritic eutectic generated (f_E) increases with solute content (C_o), as can be approximated using the Scheil Equation for non-equilibrium solidification [8]:

$$f_E = [C_o / C_E]^{1/(1-k)}, \tag{1}$$

where C_E is the eutectic composition and k is the partition ratio. It has been suggested that alloys with large solute content, and hence large quantities of eutectic, will be less susceptible to cracking [1]. This is because: 1) there is a less extensive coherent dendrite structure and 2) shrinkage can be more readily fed by means of back-filling due to a more open dendrite array.

The above reasoning can in some cases be used to explain the peak susceptibility behavior typically observed in aluminum alloys, where high alloy content in base metal or filler metal often results in improved weldability [9]. Filler metal alloys are typically high in alloy content for this reason. Al-Si filler alloys in particular, which generate large quantities of eutectic, are known for their exceptional weldability. At the opposite extreme, however, are low eutectic Al-Mg alloys that also experience good weldability. Al-Mg alloys are expected to have a high degree of coherency (i.e. large solidification range plus small amount of eutectic) and, thus, can resist thermal strain.

Surface Tension. Borland defined the effect of surface tension on cracking in his "Generalized Theory" of cracking [10], combining some aspects of the shrinkage-brittleness theory of Pumphrey and Jennings [2] with the strain theory of Pellini [3] (see *Strain* below). Central to Borland's theory is the continuity of liquid at the base of the dendrites, during the last stage of solidification.

If the last liquid to solidify wets the dendrites (i.e. low $\gamma_{L/S}$), there will be a higher likelihood that a continuous network of liquid can provide back-filling. At the opposite extreme where no wetting occurs (i.e. high $\gamma_{L/S}$), bridging between dendrite arms is promoted, resisting strain, and thus avoiding cracking. It is at intermediate values of wettability, between these two extremes, where cracking is encountered. Experimental evidence exists to support this theory, comparing Al-Sn, Al-Cd, and Al-In alloys [10]. Two alloys with poor wetting characteristics (Al-Cd and Al-In: 90° dihedral angle) were found to be more weldable than the alloy experiencing moderate wetting (Al-Sn: 65° dihedral angle).

Surface tension must also play a role in the rupture of a liquid film (i.e. creation of new vapor/liquid interface). The force (F) required to separate a liquid film wetting two parallel plates has been predicted by Seveiko [11] to be:

$$F = c_1 \frac{\gamma A}{t}, \tag{2}$$

where A represents surface area, γ is vapor/liquid surface tension, and t is film thickness. In essence, thin films with high surface tension should prove more resistant to cracking. However, cavitation should provide a more energetically favorable mechanism for liquid rupture (see *Liquid Rupture* below), and so Eq. 2 represents an upper bound on liquid strength.

Another possible influence of surface tension involves temperature gradient driven fluid flow, commonly known as the Marongoni effect. Considered from a theoretical standpoint, it has been proposed that variations in sulfur content may influence interdendritic fluid flow and, hence, backfilling [12]. For example, it was shown that at high sulfur concentrations (i.e. $d\gamma/dT > 0$) the temperature gradient will drive flow out of the dendritic network, inhibiting feeding of shrinkage.

Grain Boundaries. Solidification cracking normally occurs along weld metal grain boundaries, although this is not necessarily always the case. The reason for this tendency is likely tied to the preferential segregation of solute or impurity elements (e.g. sulfur in steel) to grain boundaries, even though originally partitioned between dendrites. A grain boundary provides a well-defined, high energy planar interface upon which a liquid film can wet. Therefore, it follows that grain shape, structure and size should have a profound effect on cracking susceptibility.

i. Grain Shape. General wisdom suggests that conditions resulting in columnar grains growing normal to the welding direction are most deleterious [13]. Such is the case when welding at rapid travel speeds with the resultant tear-drop shaped weld pool. It may be that this tear-drop shape causes solute or impurities to concentrate along the weld centerline, where one single grain boundary film must accommodate all of the imposed strain.

ii. Grain Structure. Brooks has shown that different grain boundary structures in stainless steel weld metal may account for the difference in weldability between austenite versus ferrite primary solidification [14]. When solidifying as primary austenite, the weld metal grain boundaries are observed to be better defined (i.e. more straight and continuous) allowing for easy crack propagation, as depicted in Fig. 4. The structure of the grain boundary is affected by the morphology of the dendrites, which will vary between alloys.

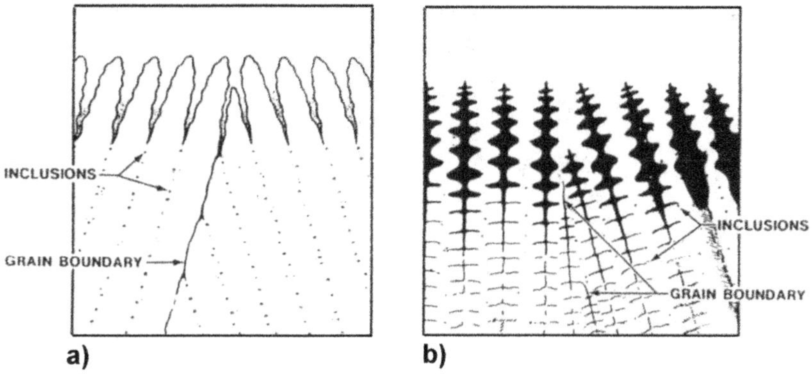

Fig. 4. Schematic comparing solidification grain boundary structure between primary austenite **(a)** and primary ferrite **(b)** stainless steel, from Brooks et al. [14]

iii. Grain Size. Numerous studies have observed dramatic improvements in weldability through means of grain refinement, principally in aluminum alloys. This can be readily explained by considering the distribution of strain (e.g. from solidification shrinkage and thermal contraction) between grain boundaries spanning the mushy zone. Smaller grains mean more grain boundaries are present, which means a smaller amount of strain is partitioned to each individual grain boundary. A high level of strain, per grain boundary, is believed to result in cracking (see *Strain* below). An example of how grain refinement can effectively reduce solidification cracking is shown in Fig. 5, where circular patch test results are given for aluminum alloy 7108 treated with varying amounts of scandium grain refiner [15].

Porosity. Porosity represents a form of liquid rupture in that it involves the formation of a liquid/vapor interface, albeit round rather than planar. Therefore, it is not unreasonable to expect that porosity might play some role in solidification crack formation. Dixon [16] has observed interdendritic porosity in close proximity with weld solidification cracks in steel, using real-time radiography. The flattening and elongation of such a pore may serve as a crack nucleus, as may the coalescence of micro-pores. A cracking model involving gas pore coalescence has been proposed [17]. On the other hand, porosity is also known to counter solidification shrinkage (e.g. killed steel versus rimmed steel), reducing or eliminating the need for back-filling. Dissolved gasses that result in porosity may also influence cavitation (see *Liquid Rupture* below).

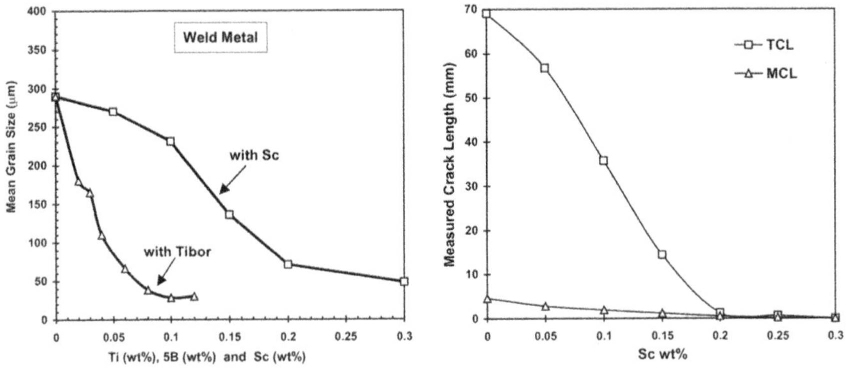

Fig. 5. Weld metal grain size versus weldability data from Circular Patch Test for aluminium alloy 7108 [15]. Grain refinement was achieved using Scandium additions

Mechanical

Strain. Originally viewed as a natural extension of concepts governing fracture in solids, strain has likewise been assumed to play an important role in controlling solidification cracking. In his "Strain Theory", Pellini [3] proposed that cracking occurs when an intergranular liquid film is strained beyond some critical value. Furthermore, he points out that the amount of strain that a liquid film will experience is determined by the film life, as determined by the solidification range and the weld cooling rate. It follows that when sulfur is added to steel, the solidification range is extended, liquid films will be exposed to more strain, and there will be a higher likelihood that a critical strain will be reached.

Building upon the strain limitation concepts of Pellini, both Prokhorov [18] and later Senda, et al. [19] established ductility curves, defining the maximum strain tolerated before cracking occurs. An example of this approach is shown schematically in Fig. 6, where the solidification range defines the upper and lower temperature bounds, and the critical ductility curve is determined experimentally (e.g. by applying controlled strain in weldabilitiy tests). Cracking will occur if the deformation curve, representing strain across the mushy zone, intersects the ductility curve.

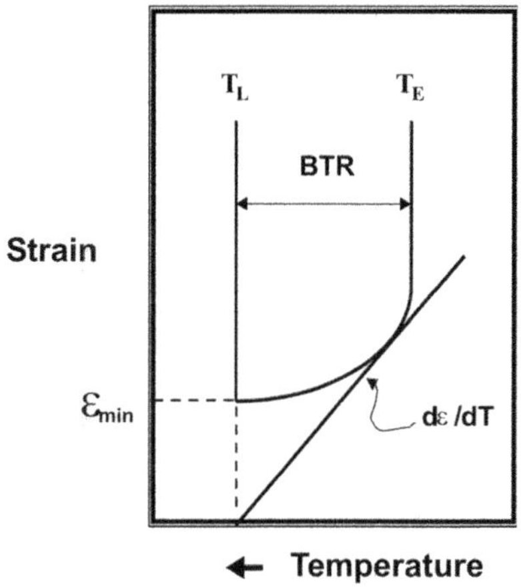

Fig. 6. Schematic showing ductility curve and Brittle Temperature Range (*BTR*) from theory of Senda et al. [19]

Stress. Because stress and strain are linked through continuum mechanics, stress must play a role in any of the cracking mechanisms discussed above involving strain. Chihoski [20] provided an early analysis of the local compression and tension stress cells that follow a moving weld pool, demonstrating how the relative size and location of these cells will vary with welding parameters.

More recently, finite element models have been used to evaluate local stresses in weldability studies [21, 22, 23]. Zacharia [21] has shown that by applying a high cross-weld stress in a Sigmajig test, the trailing end of the mushy zone will experience a tensile stress resulting in cracking, as indicated in Fig. 7.

Eq. 2 suggests that liquid films possess a critical strength dependent upon surface tension and thickness. Also, dendrite coherency should provide additional resistance to stress. Thus, it seems plausible that a weld mushy zone must possess some inherent strength. Various experimental tests have been developed in an attempt to measure the strength of liquid-solid mushy zones, including the Gleeble™ test [24].

A quartz jacket, placed around a cylindrical test specimen, can be used to support a molten zone while applying a uni-axial tensile stress. Results from these tests have typically been inconclusive, reflecting the complex nature of deformation in the mushy zone.

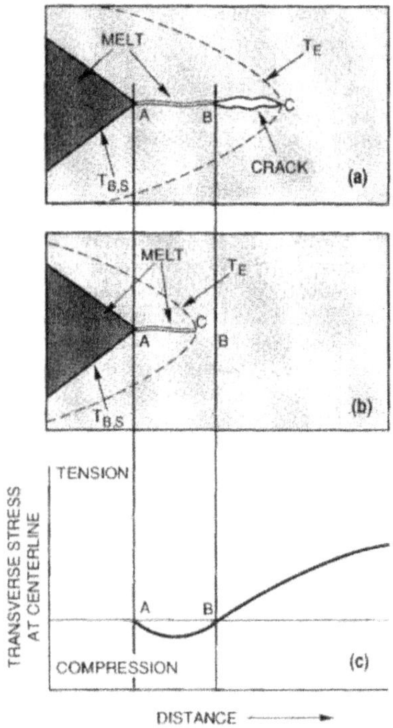

Fig. 7. Schematic showing two different welds **(a)** and **(b)**, where the mushy zone of weld (b) avoids solidification crack by remaining in compression, from Zacharia [21]

Strain Rate. The importance of strain rate on cracking has been appreciated from early analyses. From the work of Prokhorov [18] and Senda [19] discussed earlier, it is clear that the rate of deformation, normal to the weld, serves to determine whether the critical strain for cracking is achieved. The slope of the deformation curve in Fig. 6 can be related to strain rate (dε/dt) and cooling rate (dT/dt) as follows:

$$d\varepsilon/dT = (d\varepsilon/dt) / (dT/dt). \quad (3)$$

Several weldability tests have been specifically developed to measure the critical strain rate required for cracking: e.g. VDR test [25] and PVR test [26]. Another explanation for the role of strain rate will be provided in the section below in *Models for Crack Initiation.*

Restraint. It has generally been assumed that high levels of weld restraint result in higher susceptibility to solidification cracking. This belief may have evolved as an extension of the well-established relationship between restraint, residual stress, and cold cracking. Numerous different self-restrained weldability tests have evolved based upon this belief; e.g. Houldcroft [27] and Lehigh [28] "fishbone" tests. In one particular weldability test specifically developed for aluminum (a "window" restraint test), a slotted weld coupon is welded around its perimeter to a massive 5 cm thick plate (1.2m x 1.4m) to provide excessive restraint [29].

However, recent work by Kannengiesser, et al. [30] has demonstrated that high restraint does not always lead to higher cracking susceptibility. What appears to be more important here is the interaction between restraining forces and local weld strains, and specifically how this affects strain and strain rate in the vicinity of the mushy zone.

Mechanism for Crack Initiation

Based upon an accumulation of observations made over a 50-year period, as briefly outlined in the discussions above, a unified model for solidification cracking is evolving which will eventually address all of the complex aspects of this problem. Absent in all of the mechanisms discussed above, however, is a theoretical consideration of how liquid is ruptured. This would appear to be a rather serious omission, when considering that solidification cracking involves liquid rupture. Such a consideration has recently been applied to solidification cracking in castings [7] and has, in a similar manner, been applied to welding [31]. It is assumed in this model that solidification cracking is associated with a pressure drop in the interdendritic liquid, caused by solidification shrinkage, compounded by thermal contraction strains in the coherent region. When the liquid pressure reaches a critical negative value, cavitation will occur and result in liquid rupture.

Liquid Rupture

When liquid is placed in a state of hydrostatic tension (i.e. negative pressure), it becomes metastable. When a sufficiently negative pressure is met, many fine pores will nucleate spontaneously (i.e. cavitate) in a manner described by Fischer [32]. This cavitation mode of failure for liquids differs significantly from fracture in solids, in that liquid rupture is defined by the critical pressure required to nucleate one single pore. Compare this, for example, with ductile fracture in solid, which requires the nucleation, growth and coalescence of many cavities over an extended period of time. The critical fracture pressure required for homogeneous nucleation of one pore was shown by Fisher to be:

$$p_c = -\left[\frac{16\pi}{3kT}\frac{\gamma^3}{\ln(NkT/h)}\right]^{1/2}, \tag{4}$$

where γ is surface tension, k is Boltzmann's constant, T is absolute temperature, N is Avogadro's number, and h is Planck's constant.

Campbell [33] took a closer look at Fisher's liquid fracture pressure, examining both homogeneous and heterogeneous nucleation. Values of p_c required for homogeneous nucleation are compared in Table I for four different liquid metals and water.

Table 1. Fracture Pressures [33]

Liquid	Surface Tension (erg/cm)	Temperature (°K)	p_C (atm)	Experimental (atm)
Water	72	300	-1,380	-270
Mercury	490	300	-23,100	-425
Aluminium	850	933	-30,500	–
Copper	1,300	1,356	-48,000	–
Iron	1,850	1,800	-70,800	–

Experimental measurements for fracture pressures are typically orders of magnitude less than predicted values, indicating that additional factors must be at play here. One such factor involves dissolved gasses in the melt, making it easier to initiate fracture by lowering the required pressure by an amount equal to the internal gas pressure p_g (i.e. p_c-p_g). Another factor lowering the critical fracture pressure involves heterogeneous nucleation on substrates (e.g. inclusion particles or oxide films). Gas bubbles entrapped in grooves on inclusions or oxide films would provide an even more effective means of initiating failure.

Models for Crack Initiation

Solidification cracking in castings has been related to the pressure drop in inderdendritic liquid originating from 1) solidification shrinkage, 2) an inability to properly feed this shrinkage through back-filling, and 3) thermal strain resulting from cooling. In an early model by Feurer [34], this problem was partially addressed by assuming that cracking can only occur if the rate of shrinkage becomes greater than the rate of back-filling. Rappaz, et al. [7] built upon this model, adding in the strain associated with thermal contraction, and incorporating Fisher's criterion for liquid rupture. Thermal strain is limited to the coherent dendrite region, and crack initiation was assumed to occur in the thin film region immediately ahead of the coherent dendrite network (Fig. 8).

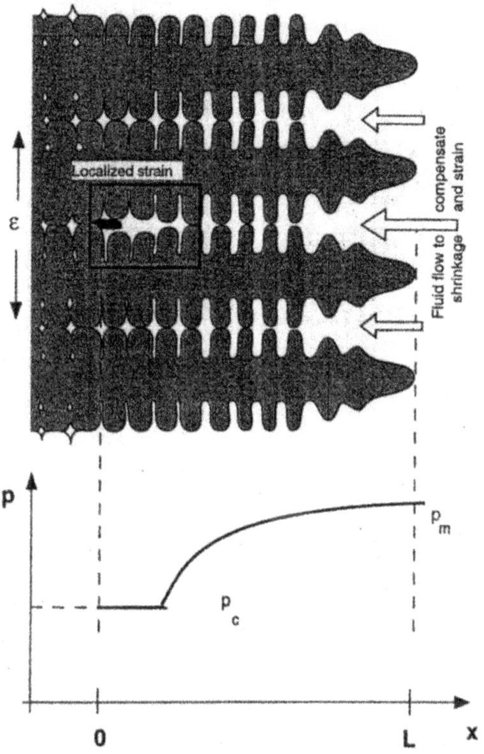

Fig. 8. Schematic showing location of solidification crack initiation relative to interdendritic pressure drop and dendrite coherency, from Rappaz et al. [7]

What remains to be done in applying the casting model of Rappaz, et al. to welding, is to account for the local transverse strain unique to welding. One might consider this to be 'solid feeding' of shrinkage. If there is sufficient inward movement of base metal behind the weld pool, the interdendritic pressure cannot drop low enough to initiate rupture. On the other hand, if there is outward movement of base metal, the cracking problem will be exacerbated. This would account for the importance of having compressive rather than tensile local stress fields to avoid cracking, as has been reported [21]. In a recent study relating local weld strain to cracking [30], transverse displacement was measured for different conditions of restraint and joint gap. It was determined that cracking was favored by conditions producing minimal inward displacement, such as a zero gap joint (Fig. 9).

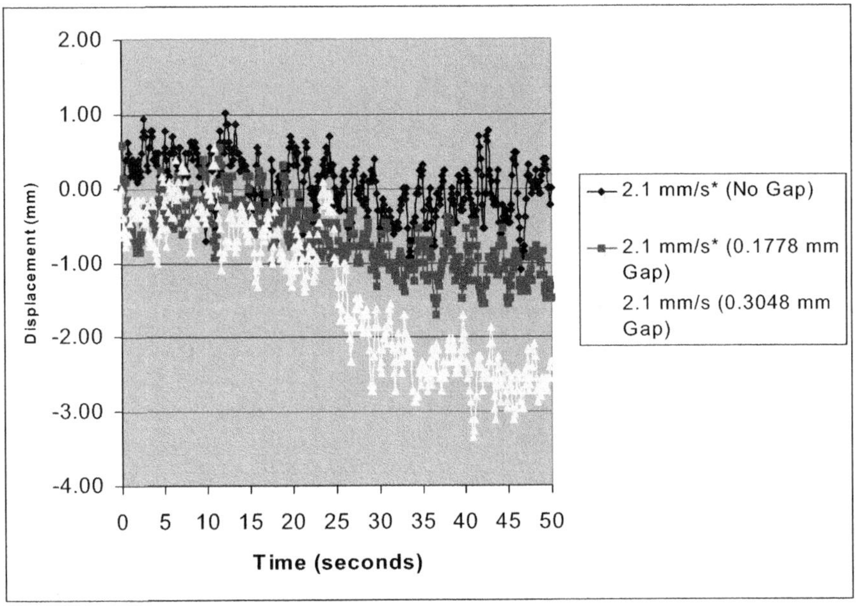

Fig. 9. Local weld displacement transverse to weld (feeding shrinkage) for three different gap conditions, from Kannengiesser et a. [30]

Summary

It has been demonstrated that solidification cracking is a many-faceted problem, with many influencing factors identified in this paper. Most attempts to model this phenomenon have concentrated on one or two of these factors (e.g. solidification range, critical strain, strain rate, etc.), but rarely have details regarding the actual failure mechanism even been considered. Models addressing liquid rupture in castings hold promise for shedding new light on weld solidification cracking, when combined with existing knowledge and an understanding of localized strain around a moving weld pool.

References

1. Cross CE, Kramer LS, Tack WT, Loechel LW (1990) Aluminum weldability and hot tearing theory. Welding of Materials. ASM Int: 275–282
2. Pumphrey WI, Jennings PH (1948) A consideration of the nature of brittleness and temperature above the solidus in castings and welds in aluminum alloys. J Inst Metals 75: 235–256
3. Pellini WS (1952) Strain theory of hot tearing. Foundry 80: 125–199
4. Cross CE, Olson DL (1986) Hot tearing model to asses aluminum weldability. Aluminum alloys-their physical and mechanical properties, Vol 3: 1869–1875
5. Lippold JC, Lin W (1996) Weldability of Al-Cu-Li alloys. Aluminum alloys – their physical and mechanical proterties. Transtec Pub: 1685–1690
6. Poirier DR, Yeum K, Maples AL (1987) A thermodynamic prediction for microporosity formation in aluminum-rich Al-Cu alloys. Met Trans 18A: 1979–1987
7. Rappaz M, Drezet JM, Gremaud M (1999) A new hot-tearing criterion. Met Mat Trans 30A: 449–455
8. Kurz W, Fisher DJ (1986) Fundamentals of solidification. Trans Tech Pub: 232
9. Mousavi MG, Cross CE, Grong Ø, Hval M (1997) Controlling weld metal dilution for optimized weld performance in aluminum. Sci Tech Weld Join 2: 275–278
10. Borland JC (1960) Generalized theory of super-solidus cracking in welds and castings – an initial development. Brit Weld J 7: 508–512
11. Saveiko VN (1961) Theory of hot tearing. Russ Cast Prod: 453–456
12. Holt M, Olson DL, Cross CE (1992) Interfacial tension driven fluid flow model for hot cracking. Scripta Met Mat 26: 1119–1124
13. Savage WF, Aronson AH (1966) Preferred orientation in the weld fusion zone. Welding J 45: 85s–89s
14. Brooks JA, Thompson AW (1991) Microstructural development and solidification cracking susceptibility of austenitic stainless steel welds. Int Met Rev: 16–43
15. Mousavi MG, Cross CE, Grong Ø (1999) Effect of scandium and titanium-boron on grain refinement and hot cracking of aluminum alloy 7108. Sci Tech Weld Join 4: 381–388
16. Dixon, BF (1988) A fractographic study of weld metal solidification cracking in steels. PhD Thesis, Univ Melbourne
17. Cross CE, Olson DL, Edwards GR (1994) The role of porosity in initiating weld metal hot cracks. In: Modeling and control of joining processes. AWS, Miami, pp 549–557
18. Prokhorov NN (1956) The problem of the strength of metals while solidifying during welding. Svar Proiz 6: 5–11
19. Senda T, Matsuda F. Takano G (1973) Studies on solidification crack susceptibility for weld metals with trans-varestraint test. J Japan Weld Soc 42: 48–56

20. Chihoski RA (1972) The character of stress fields around a weld arc moving on an aluminum sheet. Weld J 51: 9s–18s
21. Zacharia T (1994) Dynamic stresses in weld metal hot cracking. Weld J 73: 164s–172s
22. Feng Z, Zacharia T, David, SA (1997) On the thermomechanical conditions for weld metal solidification cracking. In: Mathematical modelling of weld phenomena 3. Inst Materials, London, pp 114–148
23. Dike JJ, Brooks JA, Li M (1998) Comparison of failure criteria in weld solidification cracking simulations. In: Mathematical modelling of weld phenomena 4. Inst Materials, London, pp 199–222
24. Batra R, Wilber GA, Breit HF Childs WJ (1975) Programmed In Situ Melting, Freezing , and Tensile Testing for Laboratory Study of High Temperature Properties of As-Cast Metals. Journal of Testing and Evaluation, Vol 3 (1): 68–74
25. Matsuda F, Nakagawa H, Nakata K, Okada H (1979) The VDR cracking test for solidification crack susceptibility on weld metals and its application to aluminum alloys. Trans JWRI 8: 85–95
26. Herold H, Streitenberger M, Pchennikov A (2001) Modelling of the PVR-test to examine the origin of different hot cracking types. In: Mathematical modelling of weld phenomena 5. Inst Metals, London, pp 783–792
27. Houldcroft PT (1955) A simple cracking test for use with argon-arc welding. Brit Weld J 2: 471–475
28. Stout RD, Tor SS, McGeady LJ, Doan GE (1946) Quantitative measurement of cracking tendency in welds. Weld J 25: 522s–532s
29. Robinson IB, Baysinger FR (1966) Welding aluminum alloy 7039. Weld J 45: 433s–444s
30. Kannengiesser Th, McInearney T, Forian W, Böllinghaus Th, Cross CE (2002) The influence of local weld deformation on hot cracking susceptibility. Mathematical modelling of weld phenomena 6. Manley, London, pp 803–818
31. Rindler W, Kozeschnik E, Enzinger N, Buchmayr B (2002) A modified hot tearing criterion for steels. In: Mathematical modelling of weld phenomena 6, Manley, London, pp 819–835
32. Fisher JC (1948) The fraction of liquids. J Appl Phys 19: 1062–1067
33. Campbell J (1968) Pore nucleation in solidifying metals. In: The solidification of metals. Iron Steel Inst, London, pp 18–26
34. Feurer, U (1977) Influence of alloy composition and solidification conditions on dendrite arm spacing, feeding, and hot tear properties of aluminum alloys. In: Proceedings of the international symposium on engineering alloys, Delft: 131–145

New Insight into the Mechanism of Ductility-Dip Cracking in Ni-base Weld Metals

A.J. Ramirez[1], J.C. Lippold[2]

[1] Synchrotron Light National Laboratory, Campinas-SP, Brazil
[2] The Ohio State University, USA

Abstract

The ductility dip cracking resistance of Ni-base Filler Metal 52 (AWS A5.14, ERNiCrFe-7; ISO SNi6052, 59 % Ni, 29 % Cr, 9 % Fe, 1 % Mn, 1 % Al) and Filler Metal 82 (AWS A5.14, ERNiCr-3; ISO SNi6082, 72 % Ni, 20 % Cr, 1 % Fe, 3 % Mn, 3 % Nb) has been extensively evaluated using the recently developed strain-to-fracture test in conjunction with microstructural characterization using electron microscopy. This paper provides new insight into the creep-like, grain boundary sliding mechanism that leads to elevated temperature intergranular cracking in these weld deposits. The effect of precipitation on grain boundary tortuosity and sliding, and its influence on ductility-dip cracking resistance is discussed in the context of current theories of high temperature creep. Finally, the effect of impurity and interstitial elements on ductility-dip cracking is discussed.

Introduction

Many normally ductile austenitic alloys (FCC structure) such as Ni-base alloys, Ni-Cu alloys, Cu alloys, stainless steels and steels undergo a severe ductility reduction at temperatures between 0.5 and 0.8 of their melting temperature [1, 2, 3]. The exhaustion of ductility during the high temperature processing or welding of these alloys results in intergranular cracking often referred to as ductility-dip cracking (DDC).

Previous studies with FM-52 and FM-82 addressed the factors that influence elevated temperature ductility and the DDC mechanism [4]. These factors are:

- Weld metal chemical composition including impurity and interstitial elements;
- Segregation to the grain boundaries;
- Precipitation behavior;
- Grain boundary migration and pinning;
- Grain boundary orientation (macroscopic) relative to the applied strain;
- Dynamic recrystallization.

Despite the extensive research performed on ductility dip cracking, a complete understanding of this phenomenon has not been achieved. In order to provide a better understanding of the influence of the above mentioned factors on the DDC phenomenon it has been necessary to develop new testing techniques and to perform detailed microstructural characterization.

Several test techniques have been used in the past to quantify DDC susceptibility. Among these techniques are: elevated temperature tensile test, hot ductility test, and Varestraint-based tests. However, most of these techniques have poor repeatability and/or difficulty differentiating between DDC and other high temperature cracking phenomena. For this reason, the strain-to-fracture (STF) test was used to reproduce the DDC conditions, avoiding most of the problems with the previous techniques [5, 6]. Thus, using samples prepared using the STF test it is possible to more effectively study the DDC phenomenon in the context of creep-like grain boundary sliding and analyze the effect of precipitates and impurities on this cracking process.

Materials and Sample Preparation

Two Ni-base alloys, designated Filler Metal 52 (AWS A5.14, ERNiCrFe-7; ISO SNi6052) and Filler Metal 82 (AWS A5.14, ERNiCr-3; ISO SNi6082) were used in this study. Filler Metal 82 and 52 are commonly used to weld alloys 600/625 and 690, respectively, and to perform dissimilar welds involving low alloy steels and Ni-base alloys or stainless steels. The chemical composition of the filler metals is presented in Table 1. Because of the large variability of DDC resistance from heat to heat, two different FM-82 heats were included in this study.

Multipass weld deposits were made using the automated gas tungsten arc welding (GTAW) process. The base material used to prepare this joint was A36, a low carbon structural steel. However, extreme care was taken to reduce base metal dilution in the Ni-base weld metal.

Table 1. Chemical composition (wt%) of the filler metals (balance Ni)

Alloy	C	Mn	Fe	Cr	Nb	Ti	Al	Si	Other
FM-52 NX9277	0.026	0.25	8.88	29.1	0.02	0.50	0.71	0.17	0.01 Cu 0.05 Mo 0.004 S 0.004 P
FM-82 YN6830	0.040	2.86	1.18	20.1	2.3	0.37	–	0.12	0.09 Cu 0.05 Co 0.001 S 0.007 P
FM-82 YN7355	0.040	2.75	0.70	20.1	2.6	0.47	–	0.07	0.07 Cu 0.04 Co 0.002 S 0.01 P

Dog bone-like samples, as presented in Fig. 1, with the weld deposit in the middle were cut out transversely from this joint. Subsequently, a GTAW autogenous spot weld was made at the sample midspan, within the pre-deposited weld metal. The circular geometry of this spot weld produces a radial distribution of grain boundaries such that axial straining during the STF test allows cracking to occur along the most favorably oriented grain boundaries. More details about the welding conditions and sample preparation are presented elsewhere [7, 8].

Strain-to-Fracture Testing

This Gleeble™ based test replicates the thermo-mechanical conditions experimented by the weld metal resulting in DDC. The sample is heated to the desired temperature, held for a short period of time (customary 10 s) and then subjected to a predetermined strain under controlled conditions [6]. The strain-to-fracture test makes it possible to determine the ductility-dip temperature range and the strain threshold to cause cracking. The STF data for the FM-52 and FM-82 presented in Fig. 2 is the reinterpretation of some recent data obtained by Collins et al.[7, 9]. The new approach takes into consideration both the cracking threshold and the number of cracks as a measure of cracking susceptibility.

The high susceptibility of FM-52 to DDC, which is observed during actual fabrication of heavy sections, when compared with FM-82 was verified by the STF test results.

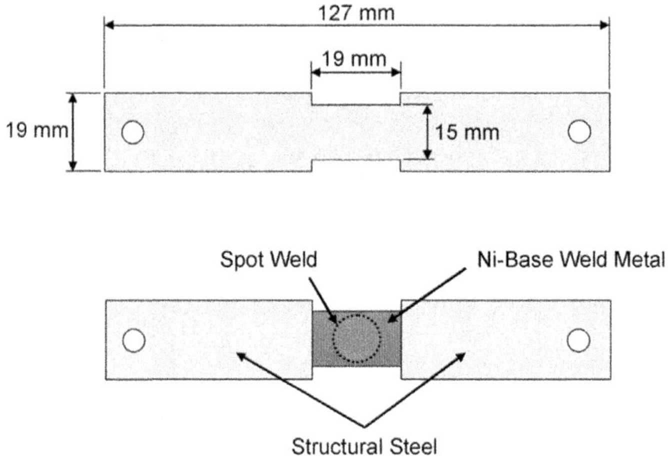

Fig. 1. Strain-to-fracture sample

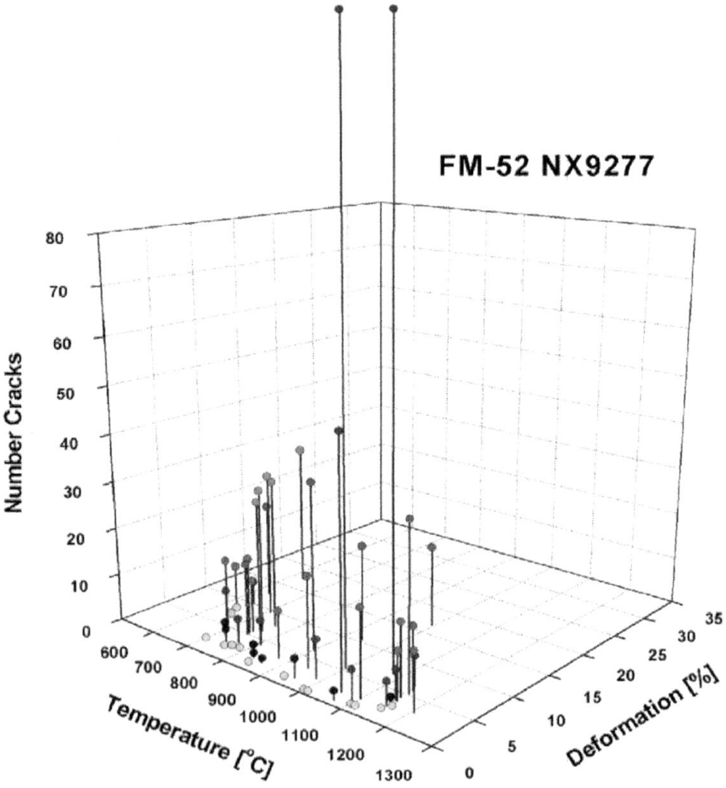

a)

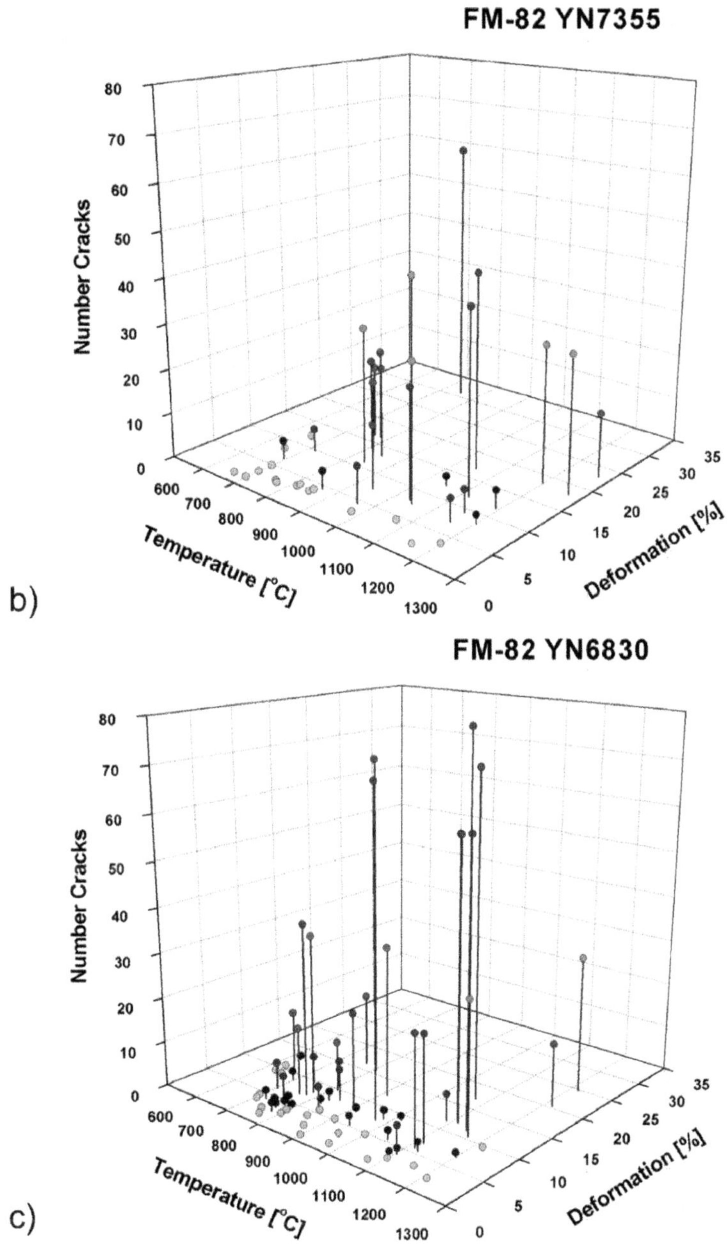

Fig. 2. Strain-to-fracture test results of **(a)** FM-52 heat NX9277, **(b)** FM-82 heat YN7355, **(c)** FM-82 heat YN6830. Adapted from "An Investigation of Ductility Dip Cracking in Nickel-Base Filler materials" [7]

FM-52 exhibited a consistently low threshold strain for cracking (between 1 and 2 %) along the entire ductility-dip temperature range (from 700 to 1200 °C), with the minimum at about 1050 °C. In addition, the number of cracks increased dramatically as the applied strain increased. On the other hand, FM-82 heats YN-6830 and YN7355 had threshold strains of approximately 3 % and 4 %, respectively, with the number of cracks increasing more gradually with the increase in applied strain when compared with FM-52. The FM-82 heat YN6830 presented a less consistent ductility-dip trough between 700 and 1200 °C, with two minimums of ductility at about 775 °C (2.5 % strain) and about 1100 °C (3 % strain). Between these two minimums the threshold ductility was about 5 %. The FM-82 heat YN7355 had a more consistent ductility-dip trough between 700 and 1200 °C, with a constant threshold of about 4 % strain throughout this temperature range.

DDC susceptibility does not depend only on the minimum threshold strain to start cracking along the whole ductility-dip trough range. In fact, DDC is a complex phenomenon, which is controlled by the available ductility along the critical temperature range where the material is subjected to severe strain and/or strain rate conditions. Finite element modeling has shown that just behind the weld pool the recently solidified material is under compressive stresses and tensile stresses start to build just up when this region reaches lower temperatures [10]. The above explains the higher susceptibility of FM-82 heat YN7355 to DDC in actual welding conditions when compared to FM-82 heat YN6830 despite the lower overall threshold strain to start cracking presented by the second alloy.

Microstructural Characterization

The STF samples were sectioned and prepared for light and scanning electron microscopy (SEM) by grinding and then polishing, using diamond pastes of 6, 3 and 1 μm. The final polishing was done with 0.05 μm colloidal silica. Electrolytic etching was performed with a 10 % aqueous chromic acid solution at 1.5 to 2.0 V for 30 seconds.

The SEM analyses were conducted using a Phillips XL-30 ESEM FEG and FEI SIRION FEG microscopes coupled with X-ray energy-dispersive spectroscopy (XEDS) and electron backscatter diffraction (EBSD). The EBSD data acquisition and analysis were performed using CHANNEL 5 and PHASE-ID software. Transmission electron microscopy (TEM) foils were mechanically thinned until approximately 100 μm, followed by jet

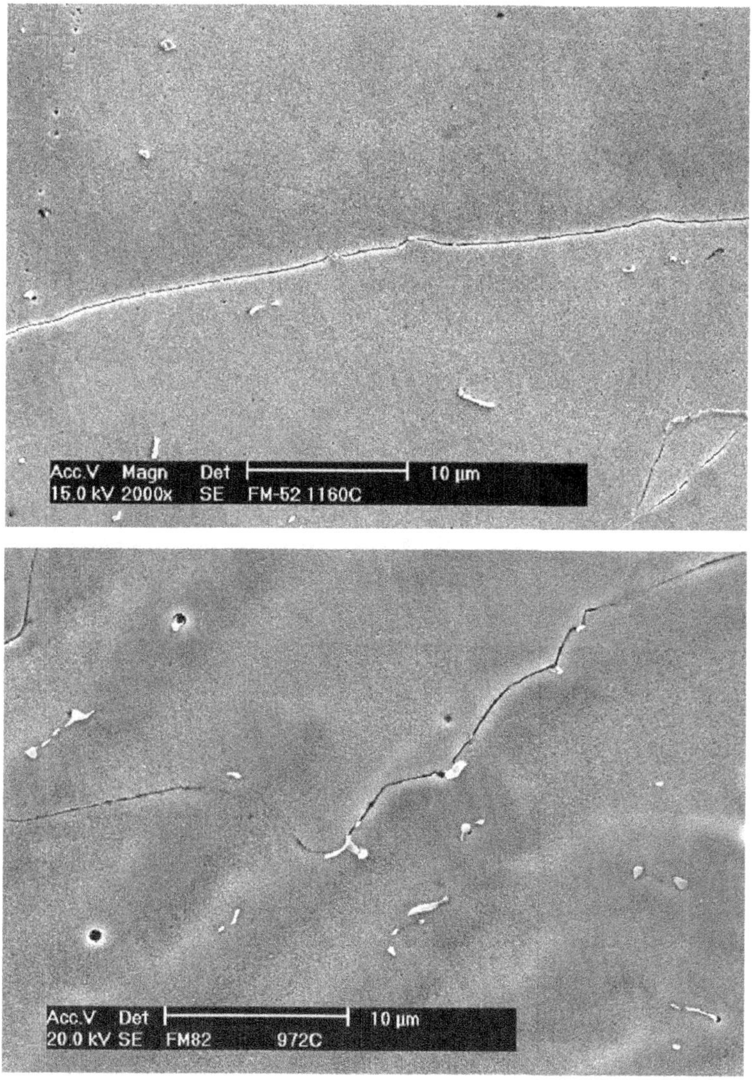

a)

b)

Fig. 3. Migrated grain boundaries in the STF samples. In **(a)** straight migrated grain boundary in FM-52 heat NX9277 (STF sample deformed 2.9% at 1160 °C) and **(b)** grain boundary pinning caused by the frequent medium size eutectic (NbTi)C precipitates in FM-82 heat YN6830 (STF sample deformed 7.5 % at 972 °C. Secondary electron SEM image

polishing using a solution of 70 %vol ethyl alcohol, 20 %vol Glycerin and 10 %vol of perchloric acid ($HClO_4$) at -20 °C applying 30 V. TEM analyses were carried out at 200 kV in a Philips CM-200 coupled with XEDS.

The summary of the optical, SEM and TEM observations on the STF samples is presented on Table 2. However, a detail characterization is presented elsewhere [11]. The most relevant aspects of FM-52 and FM-82 STF weld metal characterization are the following:

- Alloy FM-52 weld metal has straight migrated GBs and FM-82 has more tortuous GBs, as shows Fig. 3;
- FM-52 weld metal contains a low concentration of large, medium and small TiN-like nitrides and numerous small intergranular $M_{23}C_6$. Additionally, this alloy has some sulfur rich films along the GBs;
- FM-82 weld metal exhibits numerous medium and small, and sporadic very small intergranular NbC-like carbides.

Table 2. Precipitates observed in the STF samples of FM-52 and FM-82

Precipitates	Weld Metal	
	FM-52	FM-82
Large	Sporadic (Inter., Intra.) (TiCr)(CN) – 5 μm – Transferred from the wire. Fig. 4.	None
Medium	Sporadic (Inter., Intra.) (TiCr)(CN) – 1 μm. Fig. 3	Frequent (Inter., Intra.) (NbTi)C – 1 μm (Nb, Ti, Cr, Ni, Fe rich). Fig. 3,7.
Small	Sporadic (Inter.) round-like $M_{23}C_6$ – 100 nm (Cr, Fe rich) – Associated with small TiN-like particles. Sporadic (Inter.) (TiCr)(CN) – 100 nm. Numerous (Inter.) square-like $M_{23}C_6$ – 50 nm (Cr, Fe rich). Fig. 5.	Frequent (Inter.) (NbTi)C – 50 nm - Isolated, aligned or grouped. Fig. 8. Sporadic (Intra.) (NbTi)C – 50 nm – Interdendritic regions. Fig. 8.
Very small	Some (Intra.) presumably $M_{23}C_6$ – or TiN-like – 10 nm.	Sporadic (Inter.) very likely (NbTi)C – 10 nm – Associated with grouped small (NbTi)C particles. Fig. 8.
Intergranular Films	Some – Sulfur rich – Associated with medium size intergranular (TiCr)(CN) – 1 μm. Fig. 6.	None

Inter. Intergranular precipitates,
Intra. Intragranular precipitates,
Amount scale: *Numerous – Frequent – Some – Sporadic.*

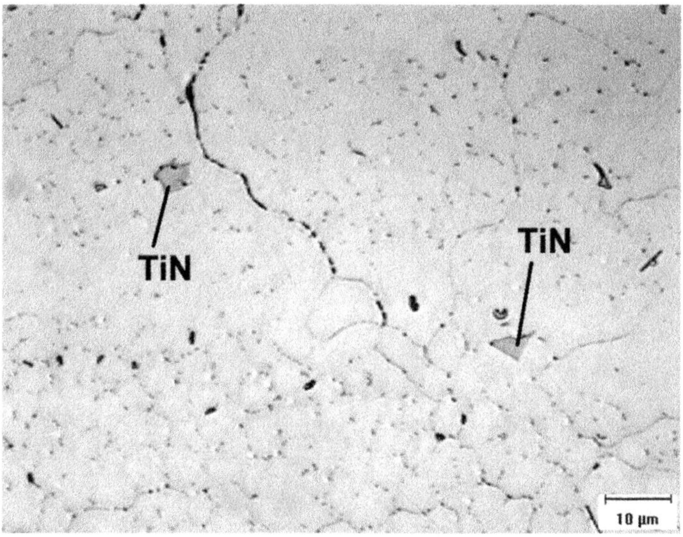

Fig. 4. Large titanium nitride (TiN) precipitates in the STF sample of FM-52 heat NX9277 (strained 1.7 % at 1090 °C). Light microscopy

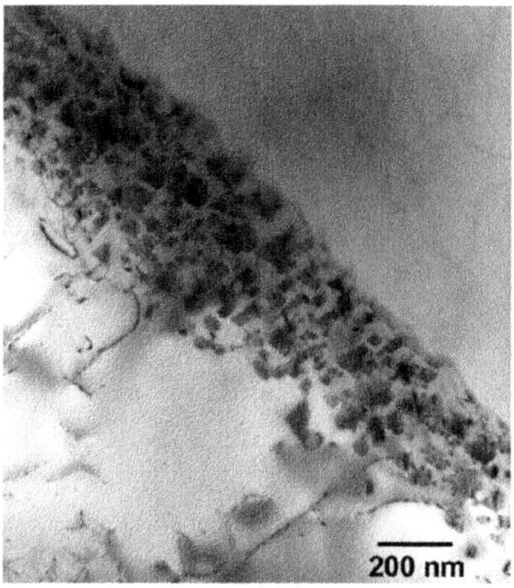

a)

b)

Fig. 5. $M_{23}C_6$ carbides precipitated along the GB of a STF sample of FM-52 heat NX9277 strained 1.6 % at 956 °C. In **(a)** detail of the carbides and **(b)** SAD pattern of the austenite grains and the carbides showing the cube-on-cube orientation relationship between the precipitates and one of the austenite grains. The SAD pattern presents both phases on the [011] zone axis

a)
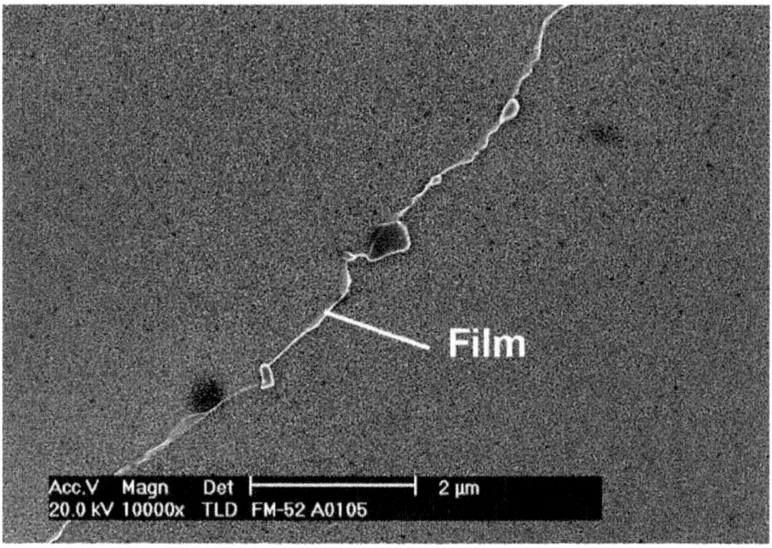

b)
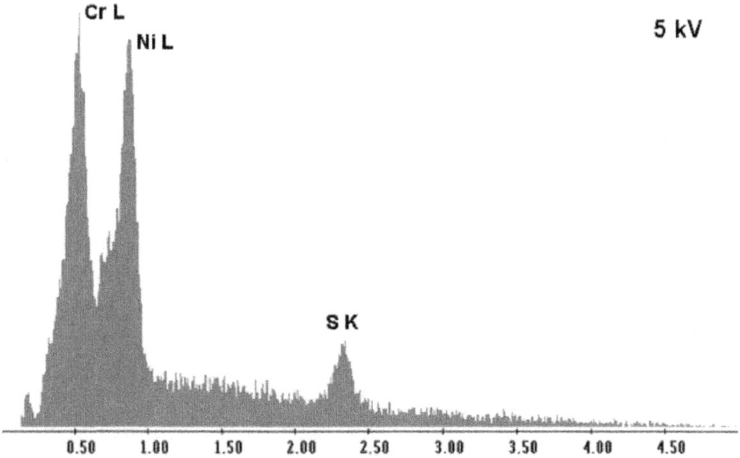

Fig. 6. Sulfur enriched film observed along grain boundaries in the FM-52 heat NX9277 weld metal. In **(a)** Secondary electron SEM detail of the film and **(b)** XEDS spectrum of this film acquired at 5 keV

Extensive GB pinning caused by the medium size eutectic NbC-like precipitates was observed in FM-82. Some evidence of GB pinning was seen in FM-52 as well. Nevertheless, the uneven distribution of TiN-like nitrides in FM-52 did not cause the efficient GB pinning effect that the NbC-like carbides did in FM-82.

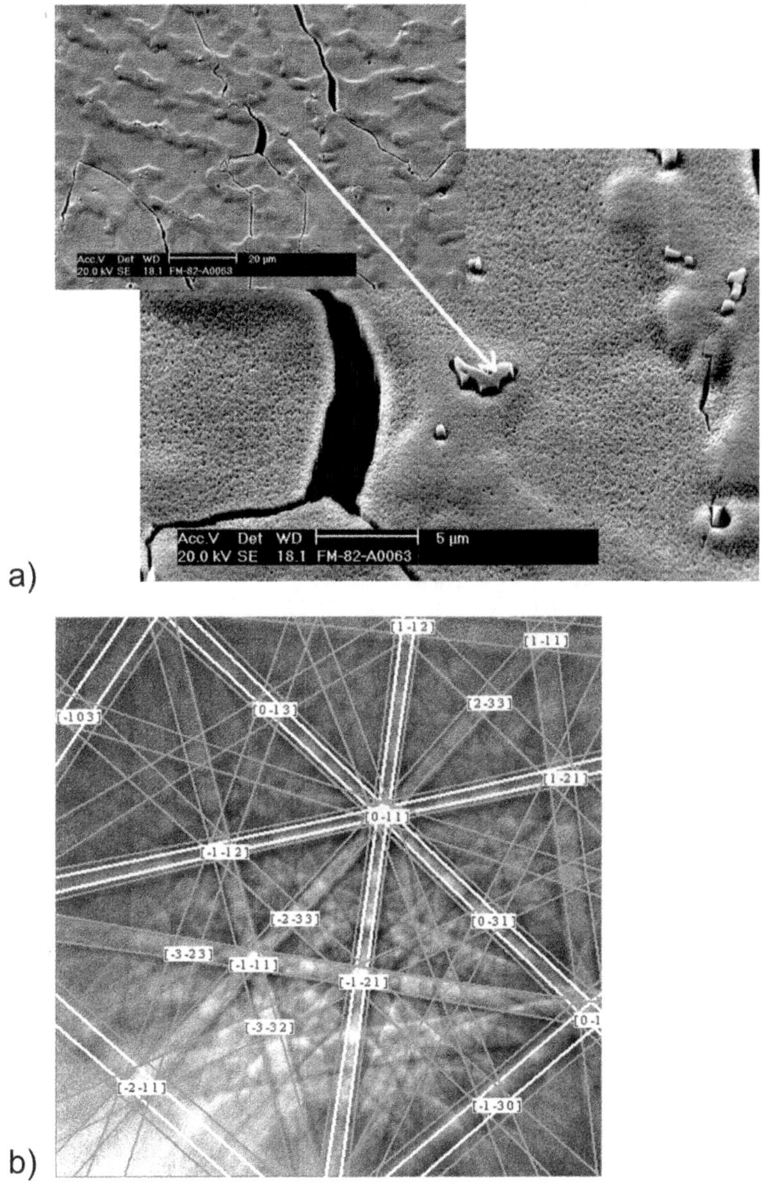

Fig. 7. DDC cracks along migrated grain boundaries and medium size (1 µm) interdendritic (NbTi)C precipitates in FM-82 heat YN6830 (STF sample strained 7.5 % at 972 °C). In **(a)** General view and detail of the intergranular cracks and precipitates, and **(b)** calculated EBSD pattern for the indicated (NbTi)C precipitate superimposed on the actual precipitate pattern

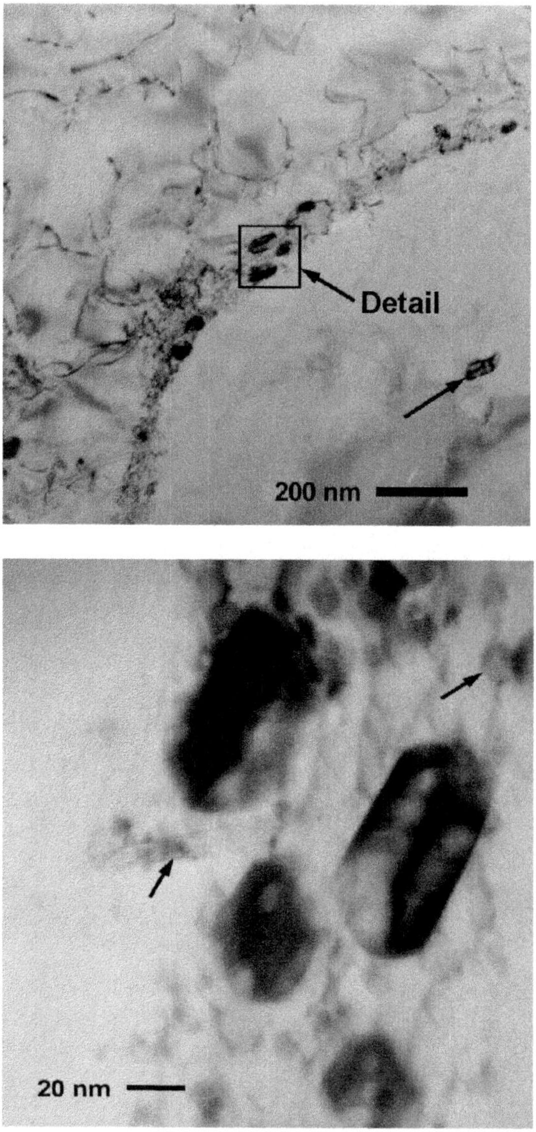

Fig. 8. Grouped small (NbTi)C and very small intergranular precipitates on FM-82 heat YN6830 (STF sample strained 3.5 % at 955 °C). In **(a)** General view of a random grain boundary with grouped intergranular small (50 nm) (NbTi)C carbides and intragranular carbides as indicated by the arrow, and **(b)** the detail of the grain boundary region showing the same small carbides and other very small (10 nm) precipitates, as indicated by the arrows. Bright field TEM images

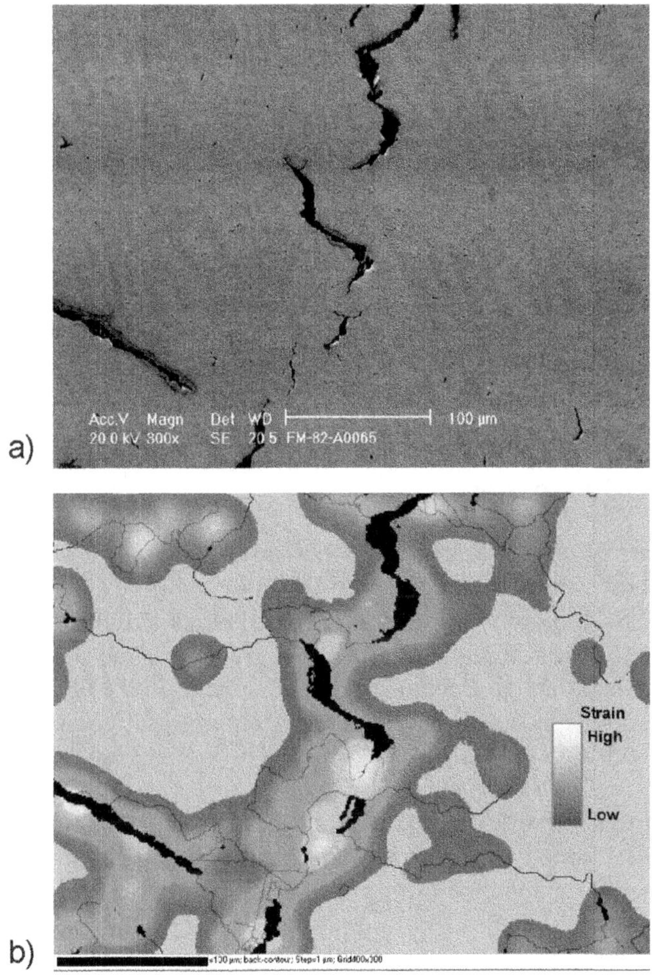

Fig. 9. FM-82 heat YN6830 strain-to-fracture sample strained 11.3 % at 1150 °C. In **(a)** Secondary electron image of the fractured sample. In **(b)** Strain distribution map from the same region obtained from EBSD measurements. The thin lines represent the high angle grain boundaries; the black regions are the open cracks; the gray scale contouring shows the strain distribution with the dark representing the lowest and the light the highest strains

The small and/or very small intergranular precipitates observed in both alloys would effectively pin the GBs. However, these precipitates nucleated heterogeneously at the GBs after most of the GB migration had occurred. Thus, the effect of the small and very small precipitates on GB migration and therefore GB tortuosity was negligible for all the studied alloys.

In addition to grain boundaries and precipitate characterization, EBSD analysis of STF samples revealed the microstrain concentration around the grain boundaries, mainly around triple points and the curved segments of the GBs.

Fig. 9 shows the microstrain map of the FM-82 YN6830 STF sample strained 11.3 % at 1150 °C. Strain concentration at the crack tips was revealed by this analysis as well. However, the strain field ahead the crack tip is a consequence of the crack propagation phenomenon and not the cracking cause. The formation of these highly strained regions resulted in dynamic recrystallization, as has been evidenced by EBSD analysis [12].

Weld Metal Grain Boundaries

Three different types of grain boundaries (GBs) can be recognized within the weld metal, solidification grain boundaries (SGBs), solidification subgrain boundaries (SSGBs), and migrated grain boundaries (MGBs). The combination of solute partitioning and the thermal gradient at the solidification front cause constitutional supercooling, which results in the instability of the solidification front itself, leading to the formation of columnar, cellular, cellular dendritic or dendritic solidification structures within the grains.

The boundaries that separate these columns or dendrites within the grains are the SSGBs, which have an important compositional component as a result of the micro-segregation during the solidification process. The SSGB are by definition low angle GBs, and possess a very small crystallographic component. On the other hand, SGBs are high angle GBs were packets of subgrains (cells and/or dendrites) intersect during the solidification process. SGBs have both a crystallographic and compositional component. Once the weld metal has solidified, the crystallographic component tends to migrate away from its original position to reduce its energy, leaving the compositional component (segregation) behind. The new GBs, which have basically a crystallographic component and formed by migration away from the segregation field, are called MGBs. These different types of boundaries are shown in Fig. 10.

The MGB carries with it the original misorientation of the SGB and is, in effect, no different than a base metal grain boundary. In the absence of grain boundary pinning agents, such as second phases or precipitates, the MGB will be relatively straight. The distance that it migrates is a function of the weld cooling rate, with slower cooling rates allowing more time for migration. Additional migration may also occur during weld metal reheating, such as during multipass welding.

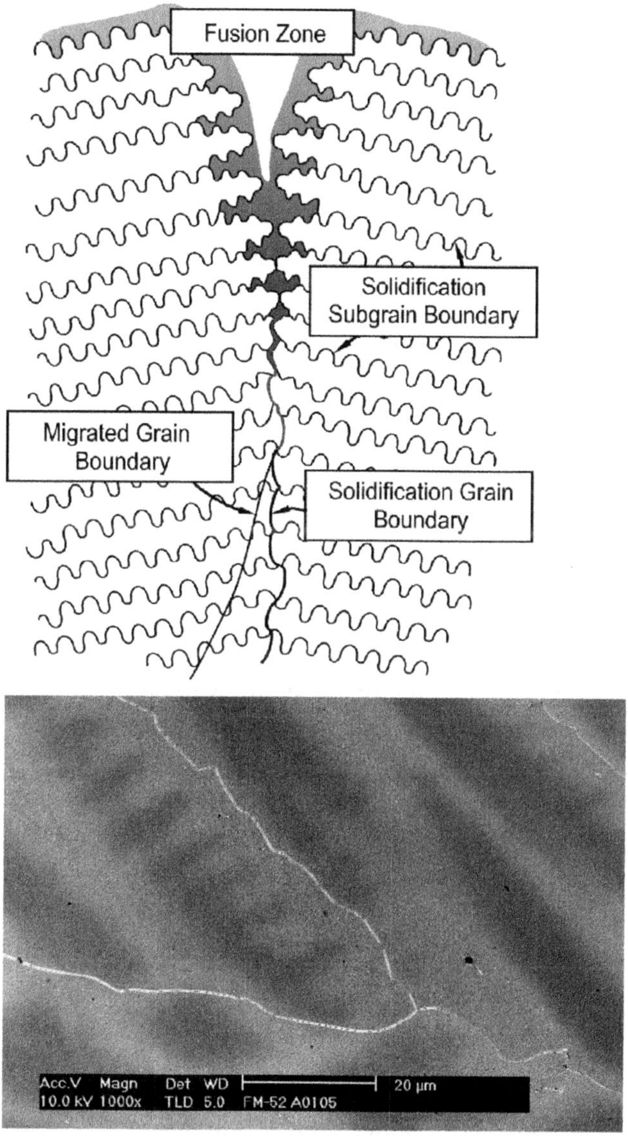

Fig. 10. Weld metal boundaries in austenitic weld metal. In **(a)** schematic illustration [13] and **(b)** migrated gain boundaries in FM-52 weld metal, which pulled away from the segregation patterns revealed by the etching

If precipitates or second phases are present, resulting either from solidification or solid-state precipitation, then the MGBs can be pinned and more tortuous boundaries will result. The microstructure characterization presented in the previous section has shown the effect of certain precipitates on GB tortuosity, with the MGBs in FM-52 weld metals being characteristically straighter, while the FM-82 boundaries are more tortuous. Examples of migrated grain boundaries in FM-52 and FM-82 are shown in Fig. 3 and Fig. 10b.

Ductility Dip Cracking Mechanism in Ni-base Alloys

Before discussing the ductility dip cracking mechanism in Ni-base weld metals, it is important to describe the principal characteristics of these cracks.

- They occur along MGBs [2, 4, 7–9];
- The orientation of these MGBs relative to the load applied to the sample has an important influence on the nature of cracking. In the STF samples, intergranular cracking normally occurred along the GBs oriented from 45 to 90° to the applied load [7];
- Evidence of GB sliding has been observed and previously associated with this type of cracking [1, 4, 14];
- The ductility recovery at the high temperature extreme of the ductility-dip trough has been associated with the onset of recrystallization [4, 9].

These characteristics suggest that DDC is a GB sliding, creep-like phenomenon, as was initially proposed by Rhines and Wray [1]. The treatment of DDC as a creep-like phenomenon, explains the following DDC characteristics.

- The ductility drop between approximately 0.5 and 0.8 of the alloy melting temperature is limited at the high temperature side by the ductility recovery due to the onset of recrystallization, which eliminates the deformation that is accumulating along the GBs and therefore precludes void formation at either triple points or GB sites. Recrystallization has been observed by the authors in actual multipass weld metal and in STF samples tested at temperatures above 1000 °C. In these cases, recrystallization occurs locally along the GBs and at triple points, where most of the deformation is concentrated. Recrystallization was also associated with the regions around the ductility-dip cracks, especially ahead the cracks. This observation suggests the possible participation of recrystallization on crack arrest at the higher temperature range. On the other hand, the low temperature ductility recovery is due to the inoperability of GB sliding at these lower temperatures [15, 16].

- The GB orientation dependency of DDC is also an important characteristic of creep, where the intergranular cavities and cracks are normally formed at about 90° to the applied load. The STF test sample design, having a spot weld with a radial orientation of migrated grain boundaries provides a spectrum of GB orientations to the applied load ranging from 0 to 90 degrees.
- GB sliding in a three dimensional array of grains causes strain concentration at triple points and at other GB irregularities. This strain concentration occurs at locations where the GB has sudden directional changes, such as at "pinning points" associated with intergranular precipitates or at GB steps. In general, any feature that inhibits GB sliding will be a preferential site for strain concentration during the deformation process. Strain concentration at these grain boundary sites can lead to intergranular cavity formation and subsequently to intergranular DDC cracking.
- The chemical composition of the alloy has an important influence on the GB sliding phenomenon. Two of the most important factors that chemical composition controls are 1) the type, size, morphology and distribution of precipitates within the microstructure and 2) the intergranular embrittlement caused by impurity and interstitial segregation. Both factors will be addressed later in this paper.

Influence of Precipitates on the Creep-Like Induced Cracking Phenomenon

The complex effect of intergranular and intragranular precipitation on the creep-like DDC phenomenon should be analyzed from two different perspectives. The first is the effect of precipitation on GB migration and the subsequent effect that GB migration and pinning has on DDC. The second is the direct effect that intragranular and mainly intergranular precipitates have on the GB sliding phenomenon. The effect that precipitates will have on GB migration and sliding depends on the nature, size, distribution and evolution of the precipitates in the microstructure during the weld thermal cycle.

The effect that different sizes and types of precipitates have on the on GB migration of FM-52 and FM-82 was discussed previously and the precipitate types summarized in Table 2. In general terms some of these precipitates pinned the GBs, reducing or preventing GB migration. As GB migration is inhibited, the average grain size remains smaller, sweeping of impurities during migration will be reduced, and, perhaps most importantly, GB tortuosity increases due to the pining effect (Fig. 3). The smaller grain size results in lower strain concentration at individual GBs and reduces the tendency for intergranular void formation and subsequent cracking [17].

When GB tortuosity is maintained or increased, sliding is restricted due a mechanical interlocking effect. This causes a reduction in strain concentration around triple points and therefore, reduces the probability of void formation at these sites, as revealed by micro-strain measurements using EBSD [12]. The restriction of GB migration imposed by the precipitates may maintain and/or promote GB tortuosity. However, the precipitates have to be present in the microstructure at the right time in order to prevent GB migration and have an effect on DDC resistance. In addition, precipitates size and distribution also play an important role in restricting GB migration.

In the case of the FM-52, the large (5 μm) and medium (1 μm) TiN-like precipitates, which were either transferred from the wire and/or formed in the liquid during solidification, were present just sporadically in the microstructure. Thus, the average GB pinning effect of this low fraction of relatively large particles was negligible, as evidenced by the long and straight GBs observed in FM-52 weld metal. On the other hand, the numerous medium size (1 μm) interdendritic (NbTi)C carbides observed in FM-82 effectively pinned the GBs during the cooling following the solidification and during the subsequent reheating, resulting in tortuous GBs. The Nb-rich precipitates form as the result of a eutectic reaction at the end of solidification [18, 19]. Tortuous MGBs make GB sliding more difficult and, as a result, the DDC resistance of FM-82 is improved relative to FM-52, as revealed by STF test results and actual welding experience with these filler metals [7–9, 20].

As previously mentioned, the small precipitates observed in both filler metals precipitated heterogeneously at the GBs. Therefore, they formed after most of the GB migration has occurred. Thus, despite the important pinning effect of these numerous small precipitates, their effect on GB tortuosity was negligible.

In addition to GB tortuosity, the degree of GB coherency plays an important role on the GB response to high temperature deformation. Work by Lim et al. [21] and Lehockey et al. [22] showed that low Σ coincidence site lattice (CSL) GBs are less prone to cavity formation during high temperature deformation of nickel. Lim et al. explained this behavior by the faster accumulation of misfit strain on the higher Σ GBs [21]. However, previous results have shown that GB coherency is not an important factor on actual

DDC formation during multipass welding of FM-52 and FM-82 due to the low fraction of low Σ CSL GBs in the weld metal [4, 11].

The effect of intergranular carbides on creep processes is more complex than just the role they play on the control of GB migration. The intergranular precipitates cause a locking effect on the GB, making GB sliding more difficult. The effect of intergranular precipitates depends on the creep stage, particle size, and particle distribution [17]. Initially the carbides extend the creep life by the reduction of creep deformation, but at the end of the creep life, void (cavity) nucleation is associated with the intergranular carbides. But even during this final stage of creep life, the carbides may improve the material response reducing the agglomeration of voids, as observed in some cast stainless steels [23]. Continuous carbides along the GB, as the numerous small $M_{23}C_6$ precipitates observed on FM-52, are easy crack propagation paths [24]. However, isolated blocky carbides along the GB may be beneficial for the high temperature cracking resistance, due to the reduction in GB sliding. Because DDC is a GB sliding process, a similar effect of the intergranular precipitates is expected under DDC conditions, which explains the beneficial effect of the numerous eutectic, medium-size carbides on the migrated grain boundaries in the FM-82 weld metal.

Intragranular carbides play a role in the DDC mechanism as well. They increase high temperature strength and reduce the creep deformation due to the dislocation pinning effect [25]. Stronger grain interiors make strain relaxation along the grain boundaries more difficult, favoring triple point void formation [26]. The amount of intragranular precipitation in the two filler metals studied here is low and its effect on the DDC phenomenon is thought to be minimal.

Effect of Impurity and Interstitial Elements on DDC

Impurities and interstitials such as S, P, and H have an important effect on intermediate temperature ductility and therefore, on DDC susceptibility. Previous results by the authors have shown that hydrogen increases the susceptibility to DDC [7–9]. The mechanism of hydrogen enhanced DDC is thought to be a combination of hydrogen-enhanced local plasticity around the grain boundaries and hydrogen-induced decohesion of GBs and/or the interfaces between the intergranular precipitates and the matrix. Although, hydrogen is generally considered to play a more important role at lower temperatures, it may also influence ductility within the DDC temperature range.

Sulfur and phosphorus are elements that segregate to the GBs during solidification, cooling and reheating and cause GB embrittlement [27, 28]. The increase in DDC susceptibility of Invar (Fe-36Ni) associated with a sulfur increase from 0.004 to 0.011 wt% has been reported to be related with the S segregation to the GBs during reheating [29]. A previous study with FM-82 showed that sulfur additions to the weld metal caused a reduction in the threshold strain to cause cracking from 6 % to 2 % during the strain-to-fracture test [9]. Thus, the presence of high sulfur levels at FM-52 migrated grain boundaries indicates that S segregation to the GBs may be playing an important role on the high susceptibility of this alloy to DDC. However, the differences in DDC susceptibility among FM-52 and two FM-82 heats cannot be simply explained based on the bulk S and P content differences, which are negligible as shown in Table 1.

Summary

The new three-dimensional presentation of the STF results data has permitted a better interpretation and understanding of the effect of strain on DDC susceptibility. Further analysis of the data has shown that the variation in DDC between FM-52 and FM-82 goes beyond the difference in strain threshold to initiate cracking. Once the cracking threshold is exceeded in FM-52, the severity of cracking (number of cracks) increases rapidly with the applied strain, revealing the higher susceptibility of this alloy to grain boundary embrittlement.

Regarding the differences in DDC susceptibility between the two heats of FM-82, the STF test results revealed differences in high temperature ductility. However, the overall threshold strain to initiate cracking is important, but not enough to characterize the DDC susceptibility of the alloy under actual production welding conditions. It is proposed that the DDC resistance is controlled by the available ductility of the weld metal within a narrow range of the ductility-dip trough where higher strains and/or strain rates are developed during weld cooling. Based on FEM analysis it is presumed that this occurs at the lower temperature extreme of the ductility-dip trough. It is anticipated that a better understanding of the DDC mechanism will allow STF test results to be a more effective predictor of DDC susceptibility and may even lead to a modification of the test procedure itself. Consequently, the better understanding of the DDC mechanism and STF and other test results will potentially make it possible to anticipate DDC problems before they occur in actual production and to develop strategies to avoid DDC.

The combination of STF results with exhaustive microstructural and mechanical (microstrain) characterization has permitted DDC to be better described as a grain boundary sliding, creep-like phenomenon. The ductility-dip trough is limited at low temperatures by the inoperability of GB sliding and at high temperatures by the onset of dynamic recrystallization. The ductility-dip crack orientation dependence to the applied load has been verified and correlated with the GB sliding phenomenon. GB triple points and any other irregularities along the GB, such as intergranular precipitates, GB directional changes or GB steps will oppose GB sliding, causing local strain concentration. These strain concentration sites along the GBs will be void initiation sites. If the temperature and strain conditions are appropriate, the growth and linking of these voids causes microcrack formation, resulting in DDC.

The effective GB pinning caused by the medium size (1 μm) (NbTi)C carbides in the two heats of FM-82, caused the formation of tortuous GBs in the weld metals of these heats. This GB tortuosity has a mechanical locking effect on the GBs, limiting GB sliding. The reduced GB sliding caused a reduction of strain accumulation at the triple points and a reduction in the void formation at these sites. The limited GB pinning effect of the sporadic TiN-like precipitates observed on FM-52 weld metal did not promote GB tortuosity and facilitated GB sliding and the consequent strain concentration and void formation at triple points. These voids progressed to form microcracks, which resulted in DDC.

The resultant beneficial or detrimental effect of the intergranular precipitates on DDC resistance is determined by the balance between their effect on preventing GB sliding and nucleating voids along the boundary. The effect of the precipitates is defined by their type, size, distribution and time of precipitation. The importance and specific effect of each one of these factors on DDC resistance is not totally clear and requires further research. However, the results presented here clearly indicate that the improvement of DDC resistance by GB tortuosity and presence of medium size intergranular (NbTi)C observed on FM-82 were more effective than the numerous small intergranular $M_{23}C_6$ carbides observed on FM-52. This DDC resistance improvement observed in the FM-82 is due to the balanced effect of GB sliding control caused by both the intergranular precipitates and the GB tortuosity. The poor DDC resistance exhibited by FM-52 is thought to be due to the rapid formation of voids and microcracks due to the straight GBs and the large number of $M_{23}C_6$ carbides along the GBs.

Another aspect that has an important effect on DDC is the segregation of impurities to the GBs. Sulfur had been shown to impair DDC resistance and the observation of S rich films along migrated grain boundaries in FM-52 is thought to be one of the factors responsible for the low DDC resistance of this alloy. Finally, hydrogen migration to the GBs and the intergranular precipitate/matrix interfaces appears to have an important influence on DDC susceptibility.

Acknowledgments

The authors would like to acknowledge BWX Technologies, Inc. for providing partial funding for this research and the members of the Welding and Joining Metallurgy Group at The Ohio State University, especially Matt Collins, Nathan Nissley, Gustavo Guaytima, and Jeff Sowards for their assistance and insightful discussions.

References

1. Rhines FN, Wray PJ (1961) Trans ASM 54: 117–128
2. Nissley NE, Collins MG, Guaytima G, Lippold JC (2002) (IIW Doc. IX-2050-02) accepted for publication in Welding the World
3. Mintz B, Yue S, Jonas JJ (1991) Inter Mater Rev 36, 5: 187–217
4. Collins MG, Ramirez AJ, Lippold JC (2003) Weld J 83 (2), Part III: 39s–49s
5. Nissley NE (2002) Development of the strain-to-fracture test to study ductility dip cracking in austenitic alloys. Masters Thesis, The Ohio State University, Columbus, OH, USA, p 104
6. Nissley NE, Lippold JC (2003) Weld J 82(12): 355s–364s
7. Collins MG (2002) An investigation of ductility dip cracking in nickel-base filler materials. Master Thesis, The Ohio State University, Columbus, OH, USA, p 240
8. Collins MG, Ramirez AJ, Lippold JC (2003) Weld J 82(12), Part II: 348s–354s
9. Collins MG, Lippold JC (2003) Weld J 82(10), Part I: 288s–295s
10. Feng Z (1993) A methodology for quantifying the thermal and mechanical conditions for weld metal solidification cracking. The Ohio state University, PhD Dissertation, Columbus, OH, USA, p 234
11. Ramirez AJ, Lippold JC (2004) Mater Sci Eng A Part I: in press
12. Ramirez AJ, Lippold JC (2002) Internal Research Report, The Ohio State University, Columbus-OH, pp 1–35
13. Lippold JC, Clark WAT, Tumuluru M (1992) In: Cieslak MJ, Perepezko JH, Kang S, Glicksman MW (Eds.) The Metal Science of Joining. The Metals, Minerals and Materials Society, Warrendale, PA, pp 141–146
14. Mintz B, Yue S, Jonas JJ (1991) Inter Mater Rev 36, 5: 187–217
15. Cottrell AH (1961) In: Structural processes in creep. Report of a symposium organized by the Iron and Steel Institute and the Institute of Metals, London, 3 and 4 May 1961, The Iron Steel Institute, London, pp 1–55

16. Evans RW, Wilshire B (1993) In: Introduction to Creep. The institute of Materials, London, UK, p 115
17. Kihara S, Newkirk JB, Ohtomo A, Saiga Y (1980) Metall Trans A, 11: 1019–1031
18. Ramirez AJ, Lippold JC (2004) Mater Sci Eng A, Part II: in press
19. DuPont JN, Robino CV, Marder AR (1998) Weld J 77: 417s–431s
20. Johnson M, Fiore S, Lippold JC (2002) Evaluation of EN82H Filler Metal and shielding Gas. EWI Project 45371CSP, EWI-OSU Final Report, Columbus-OH, USA, June, 2002, pp 1–2
21. Lim LC, Raj R (1984) Acta Metall 32: 1183–1190
22. Lehockey EM, Palumbo G (1997) Mater Sci Eng 237A: 168–172
23. Shinoda T, Zaghloul MB, Kondo Y, Tanaka R (1978) Trans ISIJ 18: 139–148
24. Kotval PS, Venables JD, Calder RW (1972) Metall Trans 3: 453–458
25. Etienne CF, Rossum OV, Roode F (1980) In: Proceedings of the International Conference Engineering Aspects of Creep, Sheffield, England, 15–19 Sept., 1980, vol. II, The Institution of Mechanical Engineers, London, pp 113–121
26. Mathew MD, Sasikala G, Mannan SL, Rodriguez P (1993) Trans of the ASME. J of Eng Mat and Technol 115: 163–170
27. Eberhart ME, Latanision RM, Johnson KH (1985) Acta Metall 33: 1769–1783
28. Tang S, Freeman AJ (1993) Physical Review B 47, 5: 2441–2445
29. Nishimoto K, Mori H, Hirata H (2001) In: Proceedings of the Today and Tomorrow in Science and Technology of Welding and Joining. 7th JWS International Symposium, Kobe, Japan, 20-22 Nov, 2001, Japan Welding Society, vol 2, pp 827–832

Influence of Welding Speed on the Hot Cracking Resistance of the Nickel-Base Alloy NiCr25FeAlY During TIG-Welding

A. Slyvinsky, H. Herold, M. Streitenberger

Institute of Joining and Beam Technology, Otto-von-Guericke-University, Magdeburg, Germany

Introduction

Efficiency, availability and operation limits of heat treating equipment and industrial furnaces are often determined by the thermal load capacity of the materials used. Within the range of nickel-base alloys, nickel-base high-temperature materials form a group especially developed to meet the ever increasing demands that arise in the construction of industrial furnaces and heat-treating equipment.

In plant construction, the joining technology most often used for such high-temperature materials is fusion welding. Here, of course, the emphasis is placed on the production of high quality welded joints which guarantee a useful life-time as well as retaining heat- and corrosion resistance.

One of the most often encountered difficulties in welding nickel-base alloys is their hot cracking sensitivity due to thermically determined and material-specific structural changes. Very often, these disadvantageous material disconnections in the weld metal structure cannot be removed without difficulties in the production process. On the other hand, as the hot cracking resistance of welded seams is clearly dependent on the crystallization character of the weld pool, there is, independently of the phase precipitation and the micro-segregation process, a chance to achieve a reduction of the proportion of directed growth of dendritic structures by optimizing the welding parameters, mainly the welding speed.

Use and Structural Characteristics of the High-Temperature Resistant Nickel-Base Alloy NiCr25FeAlY

The demand for operation temperatures up to 1200 °C with adequate life-time and corrosion resistance of the equipment is not exceptional today.

Thus, the alloy NiCr25FeAlY (Nicrofer 6025 HT, alloy 602 CA, W.-Nr. 2.4633), available in all semi-finished products since 1992, has shown an outstanding useful life-time and heat- and corrosion resistance up to 1200 °C in practical use [1, 2].

The nominal chemical composition of the material is shown in Table 1.

Table 1. Nominal chemical composition of NiCr25FeAlY, % [3]

	Ni	Cr	Fe	C	Mn	Si	Cu	Al	Ti	Y	Zr
min	bal.	24.0	8.0	0.15				1.8	0.1	0.05	0.01
max		26.0	11.0	0.25	0.1	0.5	0.1	2.4	0.2	0.12	0.10

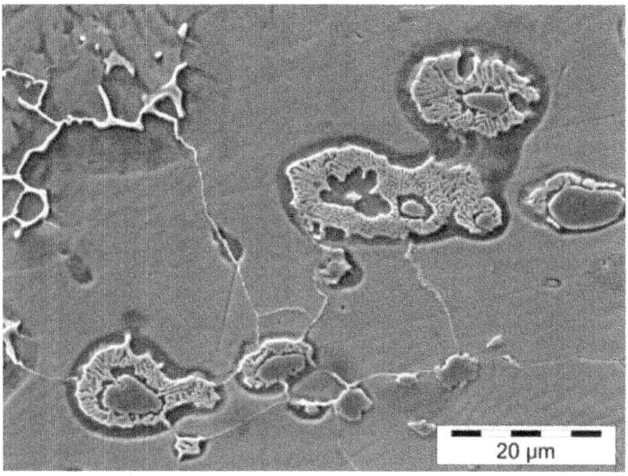

Fig. 1. Microstructure in the HAZ of a TIG -weld of the alloy NiCr25FeAlY. In the HAZ, the Cr-carbides are in solution; from them, eutectic clusters have formed. The grain boundaries towards the fusion line (top left) are greatly covered by a carbide layer; REM photo

With NiCr25FeAlY, carbide hardening with the evenly distributed primarily precipitated carbides $M_{23}C_6$ with M = (Cr, Ni, Fe) was, for the first time, put into practice on a large scale for workable nickel-base alloys [4]. The material owns a complex range of phases, covering, in the initial state, γ-solid solutions and Cr-carbides of different morphology [5, 6]. This structure is locally changed by the welding heat input. The metallographic investigations of the welded joints of NiCr25FeAlY produced by different welding procedures (TIG-, gas metal arc-, submerged arc- and manual electric welding) clearly point to a development of local eutectic fields in the overheated zone of the HAZ (Fig. 1) as well as to numerous filmy layers of carbides on grain interfacial areas in the HAZ and on dendrite interfacial areas in the weld metal (Fig. 2).

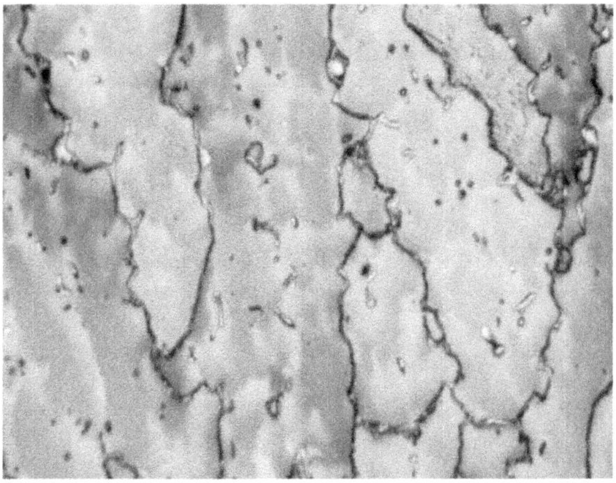

Fig. 2. Microstructure of the weld metal deposit in a TIG-welded joint of the alloy NiCr25FeAlY. Noticeable are the strong poly-crystalline layers of Cr-carbides on dendrite interfacial boundaries. Etching method: Bloch und Wedl II. Scale 625:1

The structural changes shown here cannot be avoided by optimizing the welding technology. They are typical of all welding procedures used for nickel-base alloy NiCr25FeAlY, and they are connected with hot cracking.

Determination of the Hot Cracking Sensitivity

The Institute of Joining- and Beam Technology of the Otto von Guericke University, Magdeburg, (IFST), has the equipment for a PVR-Test (Programmed Deformation Rate Cracking Test) which is used to determine in quantity the hot cracking sensitivity of the material NiCR25FeAlY and its weld using selected welding technology. The principle of the test (Fig. 3) is to produce a bead-on-plate weld on a flat tensile specimen simultaneously with a tensile load of linearly increasing speed in welding direction.

In the test, a "critical tension speed" is defined as a criterion of hot cracking resistance. This is to be understood as the speed at which the first hot crack in the raw welding bead can be detected under at least tenfold magnification.

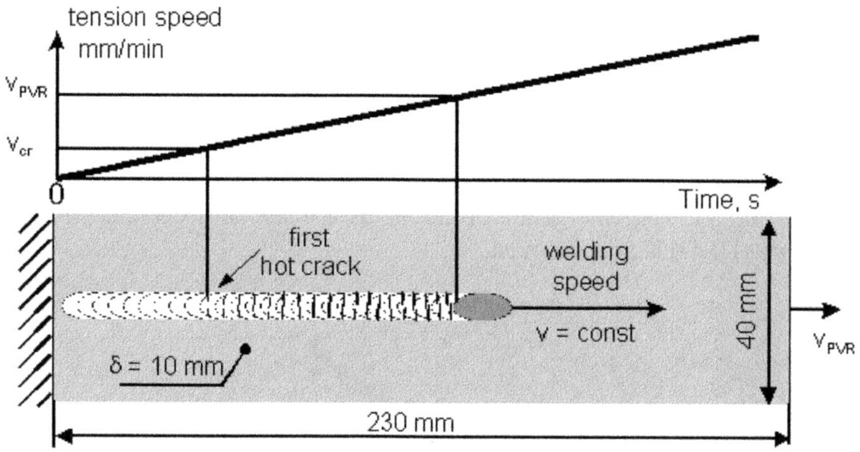

Fig. 3. Diagram of the PVR-Test

In the PVR-Test of the nickel-base alloy NiCr25FeAlY, TIG-welding without filler metal was employed. Compared to gas metal arc-, submerged arc- and manual electric welding, the TIG-welding procedure can guarantee a high weld quality with, at the same time, an as low as possible heat input, especially if, according to the regulations [3], a "hot-cracking resistant" gas mixture Ar + 1 % N_2 is used as shielding gas for the production of the test welds.

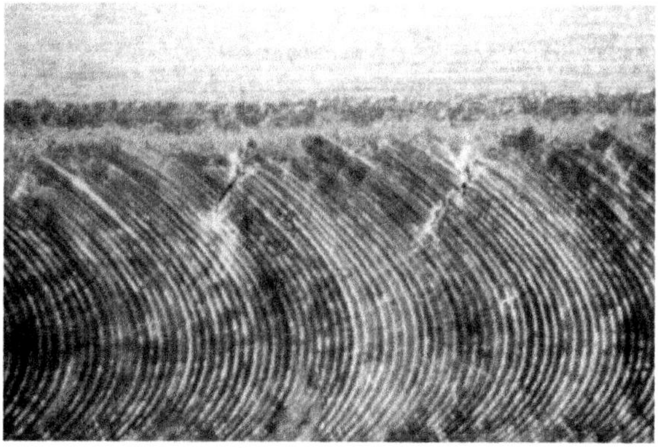

Fig. 4. First solidification cracks in the weld deposit of a PVR-specimen of the material NiCr25FeAlY. Scale 10:1

The first crack to be registered on this material is always an inter-dendritic solidification crack (SC) in the bead-on-plate weld (Figs. 4, 5). At higher tensile speeds, liquation cracks (LC) in the HAZ at the end of the specimen are to be observed (Fig. 6). Their character is remarkable as they solely form in direct combination with the solidification cracks as the continuation of which they are to be considered in the overheated zone of the HAZ (Fig. 7). The emergence of cracks as a result of a decrease in weld ductility (DDC) is not observed.

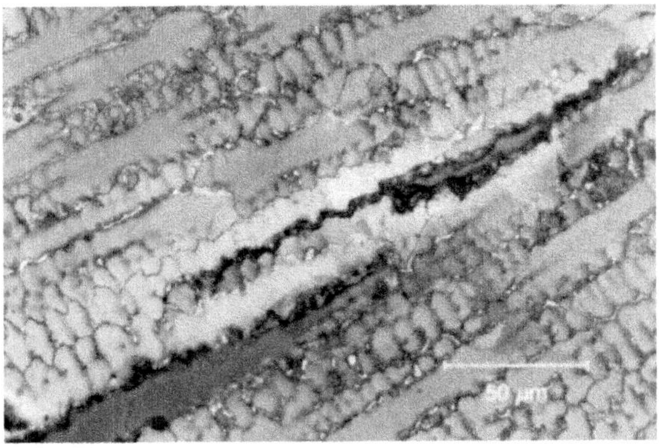

Fig. 5. Inter-dendritic solidification cracks in the weld deposit of a PVR-specimen of the material NiCr25FeAlY; Etching method: Bloech und Wedl II

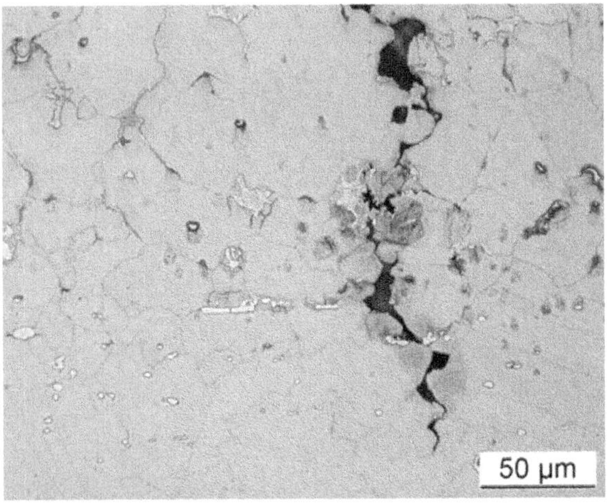

Fig. 6. Liquation crack in the HAZ of a PVR specimen of the material NICr25FeAlY

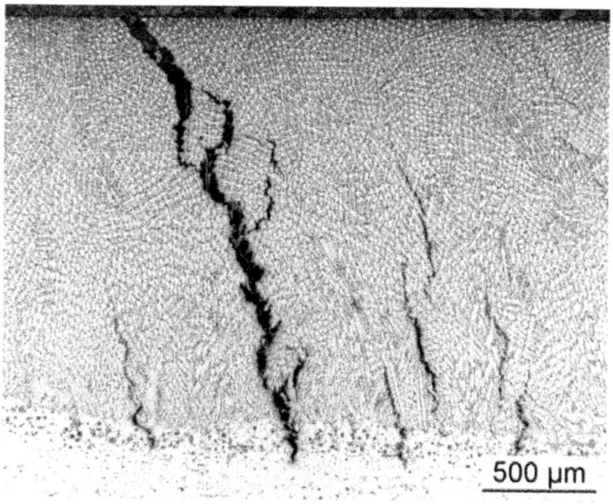

Fig. 7. Inter-dendritic solidification cracks in the weld deposit and liquation cracks in the HAZ of a PVR specimen of the material NiCr25FeAlY

From the assessment of the hot crack types produced in the PVR-Test it follows that the inter-dendritic solidification cracks in the weld deposit appear to be the most immanent danger in arc welding sheets of NiCr25FeAlY of medium thickness. This form of solidification cracks is, in general, not typical of the materials group of "nickel-base super-alloys", as these rather tend to form re-flow cracks in the HAZ, according to the mechanism of the so-called "constitutional re-flow" [7]. To assess the welding-technological influence on this phenomenon, the energy input per unit length of weld was varied by different welding speeds during a PVR-Test sequence (see Table 2).

The evaluated test results are shown in Fig. 8.

Table 2. PVR-Test. Assignment of parameters to the test variants

variant No.	v [cm/min]	I [A]	U [V]	E [kJ/cm]	L_C [mm]	$v_{cr.}$ [mm/min]
1	22.2	180	12.0	5.8	200.5	51.8
2	19.1	180	12.0	6.8	159.5	40.0
3	15.5	180	12.0	8.4	160.5	32.4
4	10.6	180	12.0	12.2	105.5	20.8

v welding speed, I welding current, U welding voltage, E energy input per unit length of weld, L_C distance up to the first hot crack, v_{cr} critical tension speed.

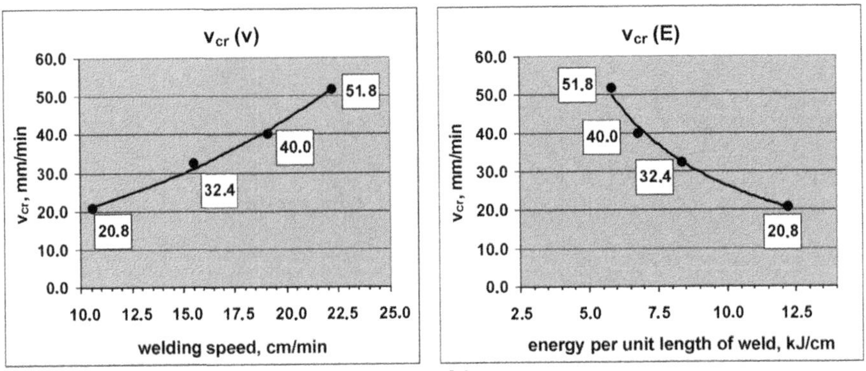

a) **b)**

Fig. 8. Dependence of hot crack resistance of the material NiCr25FeAlY on the welding speed **(a)** and the energy input per unit length of weld **(b)** in TIG-welding (determined by the PVR-Test)

The critical tension speed is directly proportional to the welding speed, and indirectly proportional to the energy input per unit length of weld, i..e. the hot cracking susceptibility decreases with the increase in welding speed under constant welding current and -voltage. In paper [8] it was stated on the basis of numerous experiments with highly alloyed Cr-Ni-materials that, with alloys susceptible to solidification cracking, the critical tension speed lies at around 30 mm/min. A high resistance to solidification cracking is shown by a weld metal deposit with a critical tension speed of around 70 mm/min and above. According to this classification, and in view of the test results, the material NiCr25FeAlY offers a resistance to hot cracking, dependent on the energy input per unit length and the welding speed, in TIG-welding with the shielding gas mixture Ar + 1 % N_2. This alloy turns out to be rather susceptible to hot cracking when welded at a welding speed of below ca. 15.4 cm/min which corresponds to an energy input per unit length of weld of about 8.4 kJ/cm.

Such an influence of welding speed on the hot cracking resistance is not equally typical of all austenitic high-alloy materials. Thus, e.g., a contrary effect of welding speed was pointed out for TIG- and submerged-arc welding of fully austenitic chromium-nickel- and chromium-nickel-molybdenum steels [9–11]. Here, low welding speeds (up to 10 cm/min), with other welding parameters being constant, have proved to be favorable for decreasing the hot cracking susceptibility , due to the type of primary crystallization. In papers [12, 13], a drastic decrease in the plasticity capacity of deposited weld metal deposits was observed in parallel with an increase of welding speed and simultaneous maintenance of a constant weld seam geometry. The influence of welding speed on the hot cracking susceptibility in laser welding is shown in [14–16] by the example of the alloy Inconel 718. Here, the hot cracking susceptibility increases with increasing welding speed. This effect is due to the cross-sectional weld geometry [14–16]. Investigations on electron-beam welded Inconel 903 showed similar results, and a decrease in the hot cracking susceptibility by decreasing the welding speed could be observed at the same time [17].

By contrast, the experiments with low-alloy steel described in [18] showed an effective influence of increased welding speed with constant welding current and welding voltage on the solidification crack resistance, which correlates with the above mentioned results of the PVR-Test.

It must be pointed out that the above mentioned papers which describe the direct as well as the indirect proportionality of welding speed and hot crack resistance always relate to the observed influence of selected welding parameters (welding speed) on the crystallization conditions of actual weld metal deposits. Therefore, the phenomenon of the influence of welding speed on the crystallization conditions in the weld metal deposit and thereby on the hot cracking resistance of the alloy NiCr25FeAlY shall be further clarified. For this reason, the solidification of the weld pool and the influence of the welding speed on the crystallization rate and the microstructure formation will be discussed in the following.

Calculation of the Mathematical Correlation Between the Crystallization Rate in the Weld Pool and the Welding Speed

Within the range of selected parameters for TIG-welding (Table 2), low values of welding current and voltage guarantee a partly constricted arc and an elliptical shape of the weld pool. At the same time, the crystallization conditions in the weld pool are influenced by changing the welding speed.

The welding speed indicates the level of thermal under-cooling of the molten metal in the tail of the weld pool and turns out to be the most important parameter that clearly influences the form and the size of the developing primary dendrites.

From theory [18] we know that, at selected points of the weld pool isotherms, the crystallization rate R can, for simple reasons of geometry, be defined by the following empirical equation:

$$R = v \ cos\beta, \qquad (1)$$

where β is the angle between the respective molten surface normal and the direction of the welding speed v.

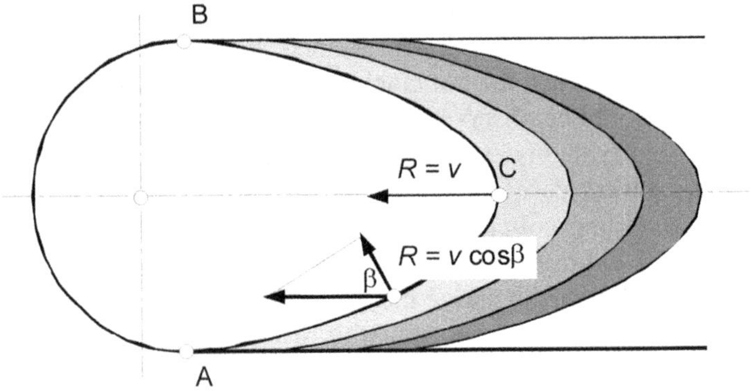

Fig. 9. Influence of the welding speed v on the crystallization rate R at selected points of the weld pool isotherms, after Savage [12]

Eq. 1 was arrived at on the basis of numerous experimental investigations and is beyond any doubt. The calculation of the crystallization rate according to Eq. 1, however, causes difficulties as it necessitates experimental finding of the angle β.

As a proportionality between the crystallization rate of the molten metal in the weld pool R and the cooling rate W assumed, another analytic term of the crystallization rate can be formulated (crystallization rate is here understood as the travel speed of the phase interface in macro-volumes):

$$R = k\, W, \tag{2}$$

where k is to be understood as a proportionality factor. For defining the cooling rate W at any point of the phase interface in the weld pool, a functional-analytic model simplification according to Rykalin can be employed for the quickly moving high-power line-source in a plate [19, 20]. The theorist takes interest in the quickly moving high-power sources for reasons of the possible simplification of the equation. The assumed quantities $q = I\,U$ (heat power) and v (speed of the heat source) are increased at possibly constant q/v (energy input per unit length):

$$q \to \infty,\ v \to \infty,\ q/v = const.$$

Apart from this, the selected scheme for the model simplification is especially suitable for calculating the temperature fields in areas near the weld [20]. With the linear heat source, the heat is evenly distributed throughout the thickness of the plate. The heat mainly spreads perpendicular to the travel direction of the heat source. The spread in travel direction is negligibly small. According to the model selected, the increase in temperature $T - T_0$ can be calculated dependent from the transverse direction y and the time t (according to Eq. 83 from [19]):

$$T - T_0 = \frac{q}{v\delta\sqrt{4\pi\lambda c\rho t}} e^{-\left(y^2/4at+bt\right)}. \tag{3}$$

In this is:

q – heat power of the source, [J/s];
v – feed rate of the heat source, [mm/s];
δ – thickness of the plate, [mm];
λ – coefficient of thermal conductivity, [J/mm s K];
c – specific heat capacity, [J/g K];
ρ – density, [g/mm³];
a – thermal conductivity, [mm²/s];
b – heat transfer coefficient at the plate, [1/s].

The heat transfer coefficient b is used for assessing the amount of heat emitted by the plate surface to the surrounding medium. In the effective range of the heat source, this coefficient is zero ($b = 0$). From this follows for the equation of the melting temperature at any point of the interface in the weld pool:

$$T_s = \frac{q}{v\delta\sqrt{4\pi\lambda c\rho t}} e^{-y^2/4at}. \tag{4}$$

In logarithmical form:

$$\ln T_s = \ln\frac{q}{v\delta\sqrt{4\pi\lambda c\rho}} - \frac{1}{2}\ln t - \frac{y^2}{4at}, \text{ or} \tag{5}$$

$$\frac{y^2}{4at} = \ln\frac{q}{v\delta T_s\sqrt{4\pi\lambda c\rho}} - \frac{1}{2}\ln t. \tag{5.1}$$

From Eq. 4 it follows that the time of existence of the weld pool ($y = 0$) is described by the following equation:

$$t_s = \frac{q^2}{v^2\delta^2 T_s^2 4\pi\lambda c\rho}. \tag{6}$$

By substitution of t_s after Eq. 6 in Eq. 5.1, the equation for the weld pool surface appears as follows:

$$y^2 = 2at \ln \frac{t_s}{t}, \tag{7}$$

with $0 \le t \le t_s$.

By differentiation of Eq. 5, the speed of temperature change at any point of the weld pool surface is arrived at:

$$\frac{\partial T}{\partial t} = \frac{T_s}{2t}\left(\frac{y^2}{2at} - 1\right). \tag{8}$$

In Eq. 8, a dependence of the speed of temperature change at the points at the weld pool surface on the interconnected parameters t and y was defined. By substitution of Eq. 7 in Eq. 8 we get:

$$\frac{\partial T}{\partial t} = \frac{T_s}{2t} \ln \frac{t_s}{te}. \tag{9}$$

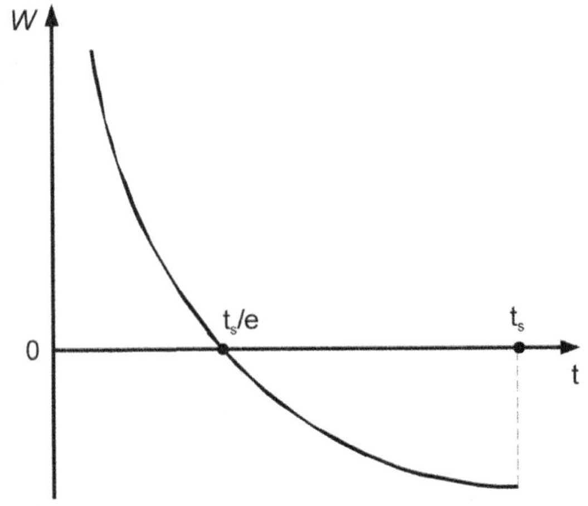

Fig. 10. Speed of temperature change (W=∂T/∂t) for points at the weld pool surface

The functional dependence of Eq. 9 is shown in Fig. 10. Obviously, with $0 \le t \le t_s/e$, the head part of the weld pool is considered ($\partial T/\partial t > 0$). For the tail part of the weld pool ($\partial T/\partial t < 0$) applies $t_s/e \le t \le t_s$, so that Eq. 9 with $t_s/e \le t \le t_s$ defines the cooling rate of the molten metal in the weld pool:

$$W = \frac{T_s}{2t} \ln \frac{t_s}{te}. \tag{10}$$

Point t_s in Fig. 10 corresponds with point C in Fig. 9. After equalization of Eqs. 2 and 10 for these boundary conditions ($t = t_s$; $R = v$), we get the new analytic term of the crystallization rate of the weld pool:

$$R = -\, v \, \frac{t_s}{t} \ln \frac{t_s}{te}, \qquad \text{with} \qquad \frac{t_s}{e} \leq t \leq t_s. \tag{11}$$

Eq. 11 is valid for any point in the tail end of the weld pool area (line B-C-A in Fig. 9).

This equation can be transformed into the Cartesian coordinate system. Under consideration of $t = x / v$ and correspondingly $t_s = L_s / v$ we arrive at:

$$R = -\, v \, \frac{L_s}{x} \ln \frac{L_s}{xe}. \tag{12}$$

In this are:

L_s – length of the weld pool;
x – linear coordinate of the point in the weld pool surface in longitudinal direction from the line-source (at the tail end of the weld pool surface $L_s/e \leq x \leq L_s$).

Discussion

In contrast to the empirical equation after Savage (1), the deduced formulae (11) and (12) consider a number of characteristic particularities of primary crystallization of the weld metal deposit which were described in technical literature many years ago. Still in the sixties of the twentieth century, practical investigations on weld seams [21, 22] showed that solidification often does not correspond with the "conditions of orthogonal growth". This means that the axes of the dendrites which develop when the weld pool solidifies are not necessarily in orthogonal direction to the isotherm interfaces. If the direction of growth deviates from the orthogonal, the crystallization rate of these dendrites increases. This will lead to several dendrites rushing ahead and with this, a local disturbance of the otherwise constant isotherm interfaces will arise [22]. For this case, the following alteration of Eq. 1 was made by Wittke [21]:

$$R = v\frac{\cos\beta}{\cos\Delta\beta}. \tag{13}$$

Here, $\Delta\beta$ is an arithmetical difference between the "real" and the "orthogonal" angle β. Unfortunately, the practical use of this equation has turned out to be problematic due to the necessity to determine both the angle β and the deviation angle $\Delta\beta$ by experiment. Besides, with very high heat input and at very high welding speed (and consequently at very high crystallization rate), a disoriented, fine-grain structure with equiaxial dendrites may emerge along the weld seam center line by heterogeneous nucleus formation [12]. In this case, the determination of the crystallization rate in the central area of the weld seam cannot be accomplished by the empirical Eq. 1. Accordingly, Eq. 1 after Savage is valid only in the special case of orthogonal and epitaxial conditions of growth.

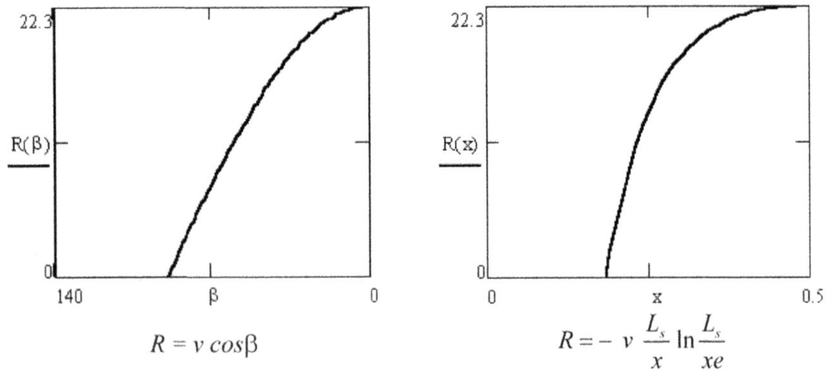

Fig. 11. Comparison of the traces of curves $R = f(\beta)$ after Eq. 1 and $R = f(x)$ after Eq. 12 (calculated for the welding speed v = 22.2 cm/min)

As in Eq. 1, a dependence of crystallization rate R from the feed rate of the heat source (welding speed) v was made evident also in the Eqs. 11 and 12. Nevertheless, here the factor of proportionality is a transcendental function. Its value is variable either from 0 with $x = L_s/e$ to –1 with $x = L_s$ (Eq. 12) or from 0 with $t = t_s/e$ to –1 with $t = t_s$ (Eq. 11). Compared to Eq. 1, the analytical Eqs. 11 and 12 consider not only the influence of welding speed v on the crystallization rate of the weld pool R, but also the influence of welding processes and welding parameters. This is achieved by considering the parameters t_s (L_s) dependent on welding procedures and welding data.

From the analysis of the Eqs. 11 and 12 follows that the flanks of the weld pool as well as the areas near the point of deepest penetration in the weld pool crystallize at a low rate. Thereby, smallest alteration of the coordinate at the considered point on the weld pool surface causes a considerable change of this rate (Fig. 11, right).

In contrast to the cosine curve after Eq. 1, Figs. 11 and 12 show that, not only in point C but also in the far end of the weld pool tail (curve C – C', Fig. 12), the amounts of crystallization rate and of the welding speed strongly approach to each other ($R \cong v$). That is the reason for special solidification conditions in the tail of the weld pool (zone C - C'), so that the dendrites can grow at any angle to the welding heat dissipation surface.

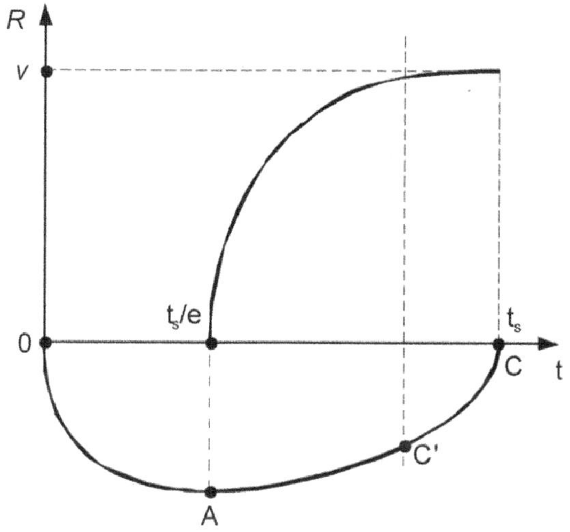

Fig. 12. Alteration in the time of the crystallization rate *R* in the tail end of the weld pool

Fig. 13 shows the macro-structure of the welded seams that correspond with the variants of welding parameters 1 (Fig. 13a) and 4 (Fig. 13b) from Table 1. The macro-structure shows a mosaic-like character which pleads for continuous nucleus formation during the dendritic growth. As follows from Fig. 13b, a three-dimensional crystallization character (dendrites growing with increasing contortion in welding direction) is typical of TIG-welding NiCr25FeAlY. With an increase in welding speed *v*, the area with fine-grain equiaxial dendrites in the center of the weld also enlarges (Fig. 13a). The building up of such a dendrite form in the center of the weld can also increase the plasticity reserve of the weld metal deposit and contribute to an accelerated formation of hot cracks.

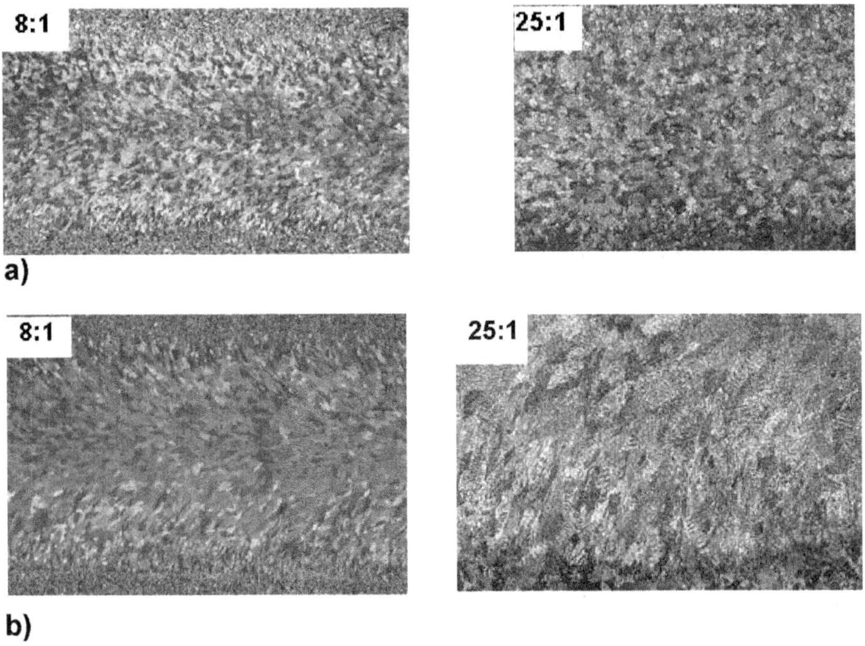

Fig. 13. Influence of welding speed on the weld microstructure in TIG-welding of nickel-base alloy NiCr25FeAlY: **(a)** $v = 22.2$ cm/min, **(b)** $v = 10.6$ cm/min

Conclusions

In the paper presented, the influence of the welding speed on the hot cracking resistance of TIG-welds of nickel-base alloy NiCr25FeAlY was investigated. Increased hot cracking resistance at a higher welding speed and constant welding current and voltage was observed.

The material under investigation shows a tendency to form interdendritic solidification cracks dependent on the energy per unit length and on the welding speed.

The experimental results of the PVR-Test allow it to select such welding speeds for TIG-welding that minimize the hot cracking sensitivity. Therefore, an energy amount of 8.4 kJ/cm at a welding speed of 15.5 cm/min should not be exceeded.

An analytic expression for the calculation of the crystallization rate for any point in the tail part of a weld pool was deduced which explains the local influence of crystallization rate and of the welding speed on the hot cracking sensitivity. This correlation is ascertained by the results of the PVR-Test.

The way in which the crystallization rate throughout the weld pool surface was changed gives reason for assuming the existence of a zone with a higher crystallization rate and, resulting from this, a quicker crystallization without diffusion in the tail zone of the weld pool. This would explain an increase in the proportion of equiaxial dendrites in the weld center area, the existence of which improves the hot cracking resistance of alloy NiCr25FeAlY due to the increase in welding speed.

References

1. Brill U (1995) Practical experience with the new material Nicrofer 6025 HT in furnace- and heat treatment plant construction (in German). Stahl 6: 37–40
2. Sölch R, Ammann T, Hoffmann T (2002) Use and welding processing of the high-temperature nickel-base material Nicrofer 6025 HT (W.-Nr. 2.4633) (in German). DVS-Berichte 219: 59–65
3. N. N. Nicrofer 6025 HT – alloy 602 CA. Material Data Sheet No. 4037, (edn) October 2002
4. N. N. High-temperature alloys from Krupp VDM for industrial engineering. Krupp VDM Report No. 25, (edn) September 1999
5. Slyvinsky A (2002) Structural characteristics and welding suitability of the high-temperature nickel-base alloy NiChr25FeAlY (in German). In: Proceedings of XXVII. Assistentenseminar „Fügetechnik/Schweißtechnik“, TU Chemnitz, Oberwiesenthal, pp 145–152
6. Slyvinsky A, Veit P (2003) Structure and properties of the welded joints of a high-temperature nickel-base alloy (in Russian). Aut Wdg 5: 7–13
7. Schulze G, Krafka H, Neumann P (1996) Welding: Materials – Design – Test (in German). VDI, Düsseldorf
8. Klug P (1980) A contribution to the testing of hot-crack resistance of high-alloy welding filler materials with the PVR-test of VEW-Kapfenberg (in German). PhD thesis, TU Graz
9. Kahovsky NI (1975) Welding of Stainless Steels (in Russian). Technika, Kiev
10. Fartushny VG, Kahovsky YuN, Savchenko VS, Demyanenko GP, Nastenko GF (1974) Increase of the hot-crack resistance of austenitic weld metals (in Russian). Aut Wdg 5: 39–43
11. Slyvinsky AM, Kahovsky YuN, Nastenko GF (1976) Influence of welding speed on the crystallisation conditions in the weld pool (in Russian). Aut Wdg 8: 6–8
12. Yakushin BF (1969) Assessment of "Technological Strength" dependent from welding parameters (in Russian). Wdg Prod 1: 7–9

13. Rybakov AA, Mandelberg SL, Shitova LG, Kireeva TS (1980) Influence of welding speed on the primary solidification of the weld deposit of low-alloy steels (in Russian). Aut Wdg 10: 15–18
14. Fontana G, Gobbi S, Rivela C, Zhang L (1999) Laser welding in the manufacture of superalloy components. Wdg International 8: 631–635
15. Shinozaki K, Kuroki H, Luo X, Ariyoshi H, Shirai M (1999) Effects of welding parameters on laser weldability of Inconel 718. Study of laser weldability of Ni-base, heat-resistant superalloys (1st Report). Wdg International 12: 945–951
16. Shinozaki K, Luo X (1999) Phenomenon and Mechanism of Weld Cracking of Ni-base Superalloys During Laser Welding. J of the Jap Wdg Soc 7: 22–26
17. Richards NL, Nakkalil R, Chaturvedi MC (1994) The Influence of Electron-Beam Welding Parameters on Heat-Affected-Zone Microfissuring in INCOLOY 903. Metall and Mater Trans 25A 8: 1733–1745
18. Rybakov AA, Mandelberg SL (1980) The influence of arc welding parameters on the formation of solidification cracks in weld seams of low-alloy pipe steels (in Russian). Aut Wdg 3: 12–15
19. Savage WF (1980) Solidification, segregation and weld imperfections. Wdg in the world 5–6: 89–113
20. Rykalin NN (1957) Calculation of thermal processes in welding (in German). VEB Technik, Berlin
21. Radaj D (1988) Heat effects of welding: temperature field, residual stresses, distortion (in German). Springer, Berlin Heidelberg New York London Paris Tokyo
22. Wittke K (1966) Characteristics of the primary crystallisation of the weld metal (in German). Schweißtechnik (Berlin) 6: 289–292
23. Wittke K (1968) Modelling of the primary crystallisation by fusion welding (in German). Schweißtechnik (Berlin) 7: 295–299

The Role of Segregation of Oxygen in Welding Alloys of the INVAR Type

K.A. Yushchenko, V.S. Savchenko, T.M. Starushchenko

E.O. Paton Electric Welding Institute, Kyiv, Ukraine

Special thermal-physical properties of INVAR alloys (Fe-36 %Ni) combined with high ductility and toughness, especially at low temperatures, enabled their extensive application in a number of engineering fields. Chemical composition of one of the alloys of the INVAR series is given in Table 1.

Table 1. Chemical composition of alloy of the 36N type

Content of elements, wt. %								
C	Si	Mn	Ni	Cr	S	P	[O]	[H]
0.05	0.1	0.35	35.6	0.45	0.004	0.006	0.002	0.00015

As shown by investigations, welds in these alloys are very sensitive to hot cracking. Reportedly [1], the cracks are formed in two temperature ranges, including temperatures below 1100 °C, i.e. at temperatures below T_s (solidus temperature). They are located along the grain boundaries and formed during cooling of the weld metal (Fig. 1).

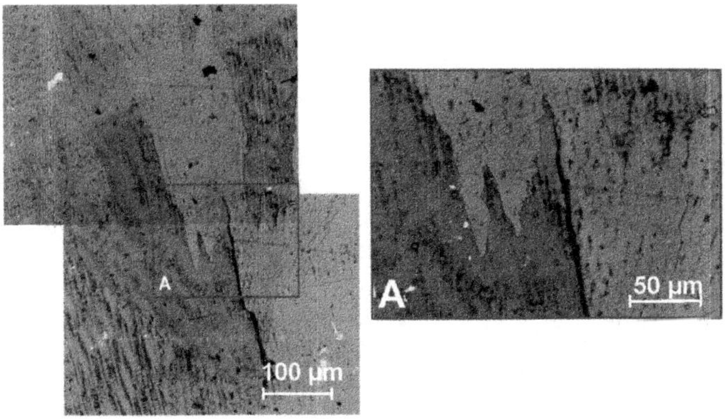

Fig. 1. Under-bead hot crack formed in multi-layer welding of alloy 36N

Low values of the concentration of sulphur and phosphorus do not prevent hot cracking of the INVAR type alloys during welding. Investigations show that formation of the said type of hot cracks can be initiated by oxygen [1, 2]. Studies of high-temperature ductility of alloys and welds of the INVAR composition indicate (Fig. 2) that metal loses ductility in a temperature range of 600–1100 °C at a selected rate of plastic deformation equal to 0.043 s^{-1}. The higher the oxygen content of a material, the lower its ductility.

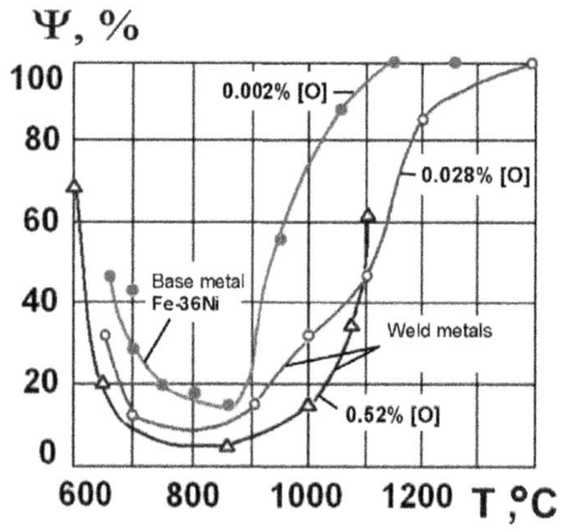

Fig. 2. Temperature dependence of ductility of base and weld metals

Presence of mostly the brittle intergranular fracture with traces of microplastic deformation can be seen in fractograph of the fracture surface of the alloy at a temperature of 725 °C (Fig. 3).

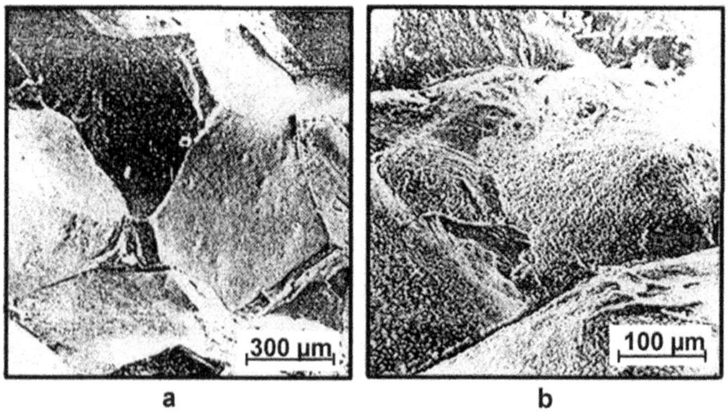

Fig. 3. Fractograph of the fracture surface of specimens in alloy 36NKh at 725 °C

It was suggested that the probable cause of hot cracking within a temperature range of 600–1100 °C (DDC) could be a non-equilibrium diffusion of atoms of the impurity elements (e.g. oxygen) contained in solid solution at the grain boundaries. A decisive factor in this case, which determines the rate of diffusion, is a high-temperature plastic deformation that always takes place in the weld and HAZ metals during welding. Sensitivity to formation of cracks in welding was quantitatively estimated by dynamically deforming the weld metal during the welding process using the testing machine of the Varestraint-Test type [3]. Plates measuring 4 x 150 x 150 mm were used as the test specimens. Welding of the specimens was performed by the TIG method under the following conditions: $I_w = 120$ A, $U_{arc} = 12$ V and $v_w = 10$ m/h, the value of deformation of the surface layers of a plate being kept constant and equal to 1.25 %.

The flow diagram of the cracking tests provides for performing two-layer welding. The first layer is deposited on the base metal to evaluate the properties of a weld. The second layer is deposited to simulate the properties of the multi-layer welds. It includes two one-layer beads, the second of them being deposited on the preliminarily made weld to overlap about 60 % of its surface area. Results of testing a series of the specimens differing in the oxygen content are shown in Fig. 4.

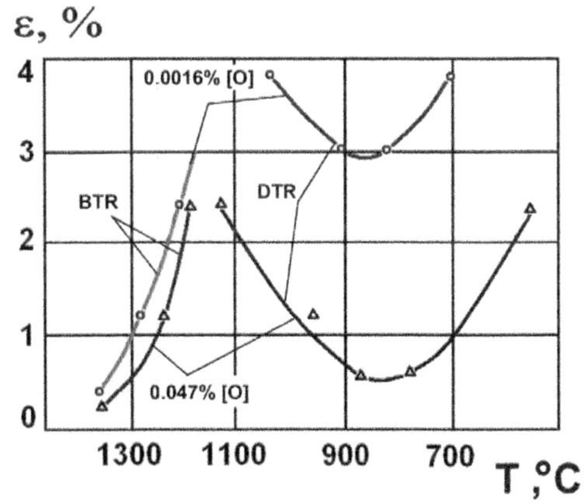

Fig. 4. Effect of oxygen on brittle temperature of alloy Fe- 36Ni

The appearance of the welds after the tests is shown in Fig. 5. Analysis of the results obtained shows that the welds in alloys of the INVAR type are characterized by two ductility-dip ranges, i.e. BTR and DTR, with oxygen affecting the ductility characteristics primarily in the DTR range (Fig. 5 a). It should be noted that the sensitivity to under-bead microcracking is higher than in the weld made during testing (Fig. 5b). It is likely that in this case the total content of oxygen affects the sensitivity to embrittlement of the first weld both during producing it outside the Varestraint-test machine and as a result of reheating, as well as during forced dynamic deformation in testing using the above machine.

The fractographs of the surfaces of the ductility-dip cracks (DDC) in Fig. 6, and the character of propagation of the cracks (Fig. 5b) show that this is the case of brittle intergranular fracture with traces of plastic microdeformation.

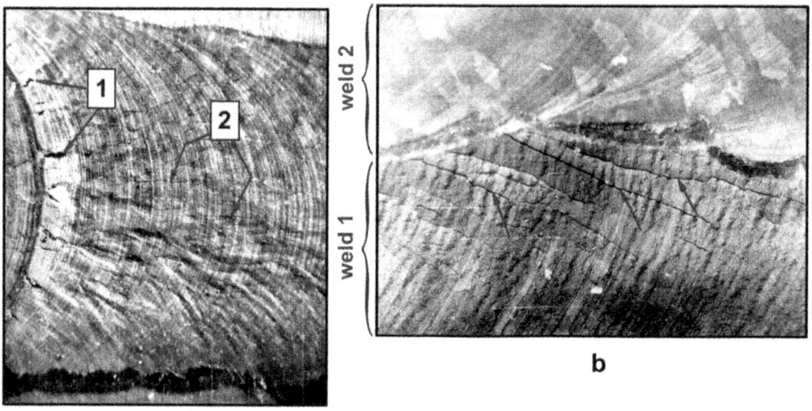

a b

Fig. 5. Appearance of the welds in alloy 36N after testing: **(a)** one-pass weld; 1 – cracks in BTR; 2 – cracks in DTR; **(b)** two-pass weld, cracks (indicated by arrows) in the first weld caused by deformation in making the second weld

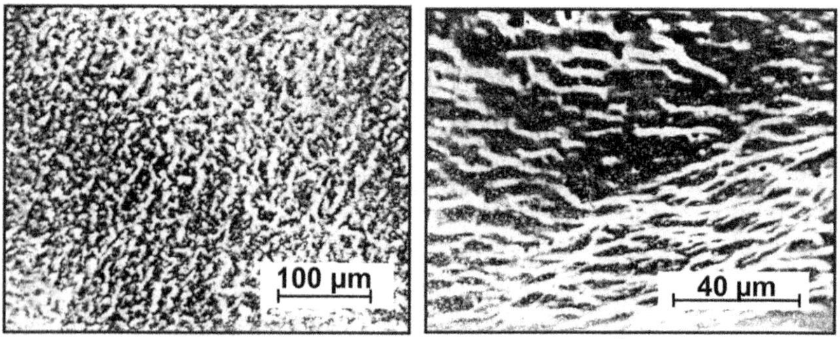

a b

Fig. 6. Fractograph of the surface of a hot crack formed in BTR

The characteristics of the crack surfaces can be estimated as type "R" (Matsuda) [4]. Element compositions of the surfaces of the ductility-dip cracks and the surface of the intact cast metal used as a reference were studied by electron Auger spectroscopy using the 3D analysis unit of the LAS-2000 model. The distribution of elements through the thickness (from the fracture surface) was studied by layer-by-layer etching with argon ions. The working vacuum during the measurement process was $1.6 \cdot 10^{-8}$ Pa. This allowed the cleanness of a specimen surface to be maintained for 16 h after removal of adsorbates and contaminants from it (by etching for up to 1 min). Profiles of distribution and variation in contents of main and impurity elements during etching are shown in Fig. 7. It follows from the presented data that the chemical composition of metal near the fracture surface remains unchanged for 4–5 min of ion etching (Fig. 7a). The chemical composition of matrix of the weld metal is revealed after 5 min of etching (Fig. 7b), whereas near the fracture surface it cannot be revealed even after 230 min of continuous ion etching.

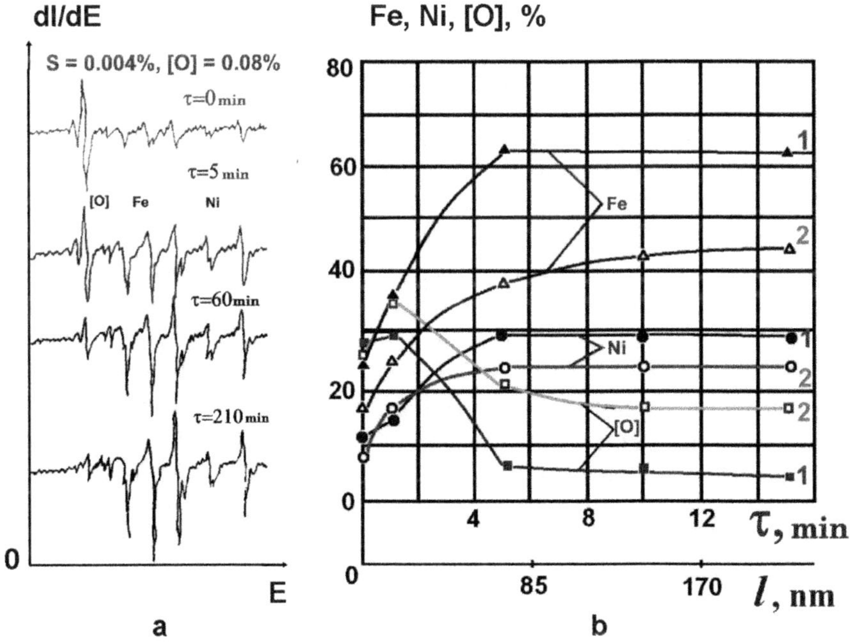

Fig. 7. Variation in the intensity of graphic display between iron, nickel and oxygen on the fracture surface **(a)** and concentration of the above elements in the surface layer **(b)** depending on the time of etching (τ) and the distance from boundary (l): 1 – intact weld metal; 2 – crack surface

It should be noted that carbon, as the interstitial impurity is for the most part removed from the surface within 30 s of etching. Its content in deeper layers of metal becomes lower than the Auger spectroscopy sensitivity limit.

A similar picture is observed also in etching the surface of a hot crack. Sulphur is removed from the surface within 15 s of etching. No nitrogen is detected in the spectrum. Hydrogen cannot be detected by Auger spectroscopy either. However, at increased temperatures (>400 °C), the mobility of hydrogen is known to be so high that it exerts no effect on the formation and growth of hot cracks in the weld metal, although the hydrogen content of the latter is a bit higher than that of the base metal (Table 1).

The oxygen peak in both cases is seen for a much longer time: within 90 min of etching for the base metal, and for more than 230 min for the fracture surface. It is likely that an oxide film about 1.3 μm thick is present on the surface of the reference specimen, and that the oxide film on the surface of a hot crack has thickness of more than 3.45 μm. Intensities of peaks of individual elements on the crack surfaces in specimens in the as-welded condition and after holding for 1.8 and 20 days in super high vacuum ($1 \cdot 10^{-8}$ Pa) were compared to determine the possibility of existence of oxygen mass transfer from the depth of the weld metal to the fracture surface. This was done proceeding from the fact that settling of adsorbates occurred uniformly and at the same intensity both on the fracture surface and on the surface of the intact weld.

As proved by the experiments, sorption of oxygen occurs most intensively during the first hours and days after surface etching. Then the rates of adsorption and desorption of oxygen level off. The chemisorbed layer is left on the surface, and the intensity of mass transfer of this impurity from the depth to the surface remains almost unchanged for 20 days. Therefore, despite the fact that the reactivity of oxygen is high and the coefficient of its diffusion in steel at room temperature is very low ($\sim 1 \cdot 10^{-16}$ cm^2/s), the surface layer of a crack has an increased concentration of oxygen caused by its transfer from the depth.

The experimental results suggest that oxygen redistribution takes place in the metal studied, resulting in increased segregation of this impurity as well as in an enrichment of boundaries and in a depletion of the bulk of grains in oxygen. It is believed that it is the peculiarities of the mechanism of intragranular deformation of the Fe-36Ni metal that can play a decisive role in changing the resistance to under-bead cracking.

According to the generally accepted concept, the process of realisation of intragranular deformation occurs either by involving a dislocation (translation) or turning (rotation) mechanism [5], depending upon the external factors and on the type of deformed metal. Moreover, characteristic regularities of the latter most often show up under complex thermal deformation conditions. It is thought that differences between the above mechanisms have a substantial effect on the resistance to plastic deformation, and thus on the formation of under-bead cracks. In this connection, it is advisable to check peculiarities of deformation (including the intragranular one) of the weld metal having a different sensitivity to under-bead cracking.

Fine weld metal structures determining the peculiarities of variations in their dislocation composition during the plastic deformation process as well as their resistance to high-temperature deformation were evaluated by the examples of two systems of alloying the welds (Table 2), i.e. with high and low sensitivity to cracking.

Comparative evaluation of variations in the fine structure was performed on specimens with a gauge part diameter of 4 mm cut from the metal of the upper bead of the multi-pass welds in the as-welded condition as well as after additional deformation (by 10 %) at a temperature of 700 °C and a rate of $5.66 \cdot 10^{-2}$ s^{-1} corresponding to the DTR temperatures.

In alloying system I (Table 2), increase in the deformation rate by 10 % leads to strong banded structure formation in the internal volumes (deformation bands), i.e. mechanical twins, and bands associated with collective forms of movement of the crystalline lattice defects propagating in the entire bulk of grains (Fig. 8a). At displacement of the deformation bands, the low-angle grain boundaries do not exert any delay effects, whereas the high-angle grain boundaries block further movement of the slip bands into the neighboring grains (Fig. 8b). Initiation of banded structures occurs in certain regions of a chaotic dislocation ensemble, which is characterised by a high density of the lattice defects ($\rho \approx 1 \cdot 10^{11}$–$1 \cdot 10^{12}$ cm^{-2}).

Table 2. Chemical composition of materials under consideration

Alloy	Content of elements, wt. %							Others	Sensitivity to under-bead cracking
	C	Mn	Si	Cr	Ni	N	Mo		
I[a]	0.05	0.35	0.1	0.45	35.6	-	-	0.004 S 0.006 P 0.002 [O] 0.0015 [H]	High
II[b]	0.030	11.2	0.50	19.6	13.5	0.25	2.6	0.004S	Low

[a] Material 36N.

[b] Material 03Kh18N19G10AM3 (Fe-Cr-Ni-Mn-N).

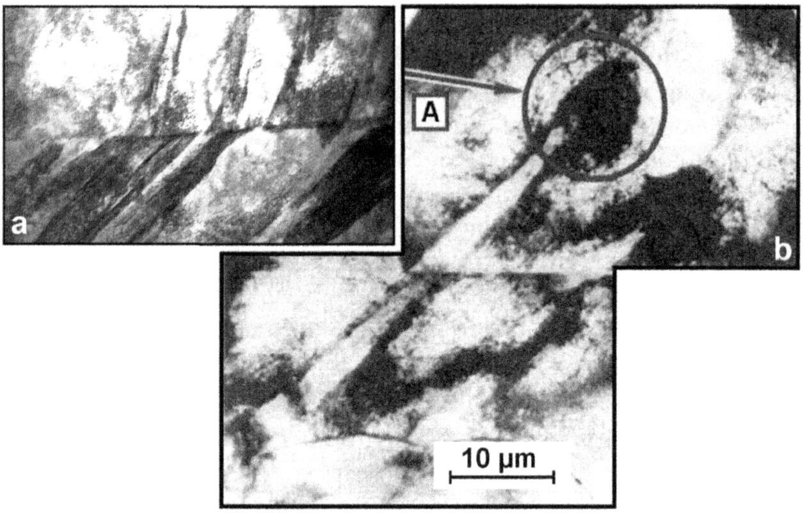

Fig. 8. Fine metal structure of the alloy of alloying system I after forced deformation at 700 °C: **(a)** intragranular **(b)** boundary zone (arrow A indicates the region of initiation of banded structures)

The location of individual dislocations along the crystallographic slip systems in the bulk of grains in a metal structure of alloying system I, in the case of absence of forced deformation and emergence of extended banded formations in grains as the external load is increased, indicates that intragranular plastic deformation in the metal occurs by the classical dislocation (translation) mechanisms [5]. In addition, the "strength" (extension) of translations and a substantial degree of mobility of the lattice defects in the stress field are caused in many respects by high values of the energy of stacking faults of the Fe-Ni system metal, which are known to serve as a weak obstacle for the deformation to propagate. It should be emphasised that strong collective displacements of the lattice defects in the bulk of grains, oriented to the boundaries, are accompanied by a "transportation" of impurity elements and their clustering in a region of intergranular boundaries, which was experimentally observed in the dark-field structural and microdiffraction images. It is likely that the specific character of displacement of the lattice defects during the process of plastic deformation results in the presence of segregation clusters and new phases along the intergranular boundaries, as well as their absence in the bulk of grains.

Therefore, as shown by the detailed analysis of structural elements, distribution of phases and peculiarities of the plastic deformation mechanism, a very heterogeneous structure with a considerable gradient of dislocation density and presence of phase precipitates and segregations between the bulk of grains and high-angle boundaries is formed in the weld metal of the Fe-Ni alloying system.

The weld metal structure of alloying system II (Fe-Cr-Ni-Mn-N) with a high cracking resistance is characterised first of all by the presence of phase precipitates with a high dispersion degree ($d_{ph.p} \approx 0.04–0.21$ μm) and by their heterogeneous distribution in the entire volume of the weld metal (particle spacing $l_p \approx 0.25–0.3$ μm).

The uniformly increasing degree of deformation in the bulk of grains entails a higher dislocation density and a greater delay of displacements in the crystallographic slip planes. The split dislocations are often seen. This points to a total decrease in the energy of stacking faults, $\gamma_{s.f.}$, of metal at the transition from alloying system I to system II.

The delay of displacements (translations) of individual dislocations within the slip systems most probably results from a total decrease in $\gamma_{s.f.}$, based on alloying. Naturally, this makes the plastic deformation more difficult to occur by the classical dislocation mechanism, of which a characteristic element is a transverse slip of the dislocations.

In this case, for metal with a low value of the stacking fault energy, dislocation transition to the other slip plane may take place, provided that particular dislocations combine ("draw together") into the total one. The wider the stacking fault, (i.e. the lower the stacking fault energy), which is characteristic of metals of alloying system II, the higher the energy intensity of the process.

It is presumed that complex dislocation rearrangements of the above type and interactions of dislocations of different slip systems lead to the formation of a reticular structure [5] (Fig. 9).

One of the consequences of a change in the deformation mechanism is the formation, on the basis of the reticular structure, of dispersed fragments approximately 0.25-0.30 μm in size, framed by the zones with a high dislocation density (Fig. 9d). Because of this, the neighbouring microvolumes turn to finite angles. In such cases, relaxation of internal stresses occurs by plastic turns (rotations).

Therefore, the conducted experiments show that under similar thermal deformation conditions the weld metals of different alloying systems are characterised by the formation of structures greatly differing in presence and distribution of impurities, morphology of phase precipitates, sizes of grains and substructural elements, and character of displacement of the crystalline lattice defects within the effective stress field. The latter points to the fact that the plastic deformation mechanism occurring in certain microvolumes of the weld metals with different alloying systems is also different. In the case of alloying system I, the deformation mechanism is mostly of a dislocation character, whereas in the case of alloying system II, the other mechanism, i.e. of a rotation character, is involved (Figs. 8 and 9). It can be suggested that a change in the mechanism exerts a decisive effect on the processes of mass transfer of impurity elements causing embrittlement along the grain boundaries. In the first case (occurrence of the translation mechanism, Fig. 8a), an intensive intragranular slip to distances comparable with the grain size (where dislocations move in the bulk of grain to its boundaries) favours the dislocation transfer of these elements to the intragranular boundaries. In the second case (with involvement of the rotation mechanism, Fig. 9), despite an identical deformation of metal, there is no explicit trend to movement of dislocations directed to the grain boundaries. Here, the directed macrodeformation of a specimen at a microlevel occurs primarily through the rotation movement of elements of an intragranular structure. As applied to the INVAR type alloys, the noted specific character of the plastic intragranular deformation mechanism leads finally to an enrichment of the intergranular boundaries with oxygen, which causes the formation of under-bead cracks in multi-layer welds (Fig. 10).

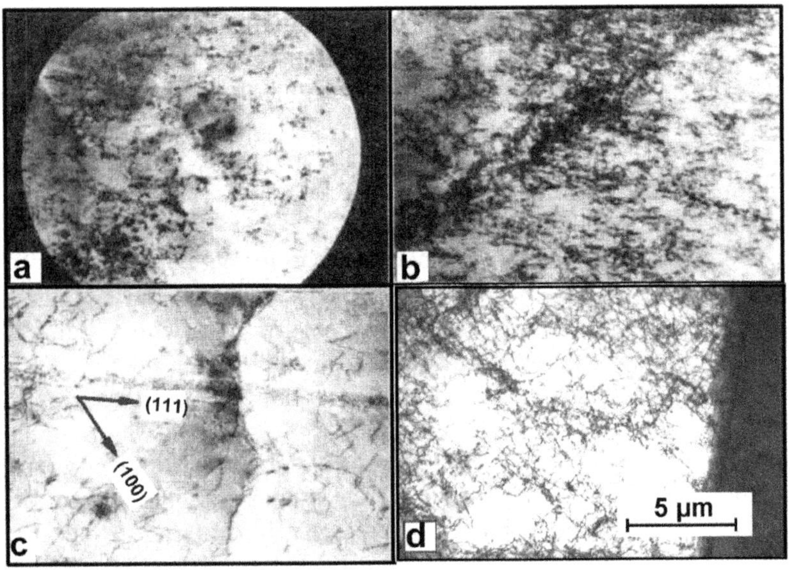

Fig. 9. Fine structure of the weld metal of alloying system II: **(a)** dispersed precipitates of redundant phases in the bulk of grains **(b)** dislocation composition of the boundary zone **(c)** split dislocations in weld structure **(d)** fragmentation of metal structure after forced deformation at 700 °C

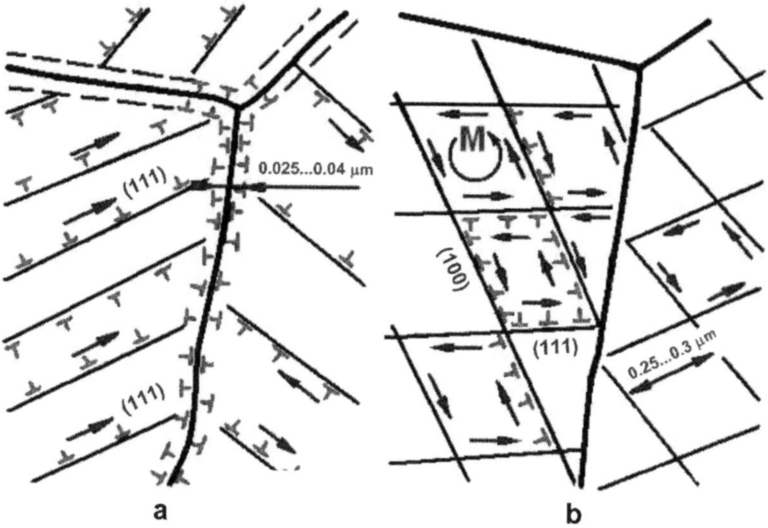

Fig. 10. Structures formed in the deformed metal of **(a)** alloying system I and **(b)** alloying system II

Conclusions

1. Under-bead hot cracks in multi-layer welds of the INVAR type are formed along the grain boundaries.
2. An increased content of impurity elements, and first of all oxygen, is observed on the surfaces of under-bead cracks in multi-layer welds of the INVAR type.
3. Enrichment of grain boundaries with oxygen during welding, leading to hot cracking of the welds, is controlled by the plastic deformation mechanism of a dislocation (translation) character.

References

1. Ющенко КА, Старущенко ТМ, Савченко ВС (1983) Механизм образования горячих трещин, вызываемых кислородом, при сварке Инвара. Автоматическая сварка. №8: с.5–7
2. Ющенко КА, Старущенко ТМ (1981) Роль кислорода в образовании трещин при сварке инвара. Автоматическая сварка. №8: с.21–24

3. Savage WF, Lundin GD (1965) The Varenstraint Test. Welding Journal, 44(10): 433–442
4. Matsuda F, Nakagawa H, Ogata S, Katayawa S (1978) Fractographic Investigation on Solidification Crack in the Varestraint Test of Fully Austentic Steel – Studies on Fractography of Welded Zone (III). Transactions of IWR I, 7(1): 59–70
5. Хирт Дж, Лоте И (1972) Теория дислокаций. М. Атомиздат: 599 с

Character of Hot Crack Formation During Welding of Cast Heat-Resistant Nickel Alloys

K.A. Yushchenko, V.S. Savchenko, N.O. Chervyakov, A.V. Zvyagintseva

E.O. Paton Electric Welding Institute, Kyiv, Ukraine

Heat-resistant nickel alloys are the main structural materials used to manufacture gas turbine engines widely applied in aircraft engineering and other industries. Engine sections for high-temperature operation are made using precipitation-hardening nickel alloys with intermetallic strengthening. To ensure the structure stability and maintain high long-term strength properties, nickel alloys are provided with complex alloying systems containing γ'-forming elements (Al, Ti, Nb), the total content of which in an alloy amounts to 6–15 % or more. Complex alloying systems and high strength properties of the alloys lead to crack formation in the weld and heat-affected zone in welding metal of even small thickness. This stipulated the interest in investigating the principles of hot crack formation first of all in the heat-affected zone during fusion welding in terms of structural transformations. Investigations were conducted on nickel alloy IN 738 used as a structural material for the manufacture of gas turbine blades. The chemical composition of the alloy is given in Table 1.

Table 1. Chemical composition of alloy IN 738

Content of elements, wt. %									
C	Cr	Co	Mo	W	Nb	Al	Ti	Ca	La
0.09	16.0	10.5	1.7	4.6	0.2	3.0	4.4	≤0.01	≤0.01

The sensitivity of the HAZ to cracking in plasma-powder welding using a powder additive with a composition identical to that of the base metal was evaluated. Welding was performed on an alloy 10 mm thick in the as-received condition. Approximate welding parameters are given in Table 2.

Table 2. Parameters used to make joints in nickel alloy by plasma powder welding

I_w [A]	U_w [V]	V_w [m/h]	Powder particle size [μm]	Argon flow rate [l/min]
100-120	25-26	4	50-150	18-20

Sections for metallographic examinations were subjected to vacuum ion etching using a high-voltage plasma discharge at a voltage of 2.5 kV and a current of 0.005 A. This treatment reveals distinct contours of cracks of different sizes. Dark- and light-field optical microscope images were used for verification revealing the zones of structural transformations in the HAZ. Sizes of structural metal components, including the γ'-phase, were estimated by scanning electron microscopy (SEM) after special etching. The character of plastic deformation on the incidence of hot crack propagation was estimated from changes in the surface profile of a welded joint by interference optical microscopy.

Metallographic analyses of welded joints in alloy IN 738 made with the weld edges at room temperature revealed microcracks propagating into the base metal (Fig. 1). A structural zone adjoining the weld, wherein the hot cracks are mainly located, was detected by dark-field image analysis (Fig. 2a, b).

Fig. 1. Crack in the HAZ of a welded joint in nickel alloy with the γ'-phase strengthening

Statistical data processing shows that the cracks are located at some distance from the fusion line. This suggests that the formation of the hot cracks is not always caused by grain boundary melting during welding. Therefore, these cracks cannot be classed as solidification cracks.

Zones with properties other than the base metal properties in the as-received conditions are formed in the base metal under the effect of the welding thermal cycle. Investigation of the formation mechanism of such zones will help to clarify the nature of hot cracking.

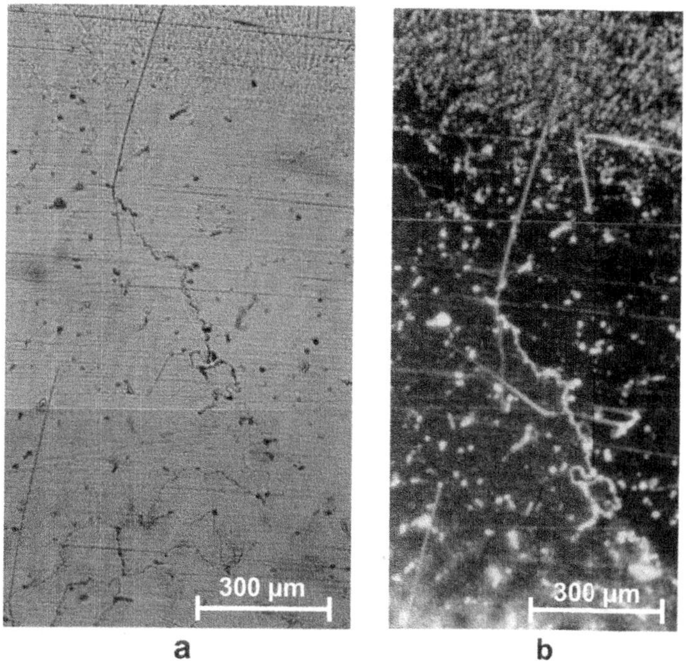

Fig. 2. Microcracks in the HAZ of heat-resistant alloy IN 738,
a – crack in the light field
b – the same crack in the dark field

Generally, structural transformations in the HAZ of welded joints depend upon the welding conditions and, first of all, upon the thermal cycles. Investigations were conducted to study the effect of the initial temperature on the formation of the structural zone under consideration, the welding parameters being kept unchanged.

Welding was performed on metal:

- preliminarily cooled to –196 °C;
- at room temperature;
- on plates heated to 1000 °C.

The HAZ-structure of welded joints in alloy IN 738 is shown in Fig. 3. It can be seen from this figure that the zone, L, undergoes substantial changes across its width. A generalization of the results obtained, shown in Fig. 4, indicates an exponential dependence of variations in the zone width upon the temperature. As confirmed by statistics, the width of the zone correlates with a mean length of the hot cracks resulting from different initial temperatures of the metal prior to welding.

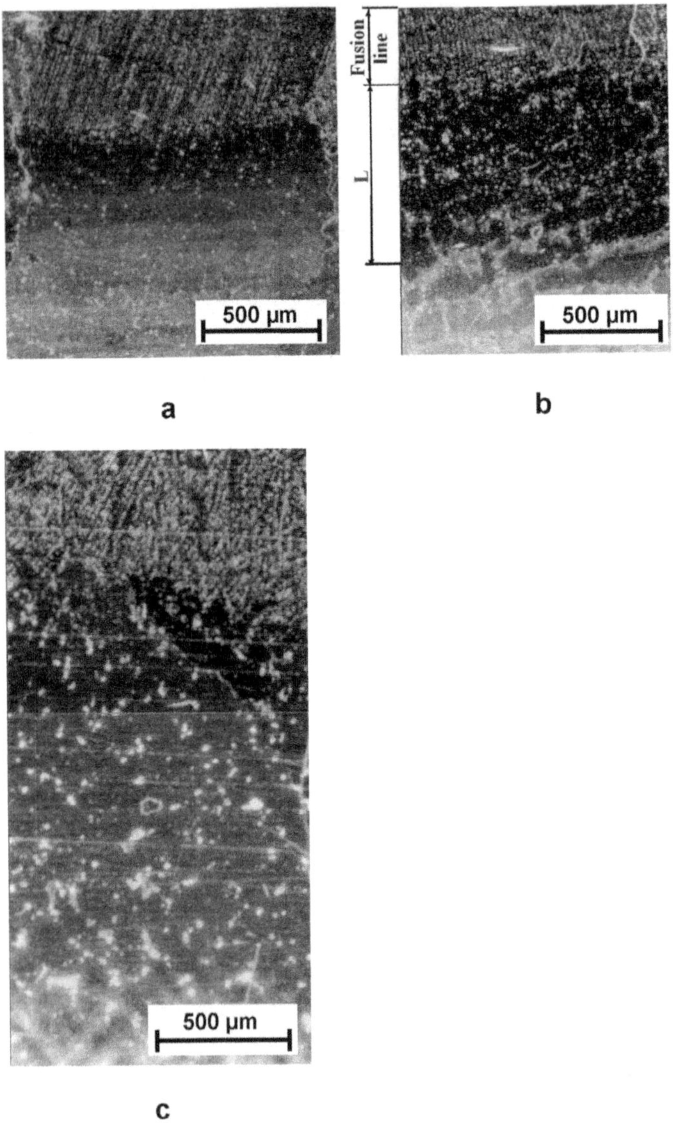

Fig. 3. Changes in width L of the zone of complete $\gamma'\rightarrow\gamma\rightarrow\gamma'$ transformations in welding of heat-resistant nickel alloy with γ'-phase strengthening
Welding was performed at the following temperatures of the alloy IN738:
a – 77 K
b – 293 K
c – 1273 K

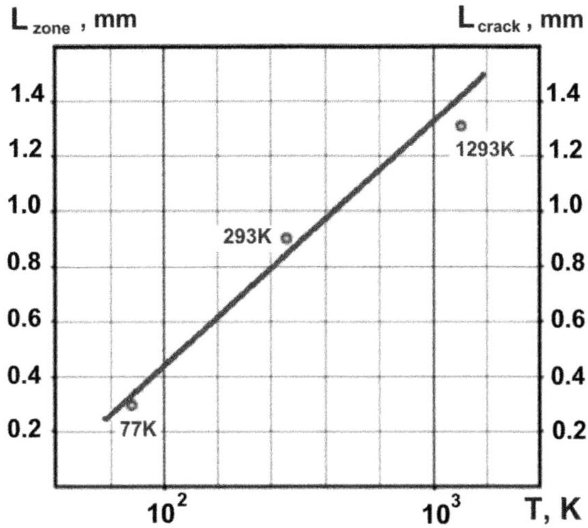

Fig. 4. Effect of temperature of alloy IN 738 prior to welding on the mean length of cracks and zone width L

The examination of the structure of the zone adjoining the weld (Fig. 5a, b) revealed precipitates of the finely dispersed γ'-phase, in contrast to coarser precipitates of the γ'-phase contained in the base metal at some distance from the fusion line (Fig. 5c).

Scanning electron microscopy proves differences in structure and sizes between the γ'-phase in the base metal (as-received conditions) (Fig. 6b) to those in the heat-affected zone adjacent to the weld (Fig. 6a).

It should be noted that changes in sizes of the γ'-phase lead to changes in strength properties of the zone examined, including the surface (Fig. 7). It can be seen from this figure that the hardness of the HAZ changes depending upon the temperature of metal prior to welding. Maximal hardness is achieved in welding of metal preliminarily cooled to a temperature of –196 °C. Metal welded after preheating has a minimal hardness. Therefore, the width of the zone with increased hardness as well as the width of the zone wherein the cracks are formed also depend upon the initial temperature conditions prior to welding.

A welded joint made on metal preliminarily cooled to –196 °C has a minimal width of the zone with an increased hardness. These differences can be explained in terms of the principles of variations in size of the strengthening γ'-phase, depending upon the cooling rate within the temperature range of the $\gamma \rightarrow \gamma'$ transformation.

It was shown [1] that an increase in the cooling rate was accompanied by a decrease in the size of the γ'-phase (Fig. 8). One might expect that a change in the strength characteristics, including hardness, takes place in this case.

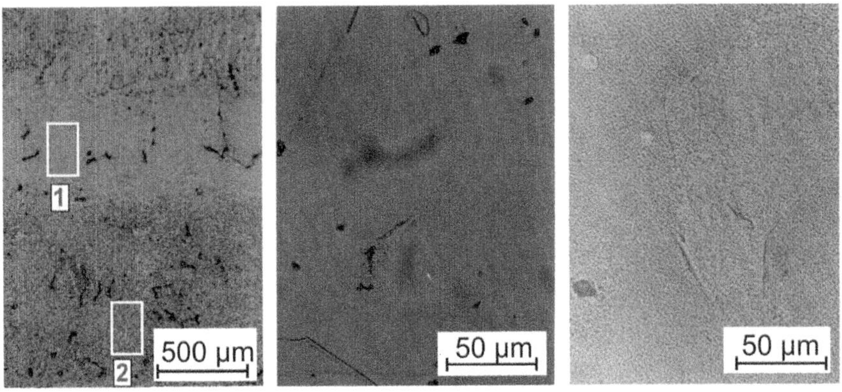

a b c

Fig. 5. HAZ structure of heat-resistant nickel alloy IN 738:
a – general view
b – fragment 1
c – fragment 2

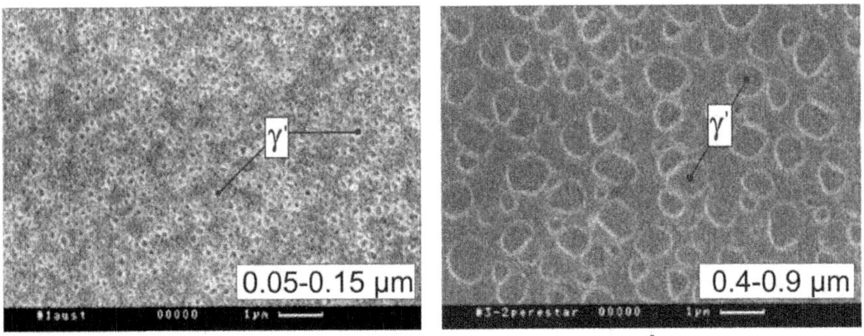

a b

Fig. 6. Size of the strengthening γ'-phase in different zones adjoining the alloy IN 738 weld:
a – zone of hot crack formation
b –base metal (as-received conditions)

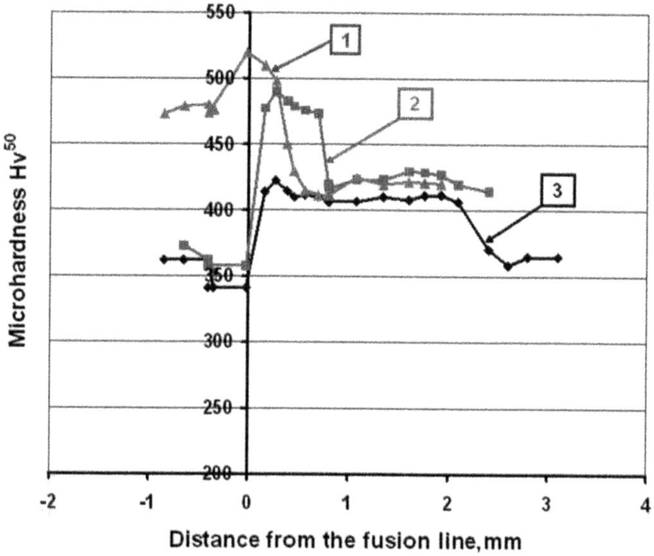

Fig. 7. Dependence of HAZ microhardness on the temperature of the metal prior to welding:
1 – welding with preliminary cooling (-196 °C)
2 – welding at room temperature (20 °C)
3 – welding with preheating (1000 °C)

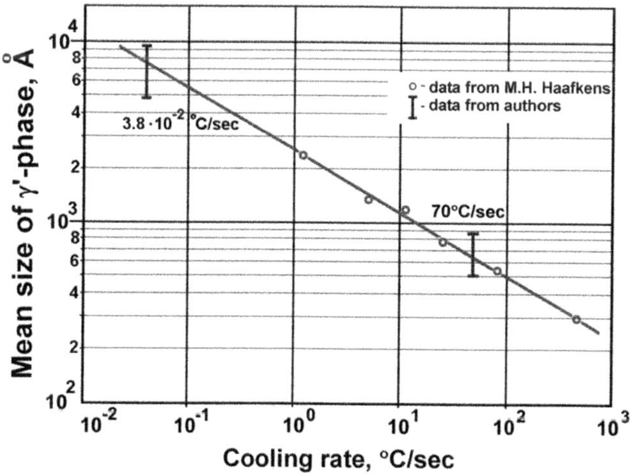

Fig. 8. Variations in mean size of the γ'-phase during cooling from temperature $T > T_{solv}$

In a general form, the strength of precipitation-hardening alloys depends upon the distance between the particles, their size and volume fraction [2], and can be determined from the following Eq. 1:

$$\sigma = \sigma_0 + \frac{c\sqrt{f}}{d}, \tag{1}$$

where

σ_0 is the tensile strength of the matrix;
c is the constant that includes the Burgers vector and shear modulus of the matrix;
f is the volume fraction of the precipitates;
d is the diameter of a particle.

This equation shows that the higher the volume fraction of the strengthening phase and the smaller the size of a particle, the stronger the alloy.

Therefore, it can be concluded that the hot cracking zone adjoining the weld metal has an increased hardness (strength) due to the $\gamma' \rightarrow \gamma \rightarrow \gamma'$ transformations (following the heating $\rightarrow$ cooling cycle) taking place under the effect of the welding thermal cycle (Fig. 9). Here, differences in hardness are caused by differences in the rate of metal cooling during welding, and thus by differences in the occurrence of the $\gamma \rightarrow \gamma'$ transformation process during cooling of the HAZ metal after the weld pool has moved in welding direction. In a general form, the diagram of preferable formation of the cracks in the weld zone of welded joints in nickel alloys with the γ'-phase strengthening and its relationship with structural transformations in the HAZ are shown in Fig. 10. Hot crack propagation along the grain boundaries from zone 1 (Fig. 10) to the base metal is retarded by relaxation of local stresses, owing to plastic deformation taking place at the crack mouth (Fig. 11). It can be seen from the diagram in Fig. 11b that this is the case for the state of plane deformation. Apparently, in this case the principles of retardation of a crack can be described by conditions of formation of the plastic deformation zones in metallic materials in terms of fracture mechanics. The results obtained are confirmed by interference optical microscopy of the surface of a hot crack mouth after welding of a nickel alloy (Fig. 12). The figure shows the interference pattern of the crack mouth as well as of a region of artificial deformation induced by a scratch on the section surface. It can be seen that the course of the interference fringes a region of the crack mouth and the scratch tends to move upward. As the scratch forms a depression in the metal, the zone of the metal ahead of the crack moves downward relative to the section plane, i.e. shrinkage of metal takes place due to realization of the plastic deformation mechanism.

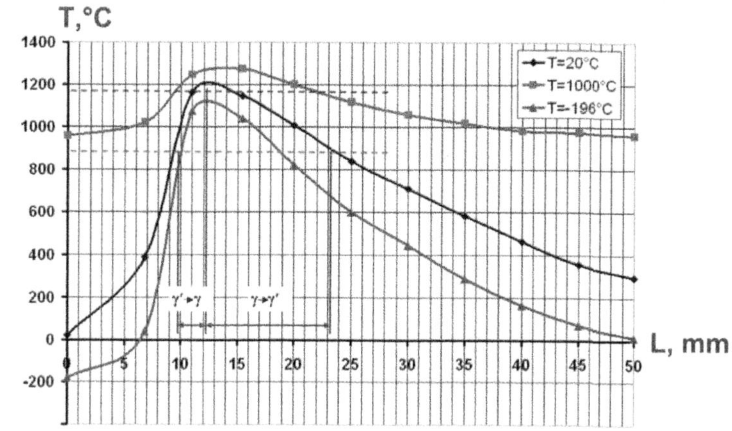

Fig. 9. Temperature distribution in the HAZ of a welded joint in alloy IN 738 and character of structural transformations in realization of the heating → cooling welding cycle

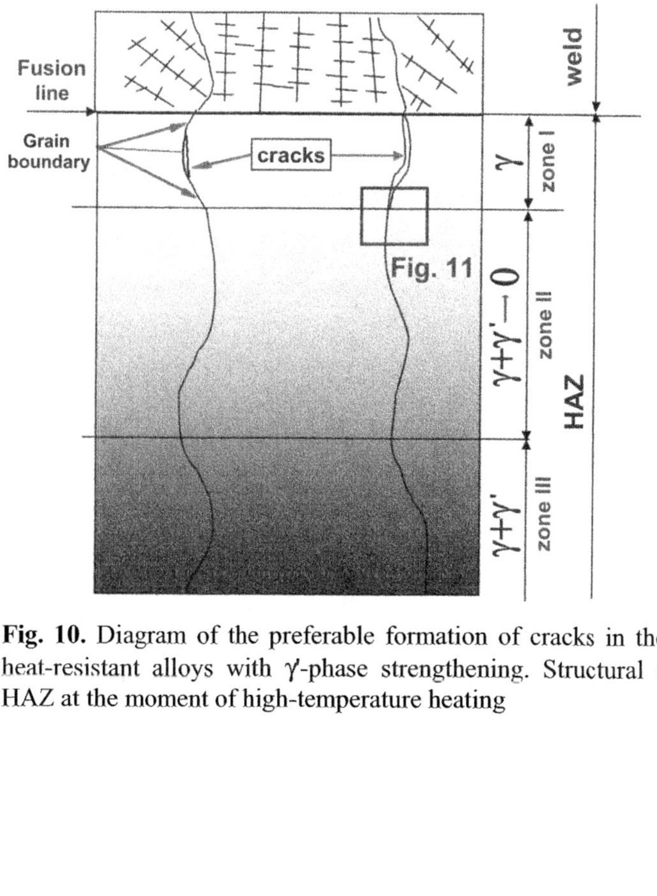

Fig. 10. Diagram of the preferable formation of cracks in the HAZ of joints in heat-resistant alloys with γ'-phase strengthening. Structural components in the HAZ at the moment of high-temperature heating

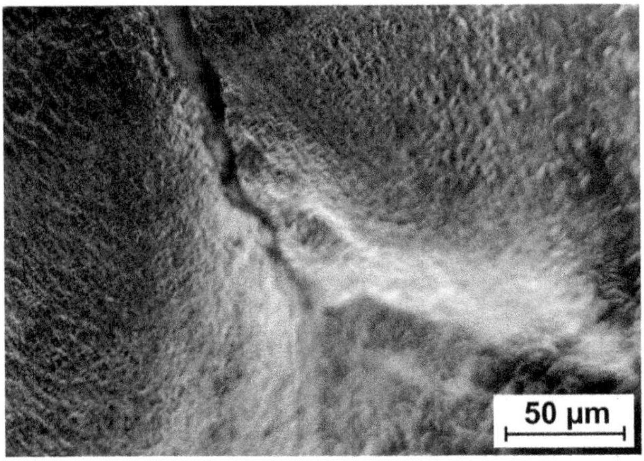

a

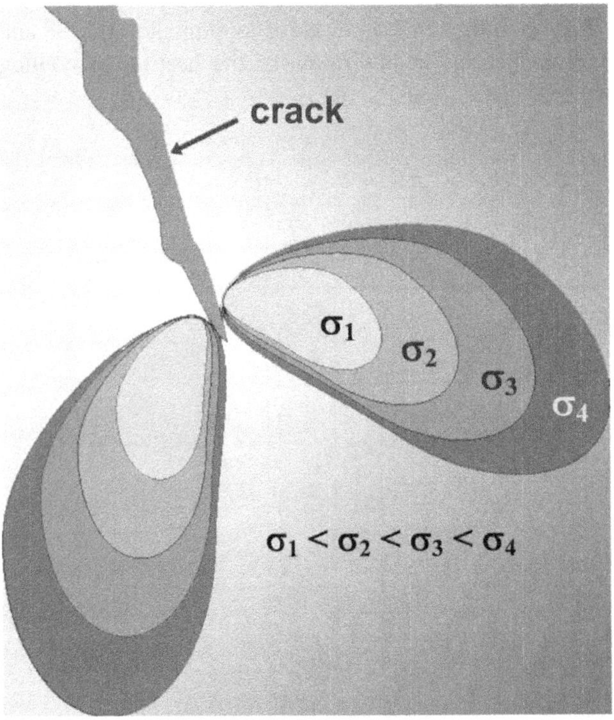

b

Fig. 11. Retardation of hot crack propagation in the HAZ due to the plastic deformation mechanism:
a – appearance of the surface of metal at the hot crack mouth
b – diagram of distribution of plastic deformation in stress fields

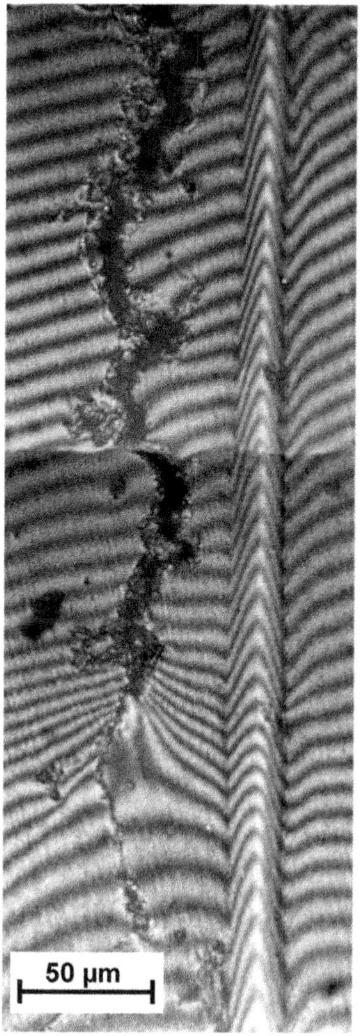

Fig. 12. Distribution of plastic deformations at the hot crack mouth

The investigation results suggest the following mechanism of hot cracking in the HAZ during welding of nickel alloys with γ'-phase strengthening:

1. In welding of an alloy, heating of the HAZ metal causes structural transformations of the strengthening γ'-phase to propagate into the γ-matrix. These transformations take place at temperatures from about 700 °C to T_{solv}, depending upon the chosen alloying system for a metal.

2. The $\gamma \rightarrow \gamma'$ transformation is accompanied by an increase in the metal volume, as at increased temperatures the crystalline lattice parameters for γ' are smaller than those for γ. This process is accompanied by the development of intensive thermal deformation processes localized in a narrow HAZ.
3. There is a high-temperature zone adjoining the weld, where the structure of a nickel alloy remains in a fully austenitic state for a certain period of time.

In this case a mechanism of enrichment of the austenite grain boundaries with impurity elements (carbon, oxygen, sulphur, phosphorus and other surface-active elements) takes place. Contribution to the embrittlement processes by plastic deformation is realized by way of interaction of mobile dislocations with impurity atoms and determined by the competing temperature processes, such as changes in the energy of interaction of atoms with dislocations (inverse temperature dependence) and the diffusion mobility of these atoms (direct temperature dependence).

The temperature range in which a sufficiently high energy of bonding of an impurity atom with dislocations is maintained and, at the same time, ensured by its mobility, owing to increase in the diffusion parameters, is the ductility-dip temperature range (DTR), as in this range an impurity element may be captured by moving dislocations and transported to the boundaries. The latter provides a change in the grain body to boundary strength ratio and leads to the formation of a crack (DDC) along the grain boundaries. As noted earlier, embrittlement of the grain boundaries takes place in a zone with austenitic structure.

As the metal cools down, the process of precipitation of the γ'-phase occurs in the bulk of grains. This is accompanied by the development and, as the amount of the γ'-phase increases, enhancement of the process of retardation of the dislocation movement as well as by an enrichment of the boundaries with impurity elements and an increase in the HAZ ductility. In this temperature range, the propagation of a crack is arrested due to realization of plastic deformation at the crack mouth.

Conclusions

1. Nickel alloys with γ'-phase strengthening are characterized by the formation of cracks along the grain boundaries in the heat-affected zone during heating and cooling in a temperature range from 700 °C to temperatures close to the melting point.

2. It is shown that phase transformations by the $\gamma+\gamma' \rightarrow \gamma \rightarrow \gamma'+\gamma$ mechanism within the base metal zone near the fusion line as well as local deformation along the grain boundaries occur during the welding process. Relaxation of stresses takes place in a crack formed intergranularly (at the crack mouth) at temperatures from 700 C to T_{melt}.
3. The probable mechanisms of formation of cracks in the HAZ include:

- a process of thermal deformation loading accompanied by segregation of impurity elements at the grain boundaries, followed by their melting and cracking;
- a process of thermal deformation loading resulting in a substantial local deformation within the ductility-dip temperature range (DTR), thus leading to cracking;
- the formation of cracks in metals with a high degree of alloying with surface-active elements, such as boron, may occur by the above two mechanisms.

References

1. Haafkens MH, Mathey HG (1982) A New Approach to the Weldability of Nickel-Base As-Cast and Powder Metallurgy Superalloys. Welding Journal, vol 61, no. 11
2. Масленков СБ (1983) Жаропрочные стали и сплавы. М.: Металлургия: 192 с

Contribution to HAZ Liquation Cracking of Austenitic Stainless Steels

P. Bernasovský

Welding Research Institute – Industrial Institute of SR,
Bratislava, Slovakia

Abstract

The paper shows results of 30 experimental stainless steels which were subjected to two laboratory liquation cracking tests. Namely the Thermorestor–W test with temperature-strain simulation of welding cycles (Japanese) and the LTP-1-6/TIG test (Russian) were employed. Preliminary criteria of liquation cracking tests (3 degrees) are proposed. Comparison of experimental results with the calculation indices L, ΔH and E_{Cr}/E_{Ni} is presented. A parametric equation $tg\alpha_K$ based on the simulation test results is also proposed. Thanks to a vacuum chamber of the Thermorestor-W simulator, the fracture surfaces of liquation cracks are suitable for further study of metallurgical cracking. Examples of low melting eutectic phases revealed by microfractographic analysis are shown. The validity of calculation indices and the preliminary test criteria of liquation cracking have to be further verified with the results obtained on real joints.

Introduction

Austenitic steels already have a certain inherent susceptibility to hot crack formation in welded joints (WJ), given by the nature of face-centered cubic lattice. Therefore, it is important to eliminate the other possible unfavorable factors (metallurgical, structural and technological ones) during welding.

Evaluation of hot cracking susceptibility in base metals (BM), means the evaluation of their resistance to liquation cracking and to ductility-dip cracking, which both are formed in the base metal heat affected zone (HAZ) of welded joints.

The ductility-dip cracks, which are formed in the lower temperature range (850–1100 °C), are usually related to decreased cohesion strength of grain boundaries during their migration. However, ductility-dip cracking is relatively uncommon. So this presentation is preferably devoted to liquation cracks in austenitic stainless steels.

Liquation cracks are usually short, they reach only several grain diameters or at maximum several millimeters, and in general they are of subsurface character. Employing old conventional non-destructive techniques, they could be sometimes hardly identified.

But, in applications where there is low defect tolerance (e.g. nuclear power plant components), their occurrence is inadmissible. Even though it is known that liquation fissures in austenitic steels affect only a little their strength properties and toughness, they may under certain circumstances reduce the fatigue, corrosion and creep properties of a welded joint. Their effect on reheat crack initiation and stress corrosion cracking is also reported.

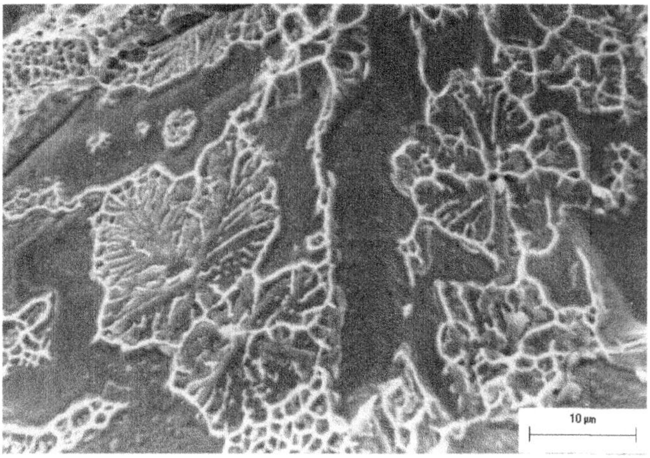

Fig. 1. Eutetic NbX on liquation crack surface in AKOR 2 steel

Liquation cracks are always intercrystalline. On fracture surfaces of liquation cracks, we can find residues of solidified molten metal in the form of the eutectic secondary phases (Fig. 1) or the round grains with typical bridges between them (Fig. 2) or thermal facets (Fig. 3) which enable distinction of liquation cracks from other lower temperature types.

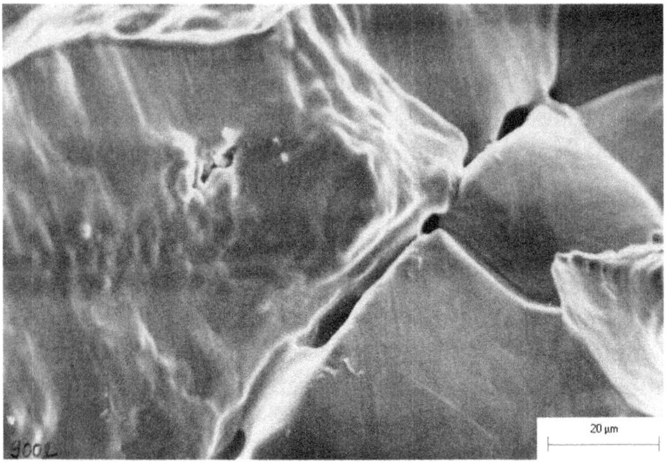

Fig. 2. Typical bridges between grains on the liquation fracture surface of 17 259 steel

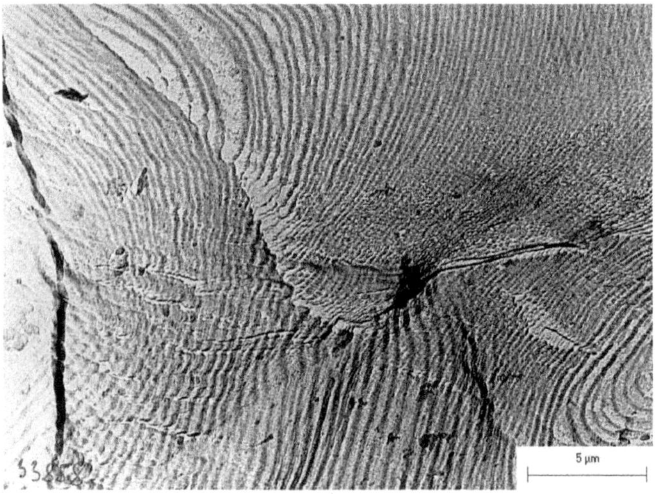

Fig. 3. Thermal facets on the hot tensile fractured surface of 17 349 (K) steel

Mechanism of Liquation Cracking

The formation of liquation cracks is conditioned by the presence of a liquid film on grain boundaries during the cooling phase of the welding thermal cycle when the thermal tensile stresses are induced. The height of temperature and strain gradient the so-called temperature strain gradient [1] depends mainly on technological and structural factors.

The formation of a liquid film on grain boundaries is explained by several processes:

a) Preferential melting of grain boundaries which can be formed also in relatively pure alloys, especially when the dendritic segregation extends the solidification interval. This is the case with some nickel alloys and some austenitic stainless steels;
b) The segregation (diffusion) of impurities or deliberately added alloying additions on the grain boundaries where they decrease the melting temperature. These segregates can precipitate as secondary phases (e.g. see Fig. 1) or they can remain in solid solution (Fig. 2);
c) The penetration of a liquid film formed from low melting precipitates entrapped on the migrating grain boundaries in the overheated HAZ [2];
d) Migration of surface-active elements from the weld metal (WM) along grain boundaries into the adjacent HAZ where they form a liquid film [3]. This mechanism does not depend on the chemical composition of base metal.

The formation of liquation cracks depends on the deformability (hot ductility) of metal and on the intensity of high-temperature strain of metal in welding. Mathematical processing of a physical module of the liquation cracking susceptibility is attractive, it was however hitherto impossible to plot a suitable model including both high-temperature strain and hot-deformability of metal. Problems arise especially with the expression of deformability of metal in intricate non-equilibrium conditions which occur in welding.

Effect of δ-Ferrite and of the Chemical Composition of Steel on Liquation Cracking

δ-ferrite is the most significant factor for hot cracking elimination in austenitic steels. Whereas the presence of δ-ferrite in weld metal at normal temperature ought to be 2–7 %, in base metal even less, up to 1 % is sufficient [4].

However, the primary ferrite segregated during solidification is decisive from the viewpoint of cracking elimination. According to the pseudobinary Cr-Ni diagram in Fig. 4 it is segregated on the right from point A through the $L \rightarrow L + \delta \rightarrow L + \delta + \gamma$ transformation or through $L \rightarrow L + \delta$ transformation (right from point B). The steels which solidified on the left from point A with primary segregated austenite are more susceptible to cracking, because with falling temperature the solubility of nickel in δ-ferrite is decreased and ferrite becomes unstable and is transformed to austenite. At normal temperature, usually only about 1/10 of its content remains preserved during solidification. Due to this fact the suitability of the conventional Schaeffler´s diagram for assessing the cracking susceptibility becomes questionable.

By projection of solidification modes into the Schaeffler diagram (Fig. 5), the zone of austenitic and austenitic-ferritic steels can be divided into characteristic bands with the aid of a, b, c lines [5]. For instance, the band limited by a-b lines corresponds to solidification in the pseudobinary diagram between points A-B. It can be seen that also purely austenitic steels can solidify though δ-ferrite at normal temperature (hatched zone).

Knowledge of high-temperature transformations of δ-ferrite is significant from the viewpoint of liquation cracking. If the underbead zone of the heat affected zone in austenitic steel reaches again the ferrite zone, the cracking susceptibility is decreased.

Residues of the so-called high-temperature δ-ferrite can then be observed in the underbead zone, because at quick cooling-down it cannot be retransferred completely (see Fig. 6 with 13 % of δ-ferrite, whereas only 0.35 % was present in the base metal).

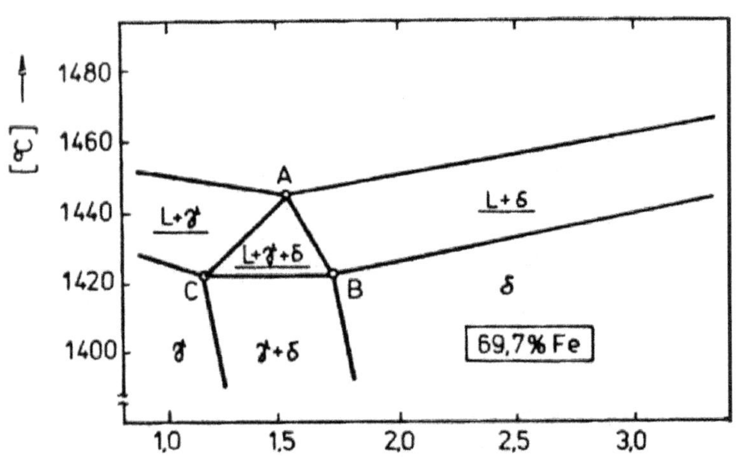

Fig. 4. Pseudobinary Cr/Ni diagram for 69.7 % Fe

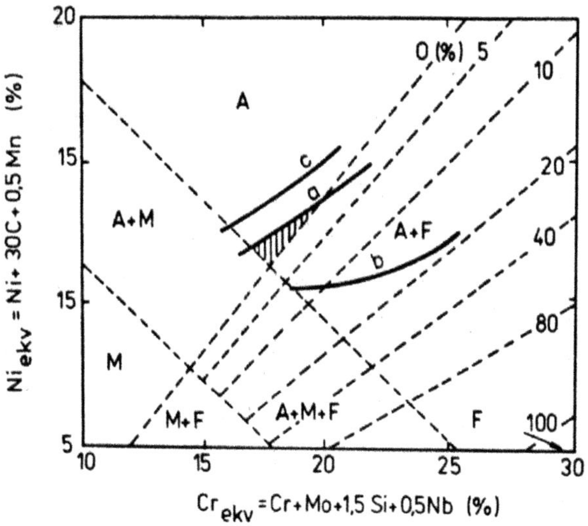

Fig. 5. Schaeffler diagram plotting the solidification zones according to Fig. 4

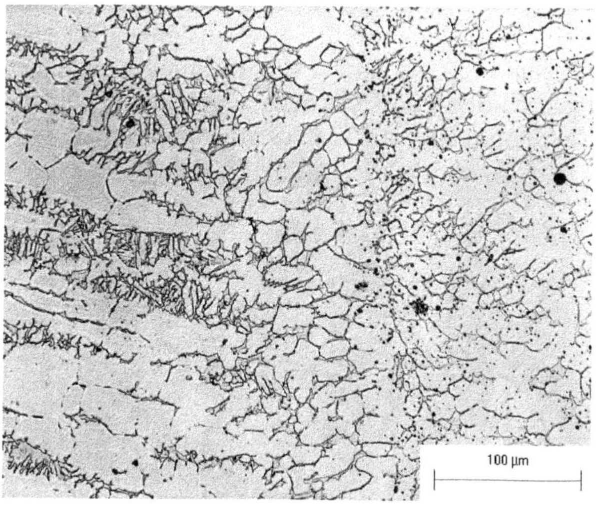

Fig. 6. High-temperature δ-ferrite in the underbead zone of 17 352 steel

The favorable effect of δ-ferrite on the elimination of hot cracks is explained by several facts:

a) Higher solubility of impurities such as S, P and harmful elements (e.g. Si) compared to that observed in austenite;

b) Enlargement of the total grain boundary area;
c) Grain size refinement due to which, similarly as in b), the covering of grain boundaries with molten phase is reduced;
d) Lowering of contraction stresses because ferrite has a lower coefficient of thermal expansion;
e) Enlargement of the dihedral angle Θ, i.e. the angle of grain boundary contact because the interphase energy $\sigma_{\delta\gamma} < \sigma_{\gamma\gamma}$ (σ_{ss} is the interphase energy between the solid-solid phase, σ_{sl} between the solid-liquid phase), [6].

$$2\cos\frac{\Theta}{2} = \frac{\sigma_{ss}}{\sigma_{s1}} \tag{1}$$

Even though the formation of liquation cracks is conditioned by the coaction of metallurgical, structural and technological factors, the primary factor is metallurgical, i.e. the proper susceptibility of steel.

Some metallurgical reasons such as impurities in steel (S, P…) are valid in general (i.e. the case of extrinsic cracks). Other, the so-called intrinsic cracks are caused by alloying elements (e.g. Nb, Ti, C, B …). Except for the fundamental chemical composition of the steel, its manufacturing process and deoxidation method etc. also affect its susceptibility to liquation cracking. The enumeration of all possible metallurgical causes for liquation cracking is as a fact impossible and each steel requires an individual approach.

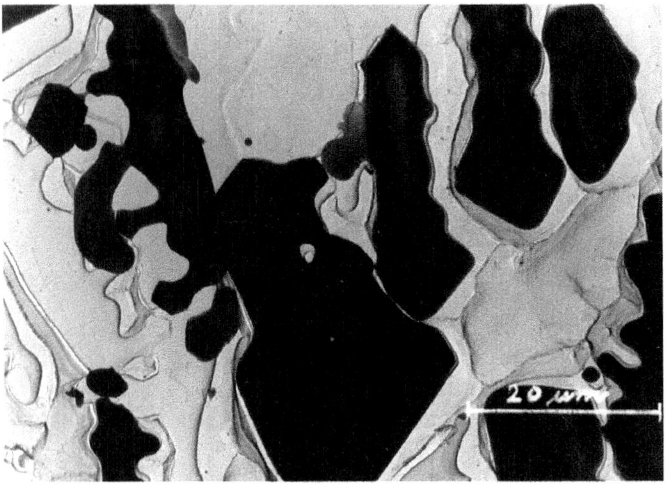

Fig. 7. Eutecticum Feγ-$M_{23}C_6$ on the liquation fractured surface in 17 255 steel

Only some of elements whose effect is already considered as more or less generally known will be mentioned below.

Carbon acts as austenitizing element which decreases the content of δ-ferrite in steel. For the creep-resistant steel type 17 255 (0.165 % C), the connection between liquation cracking and carbide eutecticum formation Fe γ - $M_{23}C_6$ (Fig. 7) was revealed. The solidus temperature of such eutecticum is decreased down to 1300–1250 °C.

S and P impurities are, in general, considered to be unfavorable. Therefore, some authors [10]require P < 0.01 % and S < 0.005 % in order to eliminate the cracks in CrNi-steel type 25/20. Phosphorus is considered to be more unfavorable than sulphur because sulphur can be eliminated by manganese. Manganese binds sulphur to sulphides of α-MnS-type which have higher melting points. These sulphides can substitute chromium up to the chemical composition of $Mn_{26}Cr_{49}$. Chromium decreases the melting point of these sulphides. A higher Mn-content in the steel is favorable because the Mn/Cr-ratio in sulphides is increased. Phosphorus forms low-melting point phosphide of M_3P-type with its eutectic temperature of 1060-1100 °C. Higher P-contents can be eliminated only by δ-ferrite which has an expressively higher solubility than austenite, e.g., for a temperature of 1365 °C: 1 %P in Fe_δ in comparison to 0.1 % P in Fe_γ [7].

Silicon is also in general considered to be undesirable because it forms a silicate eutecticum. Its content in weld metal is limited to 0.5 % and from the viewpoint of liquation cracking even a lower content of max. 0.3 % is recommended.

A low content of titanium (up to 0.1 %) is considered as favorable, since it increases the eutectic temperature of the phosphide eutecticum (~1090 °C). However, the stabilizing contents of Ti and Nb are unfavorable because they form eutectic reactions between TiX and/or NbX and Fe_γ (T_s = 1310–1350^{0}C). Moreover, titanium forms hexagonal sulphide τ - Ti_2S or $Ti_4S_2C_2$ with a melting point of ~1350 °C, which can also be the reason for liquation cracking. If the stabilizing ratio is exceeded, the niobium can also form the Laves phase Fe_2Nb with a lower melting point. From the viewpoint of liquation cracking, the excessive amount of eutecticum can, on the contrary, have a favorable effect due to crack self-healing. A similar case was encountered also with an experimental melt of 03Cr19Ni11B steels with a higher boron content (~1 %). At a temperature of 1050 °C the excess of eutectic phase $(Fe, Cr)_2$ B - Fe_γ (Fig. 8) was formed which assures liquation cracking resistance, but this steel has low toughness at room temperature. The formation of low-melting Ni_3S_2 is not likely in case of sufficient Mn-content in the steel.

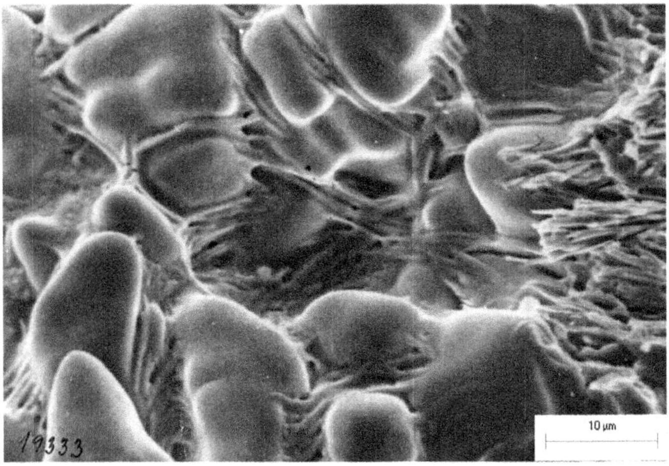

Fig. 8. Eutectic phase $(Fe,Cr)_2B$-Feγ on the liquation fracture surface in 03Cr19Ni11B steel

Only the effect of several common elements in austenitic Cr-Ni steels was mentioned, but as has already been pointed out, the problem of liquation cracking is often more complex.

Test Material and Experiments

Table 1. shows the used experimental steels. All test materials are commonly used steels made in the Czech republic. The steels designated as A, B and D are titanium-stabilized. Low-carbon steels represent another experimental group, among which G, I, J and I_1 are nitrogen-hardened and B, C and E are the strain-hardened ones. The steel designated as I has a higher manganese content (2.7 %). The steels designated as N (02Cr 20Ni 22Mo 3Nb), O (02Cr 22Ni 20) and P (05Cr 21Ni 33Al 0,3Ti) serve for demanding corrosive atmospheres. The steel designated as X is an experimental melt of austenitic manganese steel and the steel Y is a martensitic chromium steel, the only non-austenitic experimental steel. The effect of refining processes, electroslag (ESR) and vacuum remelting (VR) was verified for the steels Cl (17 249) and K (17 349). The sign of EOP means remolten in an electric arc furnace.

To verify the resistance of steel to liquation cracking, two tests were employed, namely the test with simulation of welding cycles and the VÚZ-LTP-1-6/TIG test. Both are laboratory tests. The shop tests did not prove to be reliable for the evaluation of liquation cracking.

Table 1. Chemical composition (wt%) of experimental steels

Des.	Steel	Melt	h [mm]	As-delivered Condition	C	Mn	Si	P	S	Cr	Ni	Mo	Ti	Nb	N	Cu	Al
A	17246		20	r.z.	0.068	1.6	0.46	0.032	0.007	17.7	11.1		0.97			0.08	
B	17247	E0094	20	d.s.	0.06	1.32	0.75	0.025	0.014	18.31	10.16		0.43			0.09	
C	17249	E5621	20	d.s.	0.03	1.25	0.31	0.032	0.018	18.07	11.74					0.05	
D	17347		16	r.z.	0.098	0.79	0.45	0.034	0.019	16.22	11.2	1.62	0.48			0.13	
E	17350	E7699	20	d.s.	0.03	0.66	0.31	0.037	0.021	17.59	15.0	2.60				0.13	
F	17353	9316	15	r.z.	0.067	1.15	0.54	0.031	0.016	16.77	12.84	2.29				0.11	
G	17359	E8916	25	r.z.	0.02	0.90	0.30	0.017	0.020	16.96	12.0	2.30			0.26	0.12	
I	17259	E7341	26	r.z.	0.026	2.70	0.08	0.023	0.020	18.28	12.0				0.23		
J	17360	E8882	26	r.z.	0.02	1.40	0.75	0.016	0.016	18.5	12.98	2.80			0.25	0.09	
N	AKOR	0-893	5	r.z.	0.02	0.43	8.5	0.015	0.01	20.25	22.7	3.50		0.8			
O	AKC2	0-688	5	r.z.	0.02	1.0	0.68	0.016	0.01	22.80	19.45		0.1				
P	AKR17	0-605	3	r.z.	0.05	1.2	0.6	0.021	0.015	21.0	32.80		0.5				
X	17483	2679	50	r.z.	0.025	20.23	0.47	0.016	0.015	13.68	0.5		0.34		0.014		
Y	17199	E6222	100	z.	0.2	0.7	0.09	0.014	0.006	10.45	0.15	0.94	0.08	0.07	V0.18	0.05	
E1	17350	23093	20	r.z.	0.02	1.21	0.36	0.033	0.016	17.2	13.52	2.78				0.05	
E2	17350	–	20	r.z	0.024	1.29	0.33	0.024	0.022	17.94	14.43	2.90		0.03	0.012	0.09	
H1	17352	95093E4	45 x 45	k.	0.05	1.0	0.48	0.027	0.018	17.8	12.42	3.12		0.32	0.042	0.17	
H2	17352	95019E1	45 x 45	k.	0.05	1.31	0.4	0.032	0.018	16.54	12.23	2.55		0.033	0.034	0.15	
H3	17352	94631E1	90 x 90	k	0.06	0.97	0.41	0.028	0.018	17.0	11.70	2.06		0.035	0.037	0.09	
I1	17259	E7162	25	r.z.	0.034	1.26	0.26	0.020	0.012	17.45	10.35				0.16		
C2	17249	–	14	r.z.	0.03	1.00	0.39	0.019	0.011	19.29	11.5				0.049		
C1	17249	E9597	15	r.z.	0.03	1.24	0.38	0.020	0.014	17.32	11.6	0.05			0.048		0.05
C1O	17249	EOP	Ø20	v.t.	0.03	1.35	0.46	0.014	0.014	18.12	10.95	0.04			0.034		0.13
C1E	17249	ETP	Ø20	v.t.	0.05	1.29	0.32	0.017	0.010	17.36	11.35				0.040		0.07
C1V	17249	VOP	Ø20	v.t.	0.03	1.20	0.50	0.016	0.013	17.85	11.45				0.036		0.10
K	17349		15	r.z.	0.02	1.10	0.26	0.022	0.015	16.30	13.57	2.30			0.040		0.05
KO	17349	EOP	Ø20	v.t.	0.033	1.14	0.37	0.021	0.018	17.34	12.45	2.38			0.025		0.08
KE	17349	ETP	Ø20	v.t.	0.055	1.10	0.027	0.023	0.010	16.60	12.41	2.10			0.023		0.08
KV	17049	VOP	Ø20	v.t.	0.03	0.97	0.04	0.021	0.012	18.12	12.81	2.13			0.018		0.09
K1	17349		14	r.z.	0.02	0.88	0.21	0.021	0.018	18.15	13.8	2.32			0.044		

Des. Designation, *r.z.* solution annealing, *d.s.* strain hardening, *z* quenched and tempering, *k* forged, *v.t.* hot rolling, *h* thickness.

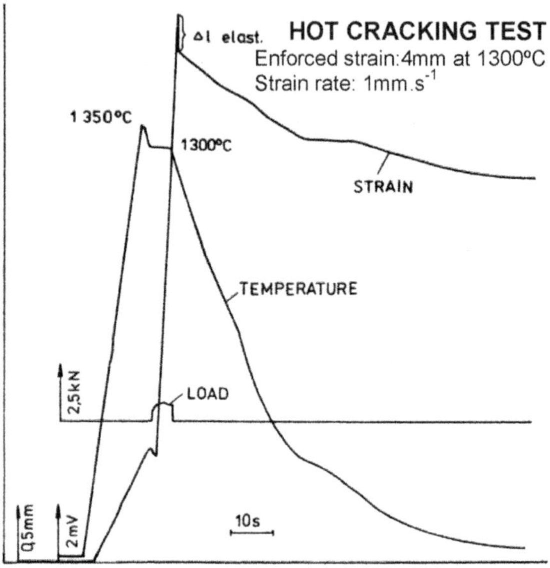

Fig. 9. Record of a hot cracking programme

The tests with simulation of welding cycles were performed on the Japanese Thermorestor-W equipment. The simulation of underbead thermal cycles with a cooling time $\Delta t_{8/5}$ = 12 s was selected. In the course of the cooling phase of the cycle the test bar 7 mm in diameter is subjected to 1–4 mm tensile strain at a strain rate of 1 mms^{-1} (Fig. 9).

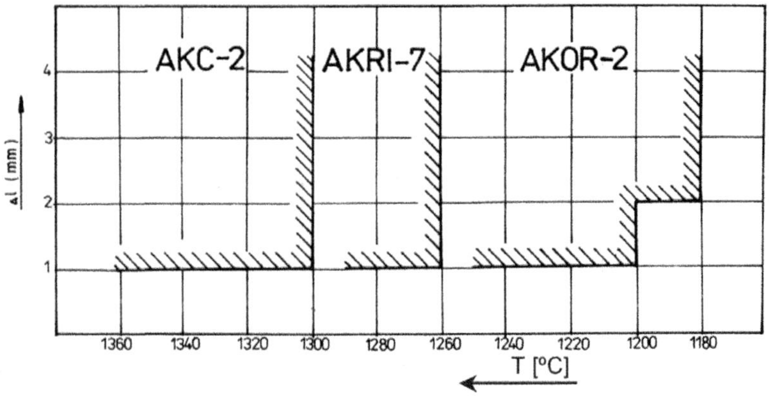

Fig. 10. Dangerous zone of liquation crack initiation in AKC-2, AKRI-7 and AKOR-2 steels evaluated by the welding cycle simulation test

The test simulates a tack welded joint in which the stitch is suddenly loosened after passing of the following run. The presence of cracks after cooling down represents boundaries of the so-called dangerous zones for liquation cracking (see an example in Fig. 10).

The calculated value of tg α_K (mm/100 °C) is the test criterion, where α_K is the angle between the directive plotted from the solidus temperature (of assessed DTA) to the boundary of the liquation crack zone and the abscissa (see an example in Fig. 11). The susceptibility to cracking is inversely proportional to the tg α_K-value.

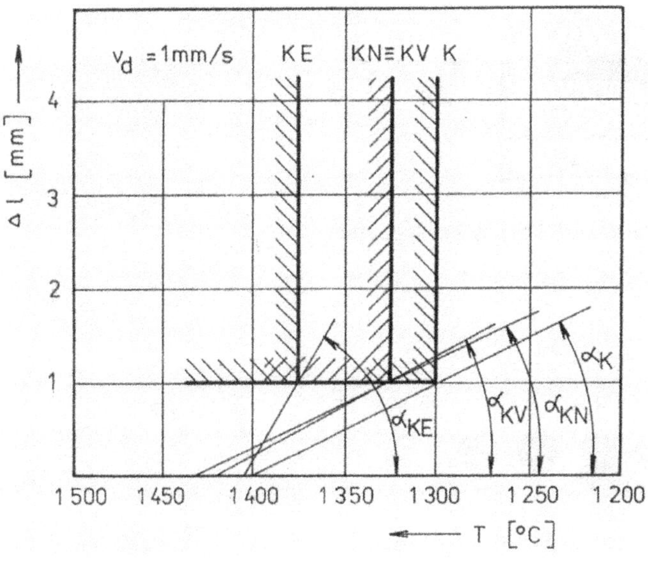

Fig. 11. An example of liquation cracking test results of four steels employing the tg α_K (mm/100 °C) criterion

In the VUZ-LTP-1-6/TIG test which is of Russian origin, specimens (Fig. 12) remelted by TIG-welding within the brittle temperature range (BTR) are subjected to tensile strain at different strain rates.

The maximum strain rate V_d (mms^{-1}) at which no hot cracks are formed in four repeated specimens is the criterion (see example in Fig. 13). The blackened part of the circles is proportional to the hot crack size. The higher the value V_d, the higher the resistance against hot crack formation. Both tests are specified in the STN 05 1143 standard.

In the course of the investigation, the preliminary criteria of both tests for evaluating high-alloy steels were determined (Table 2).

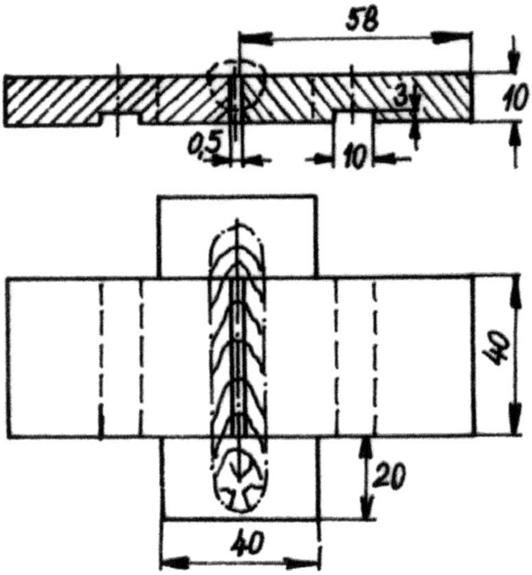

Fig. 12. Sketch of the VÚZ-LTP-1-6 test specimen

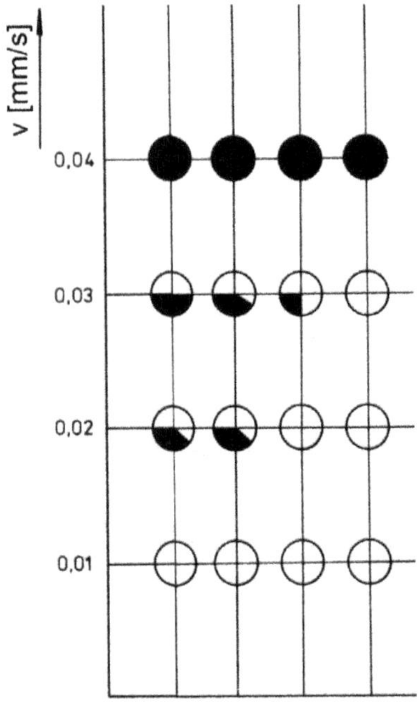

Fig. 13. Result of the VÚZ-LTP-1-6/TIG test for AKRI-7 steel ($V_d = 0.01\ mms^{-1}$)

Table 2. Preliminary criteria of hot cracking tests by simulation of welding cycles (tg α_K) and VÚZ-LTP-1-6/TIG (V_d)

tg α_K [mm/100 °C]	V_d [mms^{-1}]	Evaluation of steel
> 1.2	> 0.1	1st degree: steel resistant to hot crack initiation without special conditions during welding
0.7–1.2	0.033–0.01	2nd degree: steel partially susceptible steel to hot crack initiation which can, however, be prevented with the use of specified conditions during welding
< 0.7	<0.033	3rd degree: steel susceptible to hot crack formation

Calculation of Hot Cracking Parameters

The calculable parameters L, ΔH [8] and E_{Cr}/E_{Ni} [9] for the expression of the cracking susceptibility of austenitic steels are the most known from literature.

$$L = -299\%\,C + 8\%\,Ni + 142\%\,Nb - 5.5\%\,\delta_2 - 105, \qquad (2)$$

if L > 0, the steel is susceptible to cracking.

$$\Delta H = -700 + 17\%\,Cr - 37\%\,Ni - 117\%\,Nb + 29\%\,Mo + 188, \qquad (3)$$

if ΔH > 100, the steel is crack-resistant.

The parameter L expresses a crack length in the TIG-remelting test. The parameter ΔH was proposed based on the hot ductility test and is only valid for HAZ-cracking. The parameters are applicable to both stabilized and non-stabilized steels.

$$E_{Cr} = \%Cr + 1.37\%\,Mo + 1.5\%\,Si + 2\%\,Nb + 3\%\,Ti, \qquad (4)$$

$$E_{Ni} = \%Ni + 0.31\%Mn + 22\%C + 14.2\%N + Cu, \qquad (5)$$

if $E_{Cr}/E_{Ni} < 1.5$ and $P + S > 0.02\ \%$, the steel is susceptible to cracking
if $E_{Cr}/E_{Ni} \geq 1.5$ and $P + S \leq 0.02 - 0.05$, the steel is crack-resistant.

The term E_{Cr}/E_{Ni} contains newly proposed chromium and nickel equivalents which better express the chemical composition of the steel than the known equivalents from the Schaeffler diagram. The ratio of E_{Cr}/E_{Ni} as a fact considers the mode of primary solidification. If $E_{Cr}/E_{Ni} < 1.5$, the steel solidifies in the mode of austenite or austenite-ferrite. If $E_{Cr}/E_{Ni} \geq 1.5$, the ferritic-austenitic or ferritic mode is concerned which indicates a resistance of the steel to cracking at conventional levels of impurities.

Before using the Thermorestor-W and VÚZ-LPT-1-6 test equipment, liquation cracking tests were carried out on the Russian IMET-1 equipment which was adapted for operation in an inert atmosphere. The impact tensile test (force induced by an electromagnet) was performed in the course of the simulated thermal cycle of the overheated HAZ (controlled by resistance heating and cooling gas). The nil ductility temperature designated as T_d was evaluated in the test as a lower temperature of the brittle temperature range (BTR), see example of the evaluation in Fig. 14 (results of the IMET-1 test are not a subject of this paper).

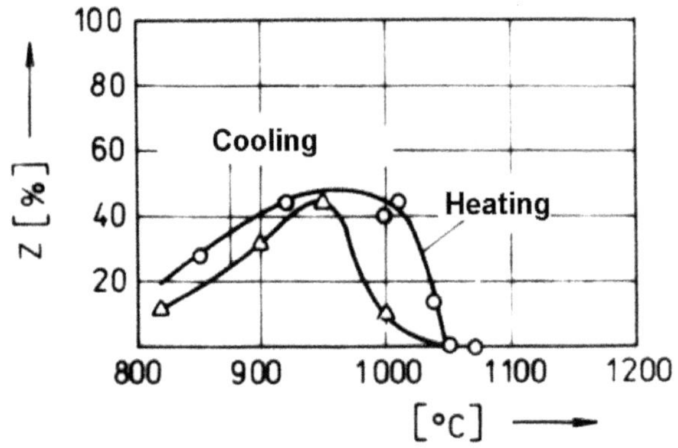

Fig. 14. IMET-1 test result of 03Cr19Ni11B steel

The solidus temperature assessed by DTA was considered as the upper BTR temperature for the case of liquation cracking. However, it has been found out that the dependence between BTR width and liquation cracking susceptibility is not generally valid. In CrNi-steel with higher boron addition (~1 %), the assessed T_d is very low (Fig. 14) because at 1050 °C, a high amount of boride eutecticum $(Fe, Cr)_2B - Fe_\gamma$ is formed as it can be seen at the fractured surface (Fig. 8). However, this material did not exhibit any cracking in real joints because a high amount of the liquid phase itself overflowed potential cracks (self-healing effect). As the IMET-1 test always results in fracture, this effect could not be considered. This was one of the reasons why we have subsequently ceased the IMET-1 test.

More objective results can be achieved on the Japan Thermorestor-W simulator where the test specimens are loaded imposing 1 up to 4 mm additional strain at a strain rate of 1 mms^{-1} which more closely resembles the real condition.

The advantage of the Thermorestor-W equipment is to be seen in the fact that the tests are carried out in the shielding atmosphere of the vacuum chamber, so that the high-temperature fractured surfaces remain free from oxides and are therefore suitable for further metallurgical analyses to identify the causes of the cracking susceptibility by SEM or TEM replicas as shown in the figures of this paper.

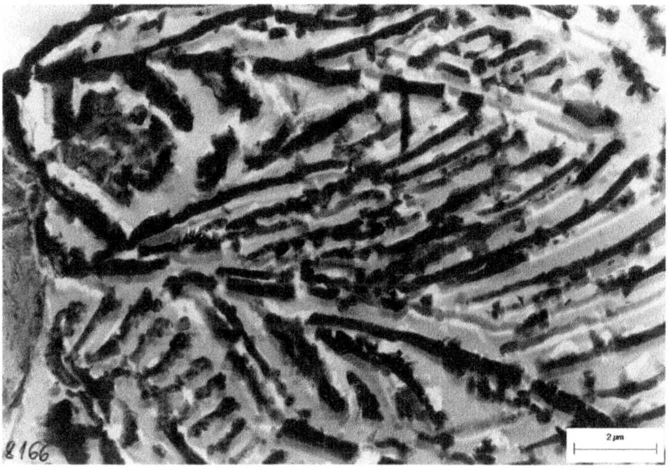

Fig. 15. Eutectic carbon nitride on the liquation fracture surface (17 347 - D steel)

Test Results

The results of liquation cracking tests obtained by simulation of welding cycles (tg α_K values) and VÚZ-LPT-1-6/TIG tests (V_d values) are summarized in Table 3. Except for a certain scatter, both methods are comparable. The complex evaluation of the test series proves that the conventional types of CrNi- and CrNiMo-steels (in Table 1 designated as class 17) can be classified by the 2nd and partially 1st degree of cracking resistance.

The worst results were found for AKOR 2 and AKRI 7 steels which have an unfavorable Cr/Ni < 1.0 ratio. Their primary solidification mode is austenitic representing the 3rd degree of cracking susceptibility in the VÚZ-LPT-1-6/TIG-remelting test. In the case of AKOR 2 steel, the liquation fracture surface was covered by eutectic particles of niobium carbonitrides (NbX).

Regardless of their purity, these steels are constitutionally susceptible to hot crack formation. Sporadic occurrence of microcracks, mainly in the remolten steel, cannot always be excluded completely and these cracks have to be evaluated by the admissibility criteria.

The melt of 17 347 steel (designated as D), which had a relatively high carbon content (0.098 %), was evaluated by the simulation test to be at the boundary of the 3rd degree. Frequent presence of eutectic constituents of coarser (TiX) and thinner (τ - $Ti_4S_2C_2$) morphology was observed on the surface of liquation cracks, as shown in Fig. 15 and Fig. 16.

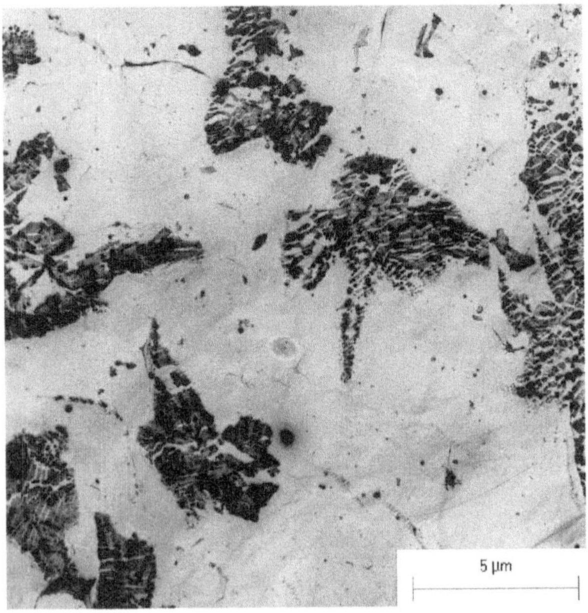

Fig. 16. Eutectic carbon sulphides $Ti_4S_2C_2$ on the liquation fracture surface of 17 347 steel

It is supposed that the decreased cracking resistance of this melt (D) is caused by the formation of a low melting point eutecticum consisting of TiX and $Ti_4S_2C_2$ phases in the underbead zone. The tests have proved that the majority of conventional austenitic steels classified by the 1st degree (crack-resistant) contain a low amount of δ-ferrite (0.3–3.5 %) and/or a lower amount of sulphur (~ 0.01%).

The calculated parameters are summarized in Table 3. The most comparable results from the verified calculation indices were attained for the parameter L including also δ-ferrite.

Table 3. Results of liquation cracking tests, calculated parameters and δ-ferrite measurements

Des.	Steel	Melt	tg α_k	V_d [mms^{-1}]	L	ΔH	E_{cr}/ E_{Ni}	P+S [%]	δ_{met} [%]
A	17246		1.5	0.15	-63.8	30	1.61	0.039	8.5
B	17247	E0094	1.25	0.15	-62.1	65	1.66	0.039	3.3
C	17249	E5621	2.0	0.13	-2.6	48	1.48	0.050	0.3
D	17347		0.7	0.06	-53.4	27	1.49	0.053	3.5
E	17350	E7699	1.18	0.08	23.9	13	1.35	0.058	0
F	17353	9316	0.95	0.07	8.4	16	1.39	0.047	1.3
G	17359	E8916	1.33	0.12	-2.0	85	1.24	0.037	0
I	17259	E7341	1.00	0.15	-1.23	36	1.16	0.04	0
J	17360	E8882	1.05	0.15	4.84	87	1.34	0.032	0
N	AKOR2	0-893	0.59	0.01	188.0	-27	1.2	0.025	0
O	AKC2	0-688	1.10	0.02	56.0	-158	1.16	0.026	0
P	AKR17	0-605	0.83	0.01	172.2	-703	0.67	0.036	0
X	17483	2679	1.66	–	-1892	384	1.91	0.036	31.1
Y	17199	E6222	0.91	–	-34.0	293	2.40	0.020	–
E1	17350	2309E	1.17	0.07	-14.9	55	1.52	0.049	1.5
E2	17350	–	1.43	–	21.9	22	1.44	0.046	0.3
H1	17352	95098E4	0.74	–	13.4	82	1.56	0.045	0.22
H2	17352	95019E1	0.72	–	11.9	52	1.46	0.050	0.25
H3	17352	94631E1	0.91	0.06	10.8	58	1.50	0.046	0.35
I1	17259	E7162	0.83	–	-12.0	78	1.29	0.032	0
C2	17249	–	1.66	–	-5.4	69	1.52	0.030	0.50
C1	17249	E9597	1.17	–	-3.2	33	1.38	0.034	0
C1O	17249	EOP	1.15	–	-8.9	71	1.50	0.028	0.30
C1E	17249	ETP	2.25	–	-4.7	28	1.32	0.027	1.00
C1V	17249	VOP	1.17	–	-4.6	46	1.43	0.029	0.20
K	17349		1.00	–	9.5	15	1.32	0.037	0
KO	17349	EOP	0.95	–	3.9	68	1.52	0.039	0.30
KE	17349	ETP	3.33	–	2.8	33	1.34	0.033	1.20
KV	17049	VOP	1.05	–	5.9	62	1.54	0.033	0.30
K1	17349		0.87	–	11.30	39	1.42	0.039	0

Des. Designation.

The parameter ΔH is the most stringent in comparison to others, no steel type was evaluated as resistant. The ratio of E_{Cr}/E_{Ni} expressing the mode of primary solidification seems to be significant, too.

The assessed scatter of results (e.g. steels D) follows from the fact that also other physical-metallurgical factors are involved which can at present hardly be defined in the parametric equations. With the help of a regression analysis of the welding cycle simulation test result, a parametric equation for $tg\alpha_K$ in dependence on the chemical composition was proposed:

$$tg\alpha_K = 4.166 + 361.30C - 4540.7C\,(P + S) \\ -117.27C\,(Cr/Ni + 90.6\,(P + S)Cr/Ni). \quad (6)$$

The equation is valid for chemical compositions of the austenitic steel within the range as follows:

C = 0.02 – 0.07 %, Mn = 0.8 – 1.4 %, P + S = 0.025 – 0.050 % and Cr/Ni = 1.12 – 1.72

The criteria of the parameter $tg\alpha_K$ are the same as those given in Table2.

Effect of Welding Parameters on Liquation Cracking

The austenitic steels are especially susceptible to grain coarsening in the overheated heat affected zone because they are monophase steels with a low thermal conductivity coefficient. The grain size in the heat affected zone depends on the size of the initial grain, on microstructural obstacles slowing-down the grain boundary migration (δ-ferrite, segregation of carbides, etc.), on the weld heat input, but also on the geometric factors (thickness). The size of the initial grain is determined by the steel making procedure. The presence of a carbide phase is mostly inadmissible from the viewpoint of the required properties (corrosion resistance).

In general, lower weld heat input is preferred to restrict the grain size, and thus to eliminate liquation cracking. Except for grain size reduction, also a narrower heat affected zone, lower segregation degree and a lower amount of the total molten phase are attained. Apart from a decrease in heat input, it is also recommended to reduce the welding speed v. Lower welding speed means lower thermal gradient of strain. In the Russian literature, the nominal heat input is denoted as:

$$\frac{q}{v} = \frac{I \cdot E}{v}, \quad (7)$$

i.e. the welding speed is considered separately; this is more practical from the viewpoint of hot cracking sensitivity evaluation. The criterion of the critical welding speed, i.e. the maximum speed which the welded joint withstands without crack initiation (while preserving the bead cross-section) is included in the Russian standard.

Preheating at low temperature, which several times proved to be suitable to provide protection against solidification cracking due to lower thermal stresses during cooling, has less effects on liquation cracking. Deposition of several smaller beads or employing a pulse process is considered to be convenient for lowering the total stresses. An other possibility of eliminating liquation cracking in the base metal is to use weld metal with a lower melting point, because, if the weld metal solidifies later, the tensile stresses cannot be transferred into the base metal.

Conclusions

Preliminary criteria of Thermorestor-W and VÚZ-LTP-1-6 tests (3 degrees) are proposed. Experimental results have been compared with the calculation indices (L, ΔH and E_{Cr}/E_{Ni}). An own parametric equation $tg\alpha_K$ based on the Thermorestor-W test are also proposed.

The majority of the examined austenitic stainless steels attained the 2nd degree of cracking resistance, and partially the 1st degree, i.e. the liquation cracking susceptibility would not restrict their weldability. The 2nd evaluation degree means only a partial cracking susceptibility which can usually be eliminated by specified welding conditions.

A resistance to liquation cracking (1st degree) of conventional austenitic steels was found at low contents of δ-ferrite (0.3–3.5 %) and at a low content of S (~0.01 %). The austenitic steels with the unfavorable ratio of Cr/Ni < 1 are susceptible to cracking (3rd degree).

The validity of calculation indices for the hot cracking susceptibility as well as the preliminary test criteria have to be further verified by results obtained from real joints.

References

1. Jakušin BF, Želev AN, Machnenko VI (1983) 2. simpozium SEV "Primenenie matematičeskich metodov pri izučeniji svarijemosti". Sozopol, Bulgaria
2. Tamura H, Watanabe T (1973) Trans of the Japan Welding Society, 2: 3–10
3. Medovar BJ (1964) Avtomatičeskaja svarka, 6: 1–13
4. Perteneder E, Rabeusteiner G, Schabereiter H, Tösch I (1979) Schweißtechnik, No 3
5. Matsuda F, Nakagawa S (1984) IIW Doc IX-1315-1984
6. Smith CS (1964) Metal Review, 9: 33
7. Kohira J (1973) Yosetsu Gijutsu, No 5: 41–47
8. Morishige N, Kuribayashi M, Okabayashi H (1979) IIW Doc IX-1114-1979
9. Kujanpää VP, Moisio T (1980) Conference Solid techn in the foundry and cast house. University of Warwick, Couventury
10. Arata Y, Matsuda F, Nakagawa H, Katayama S (1978) Transaction of Japan Welding Research Institute, No 2

Morphology of Hot Cracks in Single-Phase Weld Metal

B. Yakhushin

Baumann University, Moscow, Russia

Abstract

Solidification and liquation cracking, representing different types of hot cracking, have been investigated for austenitic stainless steels. Methods for the experimental evaluation and for the prediction of solidification and liquation cracking during single- and multi-layer arc welding are suggested.

Morphology of Solidification Cracks

Austenitic stainless steels which are often applied in safety-relevant constructions or components often exhibit a low hot cracking resistance during welding. Multiple thermal cycles affecting especially the grain boundaries of the primary weld metal are typical of multi-layer arc welding techniques and might even intensify the probability of hot crack occurrence.

Therefore investigations have been carried out concerning the testing as well as the prediction of hot cracking for single- and especially for multi-layer welding. The filler material chosen for the investigations was X5CrNiAMo 18 09 featuring a single-phase austenite microstructure up to the melting temperature. For best possible reproducibility of hot cracking during testing, the specimens were subjected to defined external loading during welding.

For hot cracking experiments concerning the root weld, pairs of specimens of 3 mm thickness were used which were welded end-to-end under simultaneous straining longitudinally to the weld direction [1]. The welding speed was varied over a wide range from 3 to 67 m/h. The welding current was adapted to the welding speed to ensure equal penetration depth of each weld. For all variations of welding parameters the detected hot cracks were directed longitudinal to the welding direction, which is especially characteristic of hot cracks in root welds. In order to minimize undefined thermomechanical loadings during overwelding the specimen's edges, slanting run-on and run-off tabs were applied (Fig. 1).

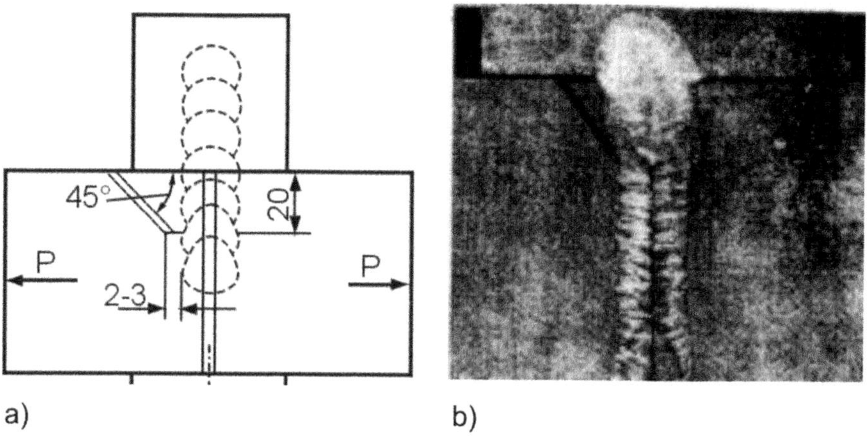

Fig. 1. (a) Outline of the specimen with a slanting run-on tab applied in the investigations and **(b)** a photograph of the topography of cracks obtained in tests

The straining of the specimens was triggered when the electrode reached the top of the run-on tab. By this special timing of the test procedure potential hot cracking could be limited to the middle part of the specimen. The period of post-weld straining was calculated as the time it took the specimen to reach a temperature between 700 °C and 800 °C, representing the lowest possible temperature of the brittle state.

Counterbalancing the opposite directed influences of arc current and welding speed on the heat input, single-layer welds of identical height but of various relative crystallite orientations in the plane of welding (Fig. 2) were produced. For obtaining a functional dependency, the general criterion R was applied:

$$R = q \cdot v_w. \tag{1}$$

The variable q physically represents the heat input per second and v_w is the weld travel speed. The usually applied criterion q/v_w varies only insignificantly when different welding speeds are counterbalanced by other welding parameters, especially if the cross-sectional area of the weld remains unchanged, but the microstructure and particularly the resistance against hot cracking vary greatly.

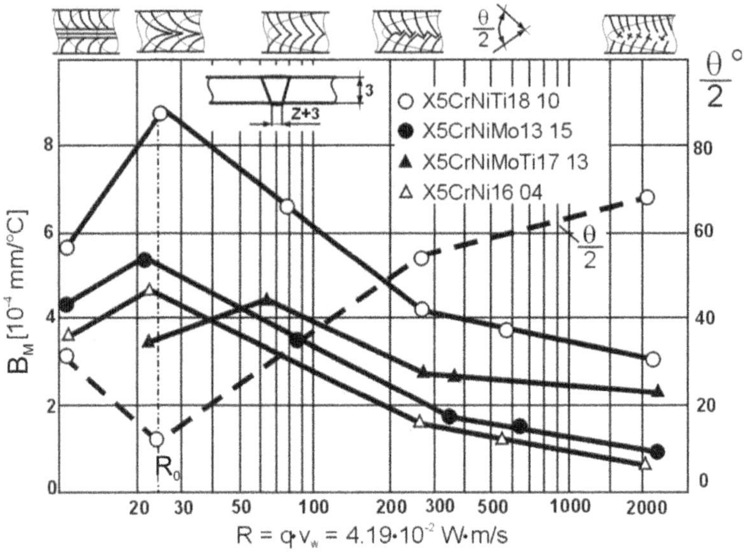

Fig. 2. Changes of the crystallite orientation «θ/2» in the weld metal center and resistance against hot cracking B_M for the $q \cdot v_w$ criterion depending on the welding speed

Fig. 2 shows the influence of welding speed on the crystallization orientation θ/2 and on the critical deformation B_M per °C. Angle θ/2 represents the direction of the solidification front measured at the surface on the weld metal centerline.

The factor B_M is calculated from the relation between the critical deformation speed v_{cr} and the cooling speed of the weld metal (v_{btr}) within the brittle temperature range (BTR), according to the following equation:

$$B_M = v_{cr}/v_{btr}. \quad (2)$$

Considering a constant heat input q, it can be seen by Fig. 2 that as the welding speed increases, the angle of crystallite orientation passes a minimum at a position where the critical strain rate B_M has its maximum.

The values of R corresponding to the maximum peak of B_M are nearly identical for various steels and range at welding speeds of $v_w = 6–12$ m/h. Such an optimal constellation of welding speed and arc current is named R_0 and means that crystals with lateral facets occur in the center of a weld. A slowdown of the welding speed, however, causes an unfavorable weld metal microstructure, since the crystals in the centerline concur under an obtuse angle. The corresponding decrease of hot cracking resistance can be measured by a lower critical strain rate B_M per °C. For values of $R > R_0$, crystals with front facets concur in the centerline of the weld leading also to a measurable reduction of the critical strain rate.

For various steels the critical strain rate B_M can be calculated from the specific R values by the following equation:

$$B_M = B_0 - k \cdot \lg(R/R_0), \tag{3}$$

whereby B_0 represents the value of B_M in the optimum mode from which the slope function B(R) for $R > R_0$ is subtracted. This means that the critical strain rate B_M is composed of a material-specific summand B_0 and a welding-specific summand represented by $B_W = k \cdot \lg (R/R_0)$.

For the calculation of the material-specific summand B_0, an equation based on statistical results of numerous hot cracking experiments was developed:

$$B_0 = [-3.779 + 2.331 \cdot E(Cr/Ni) + 0.4961 \cdot FN - 47.33 \cdot (P+S)] \cdot 10^{-7}, \tag{4}$$

whereby E(Cr/Ni) represents the chromium-nickel ratio, FN is the ferrite number and P+S represents the cumulated content of sulfur and phosphorus in the material. It follows from Eq. 4, that B_0 most significantly depends on the content of detrimental impurities but also on FN and E (Cr/Ni).

Regarding the technological and welding-related contribution to hot cracking, the value of B_W is significantly influenced by the size of the weld pool and can be calculated according to the following equation for a typical austenitic stainless steel:

$$B_W = 3.04 - 0.04L + 0.19B - 0.23b, \tag{5}$$

whereby L is the length of the weld pool, B is the width of the weld pool at the top side and b is the width of the weld pool at the bottom side.

The lengths of the weld pools L could be determined after the instant removal of the melt during welding by percussive impacts. According to the results of a statistically analyzable number of experiments, the following equation as an approximation for L was developed for steel X5CrNiTi 18 10:

$$L = 4.28 + 0.14 I_W + 0.5 V_W + 16d. \tag{6}$$

In the Eqs. (5) and (6) the fact is taken into account that the most significant parameters influencing B_W are the length of the weld pool and the width of the weld pool at the bottom side of the root weld. Under the condition of b converging to 0, a flat crystallization pattern changes to a three-dimensional pattern resulting in a reduction of unfavorable angles between crystallites and in a significant increase of the hot cracking resistance.

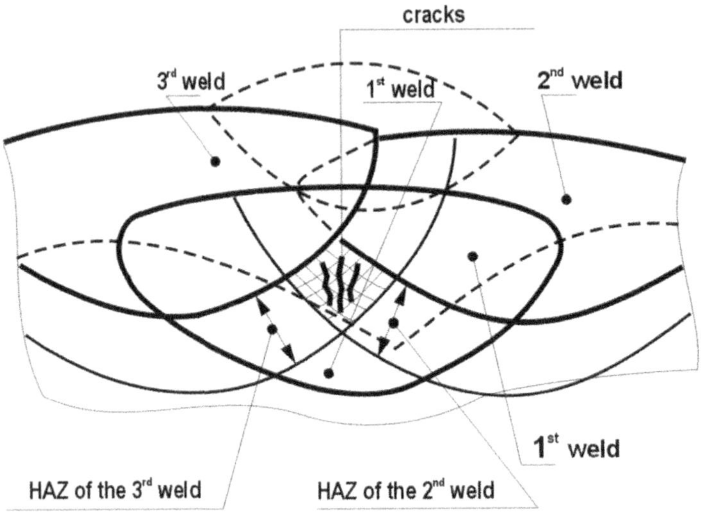

Fig. 3. Delineation of thermal interactions and the probable position of liquation cracks after three passes welded with a single-phase austenitic filler material

Morphology of Liquation Cracks

Liquation cracking has especially to be considered during multi-layer filler welding [3]. Liquation cracks usually appear as fine fissures in the previous layers of a weld and they generally do not cross the fusion line of two weld layers (Fig. 3). The histograms (Fig. 4) of 793 liquation cracks show that the liquation cracks are for the most part smaller than 0.5 mm.

They form up under the thermomechanical influence of the subsequent layer of the weld. Liquation cracks can run along migrated grain boundaries (Fig. 5).Liquation cracking as described above is assumed to result from high-temperature creep during temperature-activated relaxation of residual stresses accumulated by the previous weld pass(es).

A technological-based test procedure was developed reproducing such cracks under real welding conditions (Fig. 6). The operation principle is that the first layer of the weld is produced without any external deformation whereas during welding of the second layer the specimen with both layers is deformed by external loading. To exclude hot crack formation in the top layer, filler materials with increased technological crack resistance were applied [4].

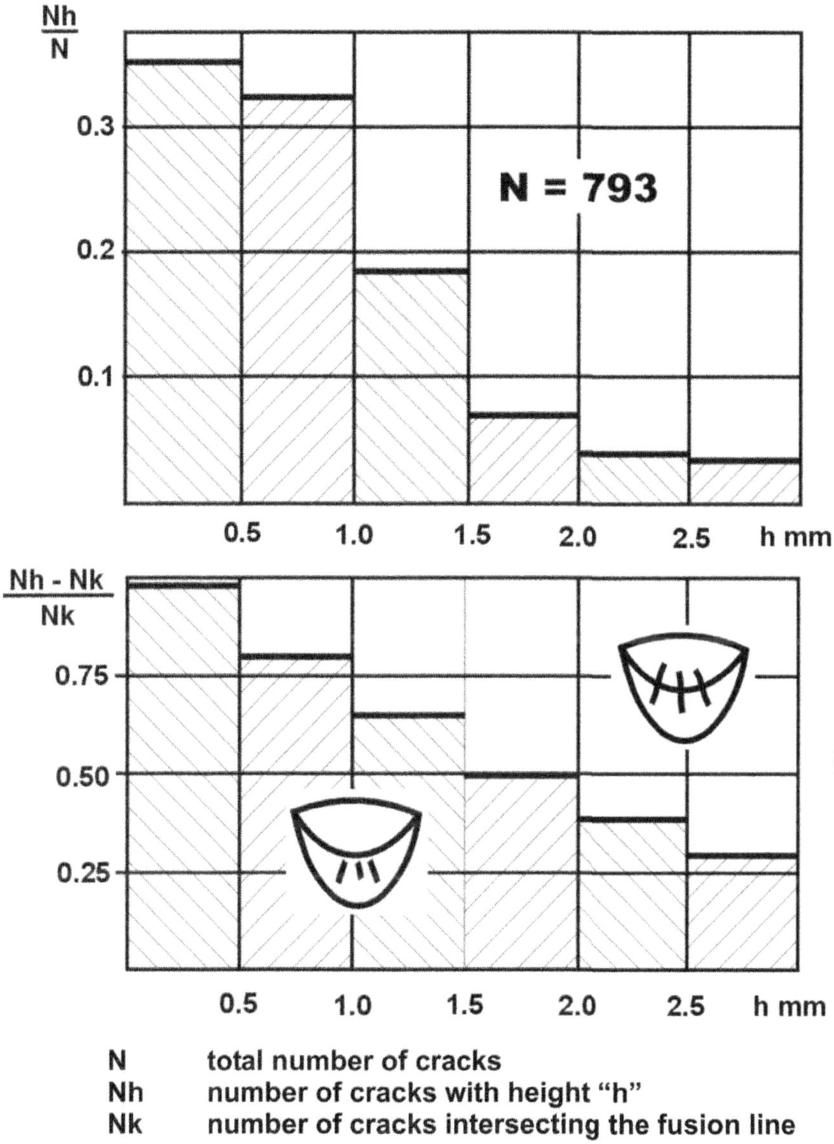

Fig. 4. Histograms of **(a)** liquation crack lengths and **(b)** frequency of cracks not crossing the fusion line; total number of analyzed hot cracks in multi-layer welds (above 100 passes): 793

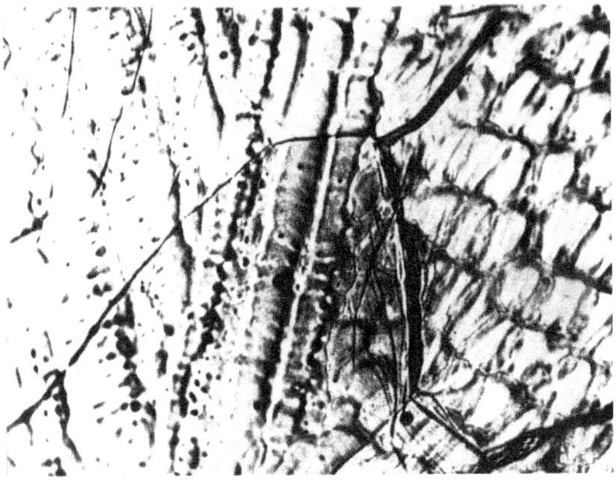

Fig. 5. A liquation crack in solid state crossing a crystallite, formed at the last stage of migration in the new position of grain boundary

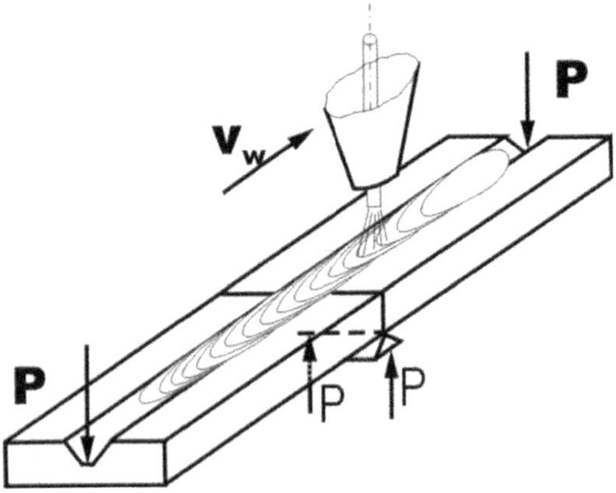

Fig. 6. Delineation of a test procedure for liquation cracking by repeated arc heating. Specimen is bent during welding of the second layer, so that the first layer represents the test weld and the second layer represents the repeated arc heating

If liquation cracking in the bottom layer of the weld is initiated, a critical deformation rate B_2 can be defined representing the hot cracking resistance of the weld metal in solid state (Fig. 7). Highest values of B_2, which means the hot cracking susceptibility is lowest, can be found in welds with 5–7 % δ-ferrite, which conforms to observations in practice.

In order to analyze the temperature range in which liquation cracks are formed, welded specimens were subjected to simulated weld cycle reheating and simultaneously exposed to different stresses (Fig. 7).

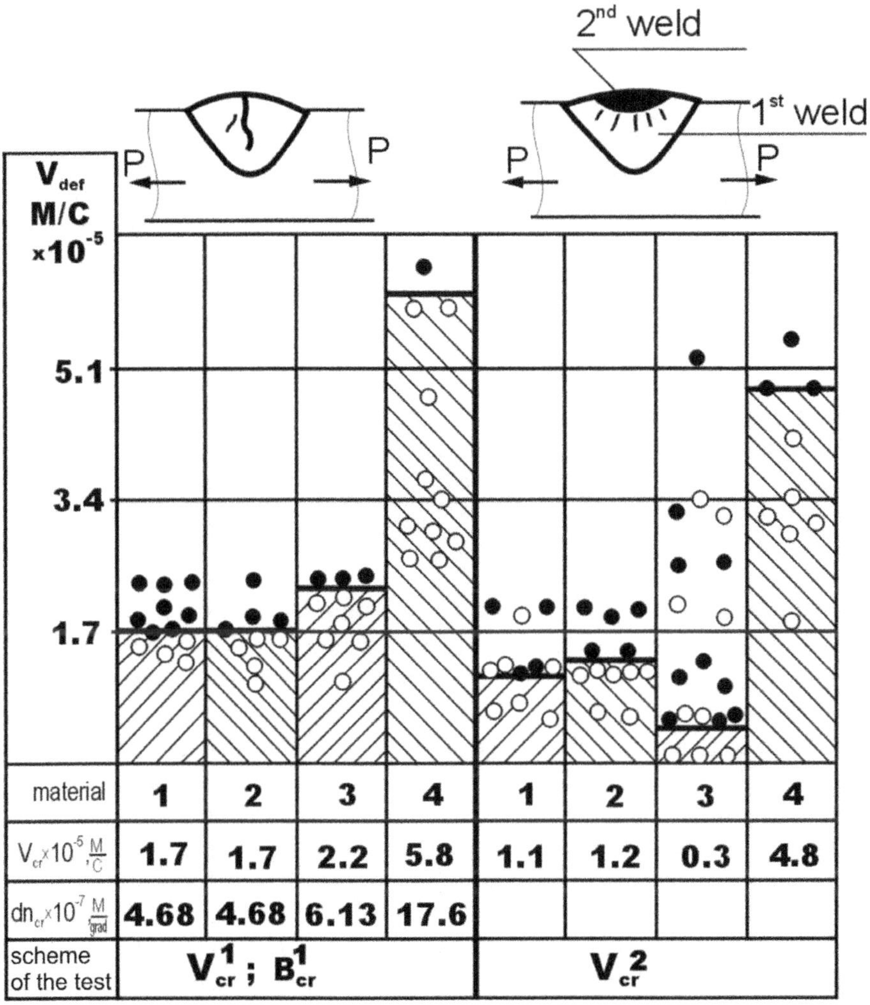

Fig. 7. Resistibility of high-alloy steels to crystallization and liquation cracking

The effects of high-temperature creep and of the temperatures at which subsequent brittle failures occurred are inversely proportional to the applied stresses.

By measuring the necking (reduction of the cross sectional area) of the specimen during heating, the brittle temperature range (BRT) was found at around 900–1100 °C (Fig. 8), i.e. in the liquation range (BRT2). When the stress loading of the specimen was further reduced, brittle failures occurred in the crystallization-liquation temperature range (BTR1) resulting from grain boundary melting.

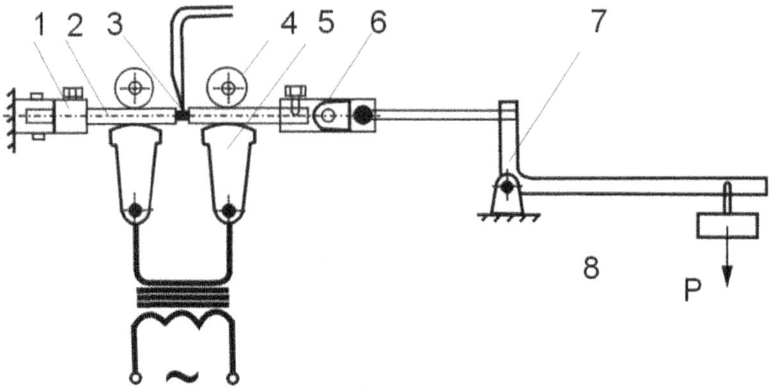

Fig. 8. Test device for the measurement of ductility and plasticity, respectively, of welded specimens *(2)* during simulated thermal weld cycle under external loading, induced by the machine parts *(8)(7)(6)(5)(4)(1)*. Part *(3)* represents the sensor for the measurement of the specimen necking

An increase of the deformation speed within BTR2 did not influence the brittle failure. For specimens containing δ-ferrite, brittle failure occurred only within BTR1. Specimens exhibiting a recrystallized base metal according to the tested grades also showed no brittle failure related to BTR2.

The results confirm the above-stated assumption that the basic reason for the existence of BTR2 is the non-equilibrium (weld metal) microstructure. The weld metal microstructure is supersaturated with vacancies which grow by deformation and which provide high speed diffusion processes during repeated arc heating.

The second reason for the existence of BRT2 is the shrinkage of the metal during cooling at speeds causing high-temperature creep. Such kind of deformation is determined by partial dislocations merging and creeping onto grain boundaries together with impurities which have a lower diffusion mobility in comparison to the elements of the base material. Segregations of this kind on grain boundaries cause grains to slide, leading to brittle failure of a solid-phase weld. By a comparison of the research results it was found that alloys exhibiting a BTR2 are characterized by a high level of the stacking fault energy, γ, responsible for their significant tendency to high-temperature creep. The stacking fault energy is calculated by the equation:

$$\gamma = -67 + 4.75\%Cr + 2\%Ni + 0.5\%Mn - 43.3\%N - 5\%Mo \qquad (7)$$

It was furthermore found that the critical deformation rate B_2 within BTR2 is proportional to stacking fault energy [5]. The stacking fault energy is recommended to be considered as a general function of the chemical composition of the weld metal.

The second factor which determines the formation of hot cracks within the BTR2 is the thermal cycle during the repeated arc welding. The period of the heating and period of the BTR2-state can be taken as the criterion for the deformation rate B_2. By adding a composition-specific term and a term considering the characteristic of the thermal cycle with linear dependencies, a basic dependence of B_2 on the chemical composition and the mode of welding was developed:

$$B_2 = n\left(\frac{1}{\gamma}\right) + m\left(\frac{t_0 - t_i}{t_0}\right), \tag{8}$$

whereby γ is the stacking fault energy, t_0 and t_i describe the time within BTR2 during welding in the optimum and in the investigated modes, respectively. m and n are factors of proportionality.

From Eq. 8 it follows that for increasing B_2 within BTR2, it is necessary to alloy a filler material with elements, as for instance nitrogen, molybdenum lowering the stacking fault energy and also increasing the cooling rate, thus reducing the period within crack-critical high-temperature states. An increase of the welding speed should also be avoided, since this might cause hot cracking within BTR1.

It is quite effective to apply low welding speeds (10–20 m/h) and to intensify the cooling of the weld metal by internal and/or external heat drainage and by electromagnetic agitation of the weld pool. It was also observed that alloying the weld metal with certain elements, for example with titanium, leads to contrary results of BTR1 and BTR2. Within BTR1, titanium leads to grain refinement, and within BTR2, it is an element, that increases the tendency to segregation. These results emphasize the need to determine B_{cr} separately for BTR1 and BTR2.

Furthermore, the prevention of solidification and liquation hot cracking can be achieved by creating the volumetric crystallization pattern in the root weld and by accelerating weld metal cooling using external or internal heat drainage [6].

Relation between Crystallization and Liquation Cracks

The described method reveals for the first time that plastic deformation of the weld metal within BTR1, which does not cause hot cracking but reduces the plasticity by over 50 %, also increases the tendency to hot cracking within BTR2 (Fig. 9). This phenomenon apparently occurred during the tests carried out with the PVR method.

Metallographic investigations showed that plastic deformation within BTR1 leads to an increase of migrations, segregations, linearity and sensibility to pickling of grain boundaries, which also indicates an increased tendency to high-temperature creep. It was found that the measured value of high-temperature deformation within BTR1, i.e. after the end of crystallization, has a significant relevance to many other characteristic material properties, for example the resistance to low-cycle fatigue, the resistance to heat loadings and to corrosion. It is proposed to explain this phenomenon, which we named technological inheritance, by the same summation of stacking defects by which the existence of two different BTRs is explained.

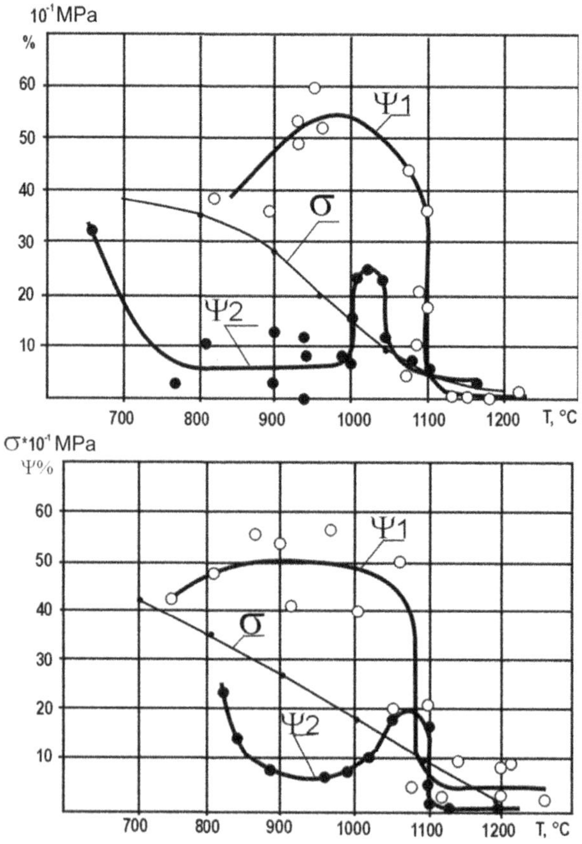

Fig. 9. Influence of plasticity exhaustion in BTR1 according to diagram (Fig.6) on the index of resistibility to liquation hot cracking within BTR2

Conclusions

The suggested test device allows it to determine the tendency to liquation hot cracking in solid-phase weld metal for repeated high-temperature cycles more precisely than other test procedures.

A high tendency of hot cracking in austenitic steels exhibiting a small temperature range of solidification results from merging of the solidification and the liquation temperature ranges of brittleness.

A higher susceptibility to hot cracking in austenitic welds at high welding speeds results from the formation of a flat crystallization pattern and the concourse of crystallites under an obtuse angle in the center of the weld, thus reducing the plasticity within both BTR1 and BTR2, respectively.

References

1. Jakushin BF (1969) Evaluation of technological durability depending on modes of welding. Welding manufacture 1
2. Jakushin BF, Gadzhiev MN, Gritsenko AI, (1991) Calculated-experimental method of determining weld metal resistibility to hot cracking. Welding manufacture 4
3. Jakushin BF, Misjurov AI (1983) Technological durability of multilayered welds with stably austenitic structure. Automatic welding 6
4. GOST of the USSR 26.389 – 84 (1985) Test methods for resistibility to hot cracking during welding by fusion. Publishing house of committee of standards
5. Vishnjakov JD (1970) Stacking defects in crystal structure. M Metallurgy: 216
6. Makarov EL (1991) Weldability of materials: Reference book. M Metallurgy
7. Rabesteiner G, Tosch J and Schabereiter H (1983) Hot cracking problems in different fully austenitic weld metals. Welding Journal, vol 62-1: 21–27

II Metallurgy and Materials

The Effect of Silicon and Iron on the Weldability of Ni-Co-Cr-Si HR-160® Alloy

I.S. Maroef[1], M.D. Rowe[2], G.R. Edwards[3]

[1]The Netherlands Institute for Metal Researches, Delft, Netherlands
[2]Haynes International Inc., Kokomo, USA
[3]CWJCR, Colorado School of Mines, USA

Abstract

Solidification cracking of HAYNES HR-160®[1] alloy was investigated with an emphasis on the interactive effects between silicon and iron concentrations. The focus on silicon was relevant to its major role in the sulfidation and oxidation at elevated temperatures, while the focus on iron was to explore the potential use for over-lay welding onto alloyed steels. Five compositions of the alloy were investigated, covering a [2 x 2] matrix of low and high concentrations of silicon and iron, in addition to the commercial alloy. To find a relationship between thermodynamic properties, microstructures, and weldability, several supporting analyses were performed. These analyses were differential thermal analysis and electron microscopy (chemical analysis), as well as fractographic investigations. Within the tested levels of alloying addition, both silicon and iron additions proved to increase the susceptibility of the alloy to solidification cracking, with silicon having the predominant influence. The detrimental effect that silicon caused was the increase in the solidification temperature range of the alloy, which in turn increased the brittle temperature range of the alloys. On the other hand, iron has the tendency to decrease the threshold strain to cracking, even though it narrowed the brittle temperature range of the alloy for a fixed concentration of silicon.

---

[1] HAYNES and HR-160 are trademarks of Haynes International Inc.

Introduction

The purpose of this study was to investigate the effect of silicon and iron on the weldability of HAYNES HR-160® alloy. HR-160 alloy is a solid solution strengthened Ni-Co-Cr-Si alloy. The alloy is designed to resist corrosion in sulfidizing and other aggressive high temperature environments. Silicon is added (~2.75 wt.pct.) to promote the formation of a protective oxide scale in environments with low oxygen activity. HR-160 alloy has found applications in waste incinerators, calciners, pulp and paper recovery boilers, coal gasification systems, and fluidized bed combustion systems.

HR-160 alloy has been successfully used in a wide range of welded applications. However, the alloy can be susceptible to solidification cracking under conditions of severe restraint. A previous study by DuPont, et al. [1] showed that silicon promoted solidification cracking in the commercial alloy. In earlier work conducted at Haynes, and also from published work by DuPont, et al., it was recognized that silicon segregates to the terminal liquid, creating low melting point liquid films on solidification grain boundaries. Solidification cracking has been encountered when using the alloy as a weld overlay on steel, and when joining HR-160 plate in a thickness greater than 19 mm with matching filler metal. The effect of silicon on the weldability of HR-160 alloy has been well documented, but the effect of iron is not well understood. Prior experience at Haynes has indicated that iron may be detrimental to the solidification cracking resistance of the alloy. Iron does not segregate to the terminal solidification product in nickel-base alloys, as does silicon [2], but iron may have an indirect or interactive influence on weldability. A set of alloys covering a range of silicon and iron contents was prepared and characterized to better understand the welding metallurgy of HR-160 alloy.

Materials and Experimental Procedures

Materials

For this study, four HR-160 type experimental alloys with different target levels of silicon (2.3 and 3.9 %) and iron (0.10 and 4.0 %) were prepared. The experimental heats represent a [2 x 2] matrix of high and low silicon and iron. The heat 8727-7-7506 was included in this study to represent the standard commercial alloy. The chemical compositions of the experimental and commercial alloys are given in Table 1.

Table 1. Chemical compositions of the HR-160 type experimental and commercial alloys (wt.pct.)

Haynes Intl. Heat #	EN2799-9-0894	EN2999-0-0896	8727-7-7506	EN2899-9-0895	EN3099-9-0897
Nominal Si & Fe	2.3% Si	2.3% Si	2.8% Si	4.0% Si	4.0% Si
Content	0.10% Fe	4.0% Fe	0.10% Fe	0.10% Fe	4.0% Fe
Source	Commercial	Exp.	Exp.	Exp.	Exp.
Element					
C	0.062	0.061	0.05	0.062	0.065
Cb	<.01	0.01	<.050	0.01	0.01
Co	30.01	29.99	30.1	29.99	29.74
Cr	27.61	27.88	27.80	27.60	27.68
Fe	0.08	4.01	<0.10	0.09	3.96
Mn	0.55	0.53	0.46	0.54	0.52
Ni	38.65	34.40	37.11	36.72	33.04
P	0.002	<.002	0.002	0.005	0.006
S	<.001	<.001	0.007	<.001	<.001
Si	2.34	2.24	2.81	4.09	4.11
Ti	0.5	0.49	0.46	0.5	0.44

The experimental alloys were vacuum melted, cast into electrodes, then electroslag remelted into 102-mm-diam. ingots. The ingots were forged to 44.5 mm, then further hot-rolled to 12.5-mm plate. The materials for varestraint testing were prepared by hot rolling at 1021 °C to 3-mm plate. The alloys were annealed at 1021 °C for 20 minutes and subsequently water quenched. Sigmajig specimens were taken from the hot rolled 12.5 mm (0.5 in.)-thick plate to be hot rolled to 2 mm (0.080 in.), then cold rolled to 1 mm (0.04 in.), followed by an anneal at 1100 °C.

Weldability Tests

Longitudinal varestraint tests were undertaken with welding parameters listed in Table 2. Each experimental alloy was tested at four augmented strain levels (nominal values of 1.0 to 3.5) in triplicate. The total crack length (TCL) and maximum crack distance (MCD) were taken as indicators of the cracking susceptibility. The maximum crack distance, MCD, is defined as the perpendicular distance from the fusion boundary to the far-end tip of the crack. Cracks were measured in the as-welded condition with a stereoscopic microscope and a filar eyepiece. An average of three time-temperature profiles were collected from a representative sample (commercial HR-160 alloy), specifically at the trailing edge of the fusion line of the weld pool center. The temperature measurements were accomplished by manually harpooning pure Pt/ Pt-13 % Re type R thermocouple wires to the weld pool during GTA welding. Before harpooning, the wires were kept trailing near the welding arc for several seconds to accumulate enough heat and thus minimize the thermal inertial problem in the temperature measurement. To approach the condition that occurs in the varestraint test, a copper block was placed underneath to support part of the length of the sample. GTA welding started at the unsupported part of the sample and thermocouple harpooning was executed when the weld pool traveled on the copper-block supported part. The average cooling rate, over the temperature range of 1400 and 1000 °C and at the trailing edge of the weld pool, was found to be 200 ± 30 °C/s.

Sigmajig testing of 50 mm × 50 mm coupons was also performed. As an indicator of cracking susceptibility, a threshold stress for crack initiation was measured. To determine the threshold stress, the sigmajig test was performed over a range of applied stress, and the threshold stress was taken as the minimum stress to produce cracking. Hot cracks that formed in the fusion zone were measured under a stereoscopic microscope.

Table 2. Welding parameters and specimen dimensions used in weldability tests

	Varestraint Test	Sigmajig Test
Specimendimensions (mm)	25 × 152 × 3	50 × 50 × 1
Current (Amp.)	70	20
Travel speed (mm/s)	1.9	14.8
Arc length (mm)	2.4	2.4
Shielding gas (L/s) argon	0.28	0.28
Type of electrode	2 % thoriated	2 % thoriated
Electrode include angle	60°	60°
Arc length (mm)	1	1
Weld configuration	Autogenous	Autogenous

Microstructural Characterization

Metallographic samples were taken from welded specimens to analyze resulting microstructures. Samples were electrolytically etched at 6 V using a solution of 5 g oxalic acid dissolved in 95 ml of hydrochloric acid. The weld microstructures were initially analyzed using light optical microscopy. The volume percent of interdendritic eutectic constituent was estimated by measuring its area fraction on a metallographic cross-section, using image analysis software. The average of 20 fields analyzed at 200x magnification was reported.

Selected fracture samples were analyzed with the scanning electron microscope for a better picture of the dendrite structures that were responsible for reduction or increase in cracking susceptibility. For chemical analysis of selected alloys, electron probe X-ray micro-analysis (EPMA-WDS) was conducted with a JEOL JXA 8900R, at electron beam energy of 15 keV and current of 50 nA, as well as a focused beam of 6 nm resolution. The measurements used Kα lines for Ni, Co, Cr, Si, Ti, Fe, and Mn. All samples for EPMA-WDS analysis were not epoxy-mounted, but were polished flat with 0.25 μm alumina slurry, then ultrasonically cleaned in acetone prior to analysis. The target outcome of the WDS measurement was an accurate assessment of the partitioning coefficient. Under an assumption of ideal condition satisfying the Scheil equation (no solid back-diffusion), the apparent partitioning coefficient ($k = kC_o/C_o = C_s/C_o$) of an element of interest can be estimated from the ratio of the concentration at the dendrite core (C_s) to that of the bulk material surrounding the core (C_o). Six-spot readings from nearby dendrite cores were collected in search of the three lowest concentrations of the segregating element. The three values were then averaged for use in the estimation of the apparent partitioning coefficient. The bulk composition was assessed in the weld metal and at the same location as that of the spots, covering an area of approximately 1600 μm^2. In average, about 20 dendrites were covered across this area of measurement. The area for both spot and bulk analyses was selected to be the weld metal adjacent to the HAZ of the sample, where primary dendrite morphology is relatively clear and where the cooling rate near solidification temperature was expected to be the highest.

Differential Thermal Analysis

Differential thermal analysis (DTA) was performed with a Netsch Thermal Analyzer STA 409, using alumina as a reference. This analysis was intended to provide some thermodynamic basis to explain the trends in solidification cracking susceptibility of the different alloys. As a parameter for solidification susceptibility, the solidification temperature range (STR) assessed from DTA was determined. Alumina crucibles and lids, designed for differential scanning calorimetry (DSC), were used to increase sensitivity of the analysis. During initial analyses (DTA1), a sample mass of 150 ± 20 mg was found sufficient for easy detection of reaction peaks. On-heating rate was determined to be 0.083 °C/s (5 °C/min.) to obtain near equilibrium solidus and liquidus temperatures. On the other hand, a rate of

0.33 °C/s (20 °C/min.) was selected for on-cooling analysis. In this way, reaction temperatures associated with non-equilibrium solidification can be measured. To detect the liquidus more accurately during cooling cycle, the analyses were repeated (DTA2) with an initial cooling rate of 0.05 °C/s to 1250 °C, followed by a final cooling rate of 0.33 °C/s. This final cooling rate was prescribed to allow easy detection of the terminal liquid reaction temperatures. The sample mass for DTA2 was determined to be 300 ± 30 mg.

Results

Weldability Data

The maximum crack distance (MCD) and the corresponding total crack length (TCL) obtained from the longitudinal varestraint test at four levels of augmented strain are shown in Fig. 1 and Fig. 2 , respectively. A decrease in the value of the cracking index for strains above 2.5 %, was evident. However, such reductions were not considered as an improvement of weldability.

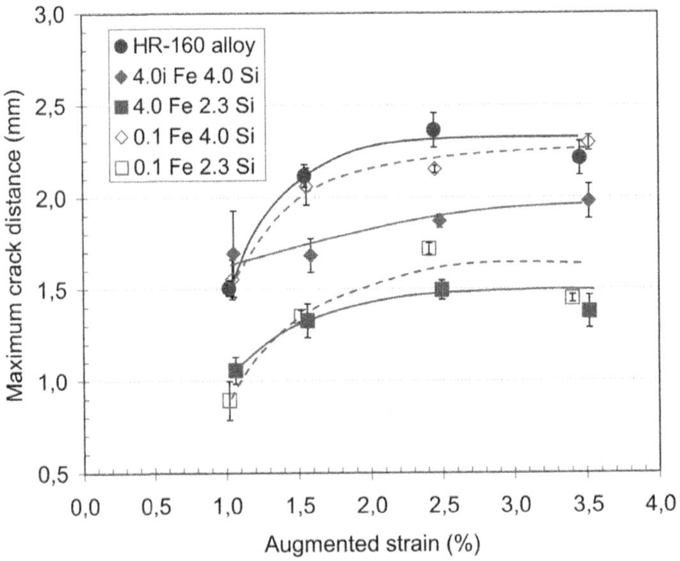

Fig. 1. Longitudinal varestraint data. Average maximum crack distance (MCD) vs. augmented strain

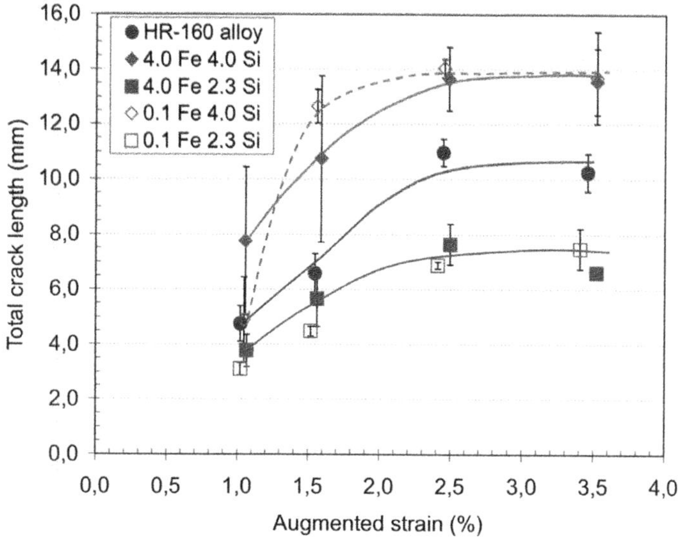

Fig. 2. Longitudinal varestraint data. Average total crack length (TCL) vs. augmented strain

Hence, an average value of crack lengths or distances was taken as the measure of susceptibility at strain levels of 2.5 % and 3.5 %. Many possible factors might have contributed to such reductions, such as change of restraint or increase in HAZ hot cracking, with increasing level of strain.

To get a better interpretation of the effect of silicon and iron addition on the weldability of the experimental HR-160 alloys, the values of MCDs and TCLs above 2.5 % are further summarized as a function of silicon concentration in Fig. 3 and Fig. 4, along with the threshold stress obtained from the sigma-jig test. Data points of iron-rich samples are filled with gray color to show the effect of iron for a given silicon concentration.

The summary plot in Fig. 3 showed that both the peak values of MCDs (at 2.5-3.5 % strain) and sigmajig threshold stresses were in a remarkably good agreement, indicating the predominantly detrimental effect of silicon over that of iron. Both types of data showed that the cracking susceptibility increased with an increase in silicon content up to approximately the commercial HR-160 alloy's concentration (2.3 wt.pct. Si).

Then, the susceptibility leveled off at higher concentrations of silicon, within this test matrix. When represented by the peak MCDs and sigma-jig threshold stresses, the weldability of the alloys appeared to be un-affected, if not improved, by the addition of iron. However, caution is necessary because detail in the varestraint data is not shown in such a summary plot.

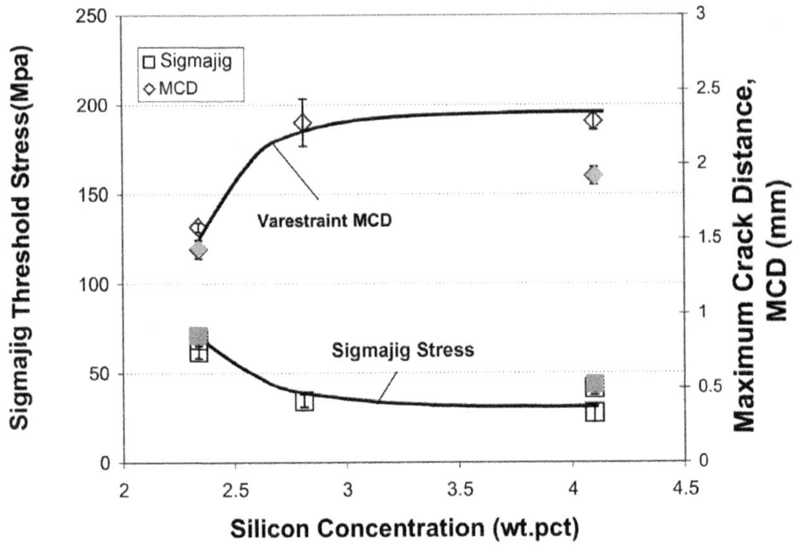

Fig. 3. The maximum crack distance (MCD) and sigmajig threshold stress as a function of silicon concentration

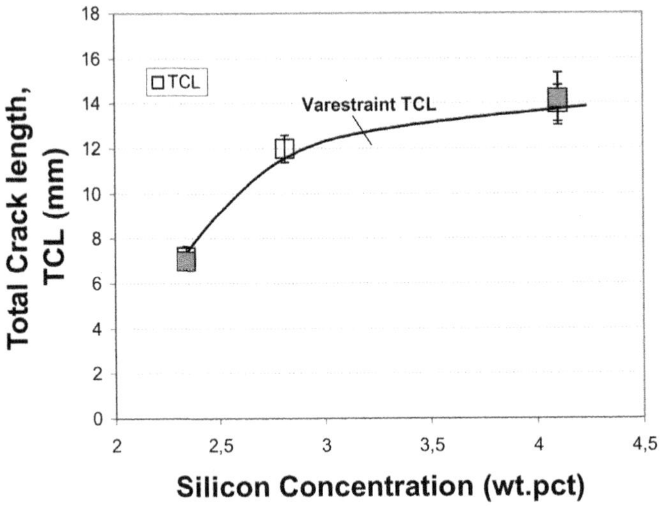

Fig. 4. The total crack length (TCL) and sigmajig threshold stress as a function of silicon concentration

Recall that for any of the high iron-containing alloy, the MCD value, and thus the cracking susceptibility, was always slightly higher at small levels of augmented strains. Testing condition at low levels of augmented strain in a varestraint test is thought to resemble an actual welding restraint severity closer than does a test at higher augmented strain level. It is worth mentioning that the local strains n front of the crack can be very large, even at low levels of augmented strain.

Despite the good agreement between MCDs and sigmajig threshold stresses, the authors considered not to rely too heavily on the sigma-jig data to support the analysis presented in the following discussions. The main reservation of using the sigma-jig was the low level of stresses measured during the testing. The sigmajig test has been shown to be marginally adequate at low values of threshold stress. Published experimental work [3], supported by finite element calculations, revealed that "free-stress" (zero preset loading) specimens could crack during welding solidification, but resisted cracking at moderate levels of pre-set loading.

It was found that, due to the peculiarity of the sigmajig set-up, the free-stress condition lacks sample restraint. During heating, the sample can easily expand in all directions, causing significant tensile stress in the transverse direction during the cooling cycle. A minimum level of preset load is required to prevent such a free expansion. Obviously, such a limitation could be overcome by using thicker samples. However, because sigma-jig testing was originally designed for thin sheet materials, production of thicker samples was not considered in the initial set-up of the study.

In contrast to MCD data, the total crack length (TCL) in Fig. 4 showed that cracking susceptibility continuously increased with an increase in silicon, within the concentration established by the test matrix. There was no measurable effect of iron, at any particular silicon concentration, to the values of TCL. The predominant effect of silicon was even more obvious when one considered these TCL data.

Quantitative Metallography

Fig. 5 shows the volume fraction of interdendritic eutectic, estimated from its area fraction, as a function of the silicon concentration, both from the varestraint and the sigmajig test samples. The non-metallic phase that forms the eutectic has been identified by Dupont as the silicon-rich G phase [$(Ni, Co)_{16}(Ti, Cr)_6(Si)_7$], which is known to form during solidification of the HR-160 alloy [1]. In this study, characterization of the eutectic phases was done only qualitatively (SEM-EDS) because, within the range of composition varied in this study, no change in eutectic solidification

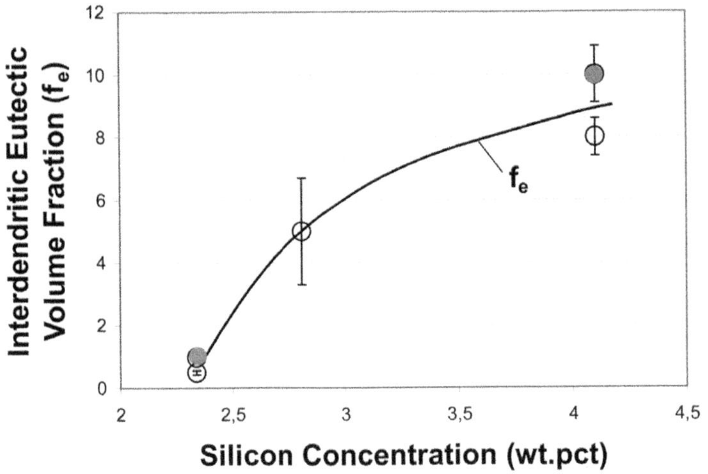

Fig. 9. The quantity of interdendritic eutectic product in the weld metal of HR-160 type alloys with different silicon and iron concentrations, as measured from varestraint and sigmajig specimens

product was anticipated. The amount of eutectic product appears to be directly proportional to the silicon content of the alloy.

The comparison of the eutectic amount and morphology of the weld metals studied can be seen from Fig. 6. The two low-Si HR-160 alloys clearly produced the lowest amount of eutectic upon welding. Their eutectic phases were droplet-shaped and completely isolated. The low-Si, high-Fe alloy produced a notably higher amount of eutectic than did the low-Si, low-Fe alloy. In the commercial HR-160 alloy, the amount and the liquid film network of interdendritic eutectic product are visibly less than in the two high-Si experimental alloys. Between the two high-Si alloys, the liquid film network of the eutectic product was roughly similar. It was recognized that the solidification structure is complex, and that more than one cross-sectional view would be necessary for one to compare the quantity and distribution of the liquid film network created by thermally cycling a given material. The ratio of TCL to MCD represents the average number of cracks occur at a given level of strain. As generally observed, the number of cracks in this study increased with increasing level of strain. One possible reason is that the MCD values did not increase linearly with strain, but even leveled off at 2.5 % strain. Therefore, as long as weak or brittle parts of the sample were still available, additional cracks would be initiated to accommodate the global strain imposed by the set-up.

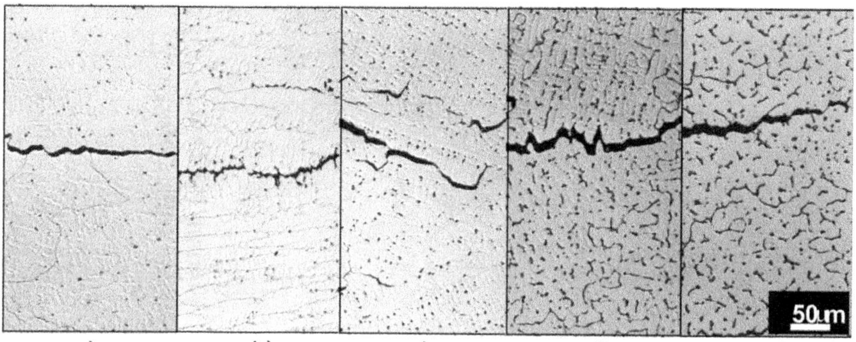

Fig. 10. Microstructures of longitudinal varestraint test specimens
(a) 2.0-wt.-pct.-Si, 0.1-wt.-pct.-Fe HR-160 experimental alloy
(b) 2.0-wt.-pct.-Si, 4.0-wt.-pct.-Fe HR-160 experimental alloy
(c) 2.8-wt.-pct.-Si, 0.1-wt.-pct.-Fe HR-160 commercial alloy
(d) 4.0-wt.-pct.-Si, 0.1-wt.-pct.-Fe HR-160 experimental alloy
(e) 4.0-wt.-pct.-Si, 4.0-wt.-pct.-Fe HR-160 experimental alloy

Accordingly, as shown in Fig. 7, the larger the volume fraction (f_e) of terminal liquid eutectic, the larger was the value of TCL. A structure with a large fraction of interdendritic eutectic phase contained many locations with poor coherency, which acted as easy crack initiation sites to accommodate a global augmented strain.

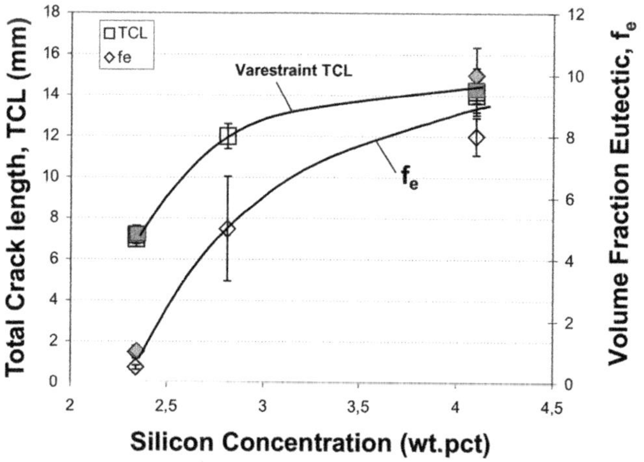

Fig. 11. The total crack length (TCL) and the volume fraction of terminal liquid eutectic (f_e), as a function of the silicon concentration

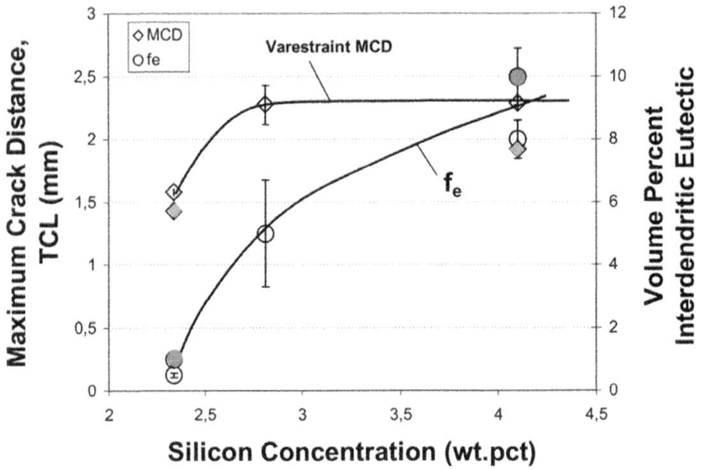

Fig. 12. The maximum crack distance (MCD) and the volume fraction of terminal liquid eutectic (f_e), as a function of the silicon concentration

In other words, larger TCL did not necessarily imply that the alloy is more brittle, but there were higher density of potential crack initiation sites. As a comparison, Fig. 8 shows that the MCD values did not depend on f_e after the silicon concentration exceeded 2.8 wt.pct.

Scanning Electron Fractography

Details of the fracture surface of the three most susceptible alloys (the commercial alloy and the two high-Si experimental HR-160 alloys) are shown from Fig. 9 to Fig. 11. The fracture surfaces shown were documented from specimens, varestraint tested at 1.5 %. strain. The fractographs shown were located immediately under the weld bead surface. The location of each surface was not necessarily nearest to the fusion line; hence the selected fracture surfaces may not represent cracking at identical temperatures.

In the fracture surface of the commercial HR-160 alloy, as shown in Fig. 9, classic dendritic structure is shown as evidence of solidification cracking. The dendrites appeared to be predominantly columnar, with moderate protrusions of secondary dendrite arms. Some plasticity, manifested as slip lines, is notable on the dendrite surfaces at high magnification. The slip lines were not necessarily involved in the cracking initiation process, but were most likely a result of crack opening displacement to accommodate the augmented strain.

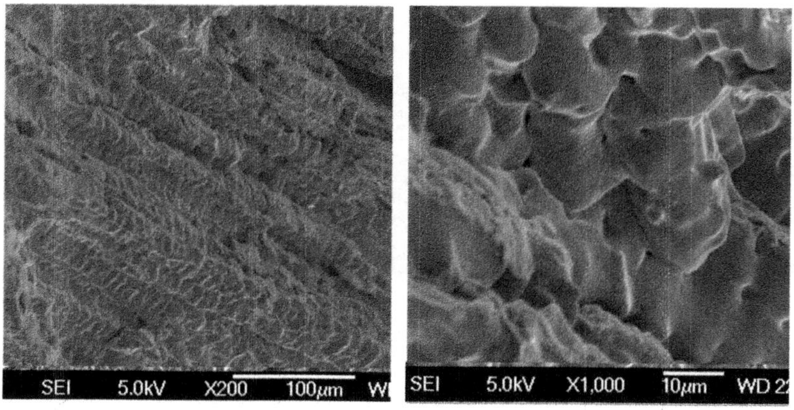

a) Low magnification b) High magnification

Fig. 17. SEM fractographs of commercial HR-160 weld metal

The fracture surface of the high-Si, low-Fe alloy is shown in Fig. 10. The primary dendritic columns are easily identified, one separated from another by a relatively long span of secondary dendrite arms. The surface revealed an apparently preferential growth of secondary arm dendrite growth in two dimensions.

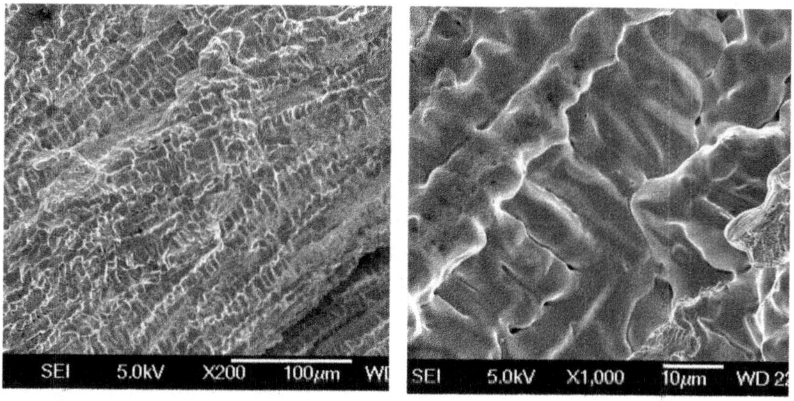

a) Low magnification b) High magnification

Fig. 18. SEM fractographs of 4.0-wt.-pct.-Si and 0.1-wt.-pct.-Fe HR-160 type experimental weld metal

a) Low magnification b) High magnification

Fig. 19. SEM fractographs of 2.3-wt. pct.-Si and 4.0-wt. pct.-Fe HR-160 type experimental weld metal

In contrast to fracture surfaces in the high-Si, low-Fe alloy, the fracture surface of the high-Si, high-Fe alloy (Fig. 11) revealed a more refined dendritic and tortuous structure. Such a three dimensional shape was not as well revealed in the cross-sectional micrographs presented previously in Fig. 6d. Some bridging, which appeared as brighter colored islands, was evident on the dendrite surfaces. At first, these islands gave an impression that they were the eutectic microconstituents, but back-scattered imaging did not distinctively show that they contained a high concentration of low atomic elements, such as silicon.

Partitioning Coefficients of Alloying Elements

Susceptibility to solidification cracking of a weld metal is frequently related to the quantity of the interdendritic eutectic product and the solidification structure itself. Both are governed by microsegregation, which itself depends on both the partitioning coefficient (k) of the particular elements in the alloy besides the solidification temperature range. Values of k for major elements can be approximated by dividing the measured dendrite core's composition (C_s) by the bulk alloy's composition (C_o); $k = kC_o/C_o = C_s/C_o$. The method assumes that dendrite tip curvature effects and solid-state diffusion are negligible, that thermodynamic equilibrium is maintained at the solid/liquid interface, and that diffusion of elements in the liquid is very fast [4]. According to this definition, an element with a low k value segregates strongly to the terminal liquid phase.

Table 3. Apparent values of initial partitioning coefficients, *k*, of selected elements in two high-Si and commercial HR-160 alloys

Element	4.0 Si – 0.1 Fe HR-160		4.0 Si – 4.0 Fe HR-160		Commercial HR-160	
	C_0	C_s	C_0	C_s	C_0	C_s
Iron	0.08	0.08	4.10	4.42	0.11	0.11
Nickel	37.92	36.98	32.90	31.75	35.75	35.10
Chromium	26.57	27.23	29.57	29.67	31.33	31.18
Cobalt	30.98	32.30	29.06	31.11	29.05	30.68
Silicon	3.56	2.74	3.73	2.52	2.81	2.11
Titanium	0.44	0.25	0.46	0.18	0.47	0.26
Manganese	0.50	0.42	0.46	0.34	0.58	0.54

Element	4.0 Si – 0.1 Fe HR-160		4.0 Si – 4.0 Fe HR-160		Commercial HR-160	
	k	(+/-)of k	k	(+/-)of k	k	(+/-)of k
Iron	1.00	0.03	1.08	0.02	1.00	0.02
Nickel	0.98	0.02	0.97	0.03	0.98	0.05
Chromium	1.02	0.05	1.00	0.05	1.00	0.03
Cobalt	1.04	0.04	1.07	0.02	1.06	0.04
Silicon	0.77	0.04	0.68	0.03	0.75	0.03
Titanium	0.57	0.02	0.39	0.04	0.55	0.02
Manganese	0.84	0.03	0.74	0.02	0.93	0.02

C_s concentration at the dendrite core (at zero fraction of solid), C_0 bulk composition (all in wt.pct).

The apparent *k* values for selected elements were calculated for the two high-Si and the commercial HR-160 alloys are listed in Table 3.

Table 4. Apparent values of initial partitioning coefficients, *k*, HR-160 commercial alloy and HR-160 overlay weld on 2.25Cr-1Mo steel (Dupont [1, 2])

Element	Commercial HR-160 (DuPont)			Weld Overlay (DuPont)		
	C_0	C_s	k	C_0	C_s	k
Iron				32.91	37.53	1.14
Nickel	37.90	36.50	0.96	25.83	23.79	0.92
Chromium	27.64	28.00	1.01	19.22	19.02	0.99
Cobalt	30.54	32.90	1.08	20.01	21.54	1.08
Silicon	2.60	1.85	0.71	1.93	1.02	0.53
Titanium	0.48	0.21	0.44	0.34	ND	NA

C_s concentration at the dendrite core (at zero fraction of solid), C_0 bulk composition (all in wt.pct.), *ND* not detected.

As a comparison, Table 4 includes the apparent *k* values assessed by DuPont [1, 2] from both the commercial HR-160 alloy and weld overlay of HR-160 superalloy on 2.5Cr-1Mo steel. The addition of iron to the high-Si HR-160 alloy increased the partitioning coefficient of silicon. With even higher iron content, as in the weld overlay data of DuPont, it was shown that the k value of silicon was further decreased.

Differential Thermal Analysis

As a typical DTA thermogram, Fig. 12 exhibits the DTA1 results for the commercial HR-160 alloy, containing both the on-heating and on-cooling thermogram curves. During heating, the sample exhibits a solidus temperature of approximately 1300 °C (determined as the onset of a detectable difference in the endothermic part of the curve) and reaches the liquidus temperature at 1370 °C (determined as the peak of the endothermic reaction). The near equilibrium melting range was 70 °C, similar to the result reported earlier by DuPont [1]. Upon solidification in alumina crucible, the alloy exhibited a large exothermic peak at an under-cooled temperature of 1340 °C, which corresponded to the formation of the primary dendrites. Such an undercooling also occurred at the analyses for the other HR-160 experimental alloys. Solidification was completed with a terminal liquid exothermic reaction at 1150 °C.

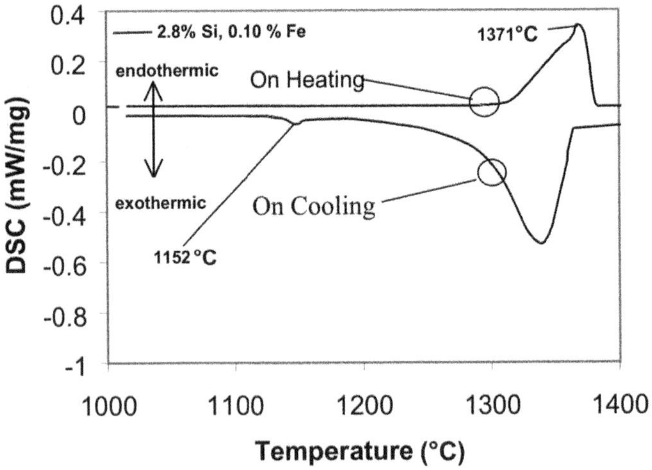

Fig. 23. DTA thermogram of the commercial HR-160 alloy during melting at 0.083 °C/s and solidification at 0.33 °C/s

Table 5. Results of Differential Thermal Analysis and Apparent Solidification Temperature Ranges

Alloy	On Heating Liquidus (°C) [a]	On Cooling Liquidus (°C) [b]	Terminal Liquid (°C) [c]	Solid. Temp. Range STR1 (°C)	Solid. Temp. Range STR2 (°C)
2.3%Si, 0.1%Fe	1378 ± 5	1350 ± 5	1280 ± 10	98± 10	70± 10
2.3%Si, 4.0%Fe	1376 ± 5	1360 ± 5	1272 ± 10	96± 10	88± 10
2.8%Si, 0.1%Fe	1371 ± 5	1364 ± 5	1152 ± 5	219 ± 5	212 ± 5
4.0%Si, 0.1%Fe	1340 ± 5	1330 ± 5	1125 ± 5	215 ± 5	205 ± 5
4.0%Si, 4.0%Fe	1341 ± 5	1330 ± 5	1130 ± 5	211 ± 5	200 ± 5

STR1 On Heating Liquidus Temp. – Terminal Liquid Temp., values from DTA1,
STR2 On Cooling Liquidus Temp. – Terminal Liquid Temp., values from DTA2.
[a] values were assessed from DTA1 with heating rate of 0.083 °C/s to 1500°C and cooling rate of 0.33 °C/s.
[b] values were assessed from DTA2 with heating rate of 0.33 °C/s to 1500 °C and initial cooling rate of 0.05 °C/s to 1250 °C, followed by final cooling rate of 0.33 °C/s for detection of terminal liquid temperature.
[c] values for terminal liquid temperature was an average of values assessed with DTA1 and DTA2.

To better approximate weld metal solidification (epitaxial growth), the solidification temperature ranges 1 (STR1) for the alloys were taken as the difference between the on-heating liquidus temperature and the on-cooling terminal liquid reaction temperature. As a complementary data, the solidification temperature ranges 2 (STR2) were also calculated, from the difference between the on-cooling liquidus and the terminal liquid temperatures, assessed during DTA2 analyses described in the experimental procedure. For the HR-160 commercial alloy, the STR1 and STR2 were 219 °C and 212 °C, respectively. Table 5 lists the reaction temperatures and the STR values of the alloys.

Discussion

The following discussion is intended not only to explore the effect of composition on the solidification cracking susceptibility, but also to relate the results to established hot tearing criteria. Factors that influence solidification cracking susceptibility have been taken into account in different theoretical formulations, although no single one has satisfactorily described the complete picture of the mechanism involved. Two main schools of thought on this subject are: 1) those of the structural considerations, such as the Clyne-Davis concept [5] and 2) those of the rheological considerations, such the Feurer's competitive processes between liquid back-filling and shrinkage [6]. The validity of each criterion appears to depend strongly on the type of process (welding or casting) and the mechanical testing applied (stress or strain based).

However, it is generally accepted that a solidifying metal starts to be vulnerable to solidification cracking when coherency is reached [7]. Coherency is defined as the condition when adjacent dendrites start to form solid/solid contacts.

A common way to analyze data obtained from a strain-based testing, such as the varestraint test, is to plot the augmented strain as a function of the cracking temperature range. This plot is usually termed the ductility curve. Such a plot for the data assessed in this study is shown in Fig. 13 below. The temperature range for cracking is deduced from the maximum crack distance, taking the temperature of the fusion line as the on-cooling liquidus of the alloy, as assessed by the DTA2 analysis. When considering the epitaxial growth during weld metal solidification, estimating the fusion line temperature from the on-cooling liquidus determined by DTA2 analysis (slow cooling rate) would be quite realistic. The temperature at the crack tip was estimated from knowledge of the cooling rate, over the temperature range of 1400 to 1000 °C, and at the trailing edge of the weld pool (an average of 200 ± 30 °C/s). Weldability of the alloy is then measured by two parameters: the brittle temperature range (BTR), and the threshold strain to initiate cracking. BTR is defined as the maximum range of cracking temperature covered by the ductility curve. An excellent weldability is then characterized by short BTR and high threshold strain to cracking.

In this study, the threshold strain to initiate cracking (an alternate ductility indicator) could not be accurately quantified because strain blocks that could give lower levels of strains were not available. To highlight the trend in the threshold strain to cracking, the cracking temperature range at 1 % augmented strain (CTR1%) of the alloys will be used for comparison purposes. It was assumed that a large value of CTR1% would tend to yield a low value of threshold strain to cracking, by extrapolation of the ductility curves.

As shown in Fig. 13, the two low-Si HR160 alloys clearly outperformed the other alloys, as signified by their short BTRs, as well as by their short CTR1%. If not for the tendency to yield a lower apparent threshold strain to cracking, the low-Si, high-Fe HR160 alloy would certainly be categorized to be as resistant to cracking as its low-Fe sister. Both the commercial and the high-Si low-Fe HR-160 alloys had similar values of BTR and CTR1%. Both alloys' BTRs were also the widest among the alloys studied. In contrast to these two later alloys, the high-Si, high-Fe HR-160 alloy yielded a BTR about 30 °C shorter, but cracked at a higher value of CTR1%. In other words, by extrapolation of the ductility curve, the high-Si, high-Fe HR-160 alloy tended to yield a lower apparent threshold strain than do the commercial and the high-Si, low Fe HR-160 alloys.

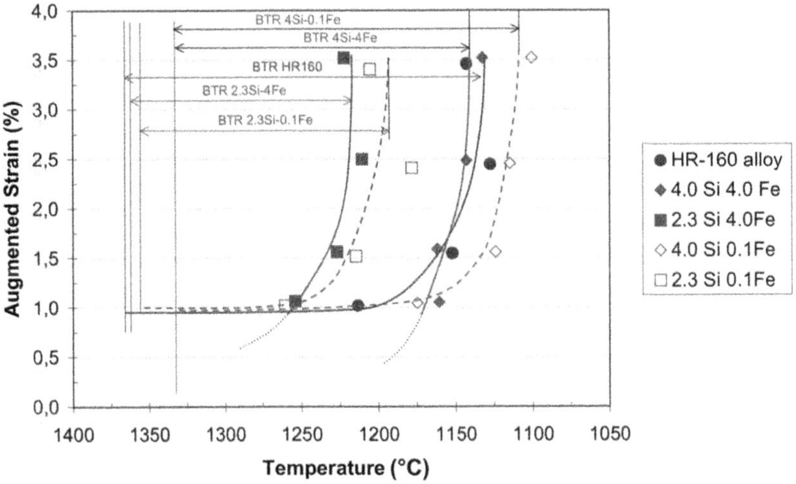

Fig. 24. Ductility curve (response) for the various HR-160 alloys upon rapid straining during the longitudinal varestraint testing

With respect to hot tearing criteria, the varestraint test may pose some limitation in contrasting the rheological process (liquid back-filling) among different alloys. As highlighted by Matsuda [8], the major part of a solidification crack in the varestraint test propagates rapidly through a rigid structure, with only dispersed and droplet-like terminal liquid.

Accordingly, the varestraint test emphasizes the structural factors of solidification cracking as opposed to the rheological factors, particularly at high strain levels. However, at low strain levels, the rheological process may play a bigger role, and may better demonstrate the beneficial effect of back-filling by a weld metal with a large volume fraction of terminal liquid (beginning from 10 to 20 pct).

The BTR values, assessed at high augmented strain levels, were predominated by the region of fast crack propagation. In this mode of fracture, the structural factors played a big role. Therefore, the material parameters that influenced the structure formation must be explored to elucidate the role of chemical composition on the solidification cracking susceptibility. Among the many materials parameters, the predominant effect of k has been well recognized [9].

To focus on the role of k, specifically k_{Si}, the two high-Si alloys were selected for a comparison study because their values of STR were similar. Comparison of the solidification structures among the alloys of interest from cross-sectional metallography was not successful, especially when the columnar growth followed the weld pool in a curved trajectory. For this purpose, the solidification structures revealed by the fracture surfaces in Fig. 10 to Fig. 11 were considered as alternative choices to assess signatures of growing dendrite structures. A revealing effect produced by the two fracture surfaces, was a decrease in k_{Si} (from 0.78 to 0.68) led to an increase in tortuosity of the dendrites. It follows that an increase in tortuosity implied an increase in coherency of the structure, as the density of solid/solid bridging during solidification should also be increased. The observation that the BTR value of the high-Si, high-Fe Hr-160 alloy (smallest k_{Si}) was considerably higher than that of the high-Si, low-Fe alloy appeared to be due to a possible increase in load bearing capacity when coherency was increased. Although the high-Si, high,-Fe HR-160 alloy was superior with respect to BTR, an explanation for why this alloy performed poorly with respect to the trend of threshold strain to cracking (higher value of CTR1%) remains elusive. Further study is needed to clarify whether it was the rheological factor (degree of back-filling) or the structural factor (degree of coherency) predominant in causing the significant reduction of the threshold strains, as the iron concentration was increased.

Fig. 14 shows the relationship between BTR and two materials parameters; namely the silicon partitioning coefficient (k_{Si}) and the solidification temperature range (STR). Without the presence of iron, it was clear that BTR increased with increasing STR, although the function was not linear. The addition of iron then made the change in the value of BTR unpredictable. There also appeared to be a relationship between k_{Si} and BTR, for the three alloys for which the data were available. The relationship was also independent of the composition of the alloys. However, more experimental data is necessary to confirm this apparent correlation between k_{Si} and BTR. The eutectic volume fraction was not included in the figure because it has been found to lack in correlation with either BTR (MCD), or with k_{Si}. This observation may change if another testing methodology is used, such as the Houldcroft or circular-patch test.

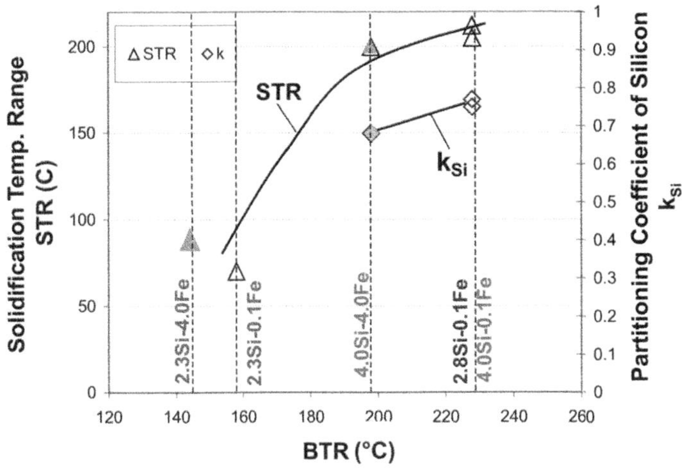

Fig. 25. Relationship between brittle temperature range (BTR) and silicon partitioning coefficient k_{si}, as well as the solidification temperature range (STR)

Conclusions

1. Within the alloying levels tested, both silicon and iron additions proved to increase the susceptibility of the alloy to solidification cracking, with silicon having the predominant influence.
2. At a fixed iron concentration, the addition of silicon increased the brittle temperature range of the HR-160 alloys.
3. The solidification temperature ranges of the HR-160 alloys depended upon the silicon content, with iron having negligible measurable effect.
4. At a fixed silicon concentration, the addition of iron decreased the brittle temperature range. However, such an apparently beneficial effect was compromised by a concurrent increase in the cracking temperature range at 1 % augmented strain. It is anticipated that iron had the tendency to yield a reduction in the threshold strain to cracking.
5. In high-Si HR-160 alloys, the addition of iron decreased the partitioning coefficient of silicon.

Acknowledgement

This research was conducted at the Colorado School of Mines and Haynes International. The financial support of the Fossil Energy Materials Program, U.S. Department of Energy, under Contract Number 19X-SW314C, is gratefully acknowledged. The authors also gratefully acknowledge the support of Dr. Stan David, Materials Division, ORNL and the technical assistance of Ms. Brenda Mulac, formerly of CSM. Ing. J. Kiersch and Ing. C. Kwaakernaak, from the Surface and Interface Research group – Delft University of Technology, the Netherlands, are also gratefully acknowledged for their help on the EPMA-WDS measurement. The support from the Netherlands Institute for Metal Researches (NIMR) that enabled the first author to complete the study was highly appreciated.

References

1. DuPont JN, Micheal JR, Newbury BD (1999) Welding Journal 81: 408–415
2. Dupont JN (1997) J. Materials Science 32: 4101–4107
3. Feng Z, Zacharia T, David SA (1997) Welding Journal 79: 470–483
4. Brody HD, Flemmings MC (1966) Trans. AIME 236: 615–623
5. Clyne TW, Davies GJ (1975) British Foundry 68: 238–244
6. Feurer U (1976) Giessereiforschung, 28: 75
7. Borland JC (1960) British Welding Journal 7: 508–512
8. Matsuda F, Nakagawa H, Sorada K (1982) Trans. JWRI, 11(2): 67–77
9. Flemming MC (1974) Solidification Processing. McGraw Hill, New York

The Influence of Different Nb-Contents on the Hot Cracking Susceptibility of Ni-Base Weld Metals Type 70/20

R. Vallant

Institute for Materials Science, Welding and Forming,
Graz University of Technology, Austria

Abstract

Metal-cored wires and flux-cored wires for GMAW of different Nb-contents have been tested concerning their hot cracking susceptibility. Using tensile test specimens and the PVR-test, an optimum content could be found, to cause low hot cracking susceptibility comparable to other welding consumables on the market. This improvement was due to the formation of NbC and by keeping the basicity index at a definite value, whereby the required mechanical-technological values could be fulfilled.

Introduction

The background of this investigation was the development of a Flux-cored wire (FCW) of type Ni-base 70/20, whereby the arc stability and the slag viscosity (weldability in horizontal position and position vertical up, bead covering and slag detachability) were optimized. However, the required fracture strain in Table 1 could not be reached, because of the appearance of hot cracks.

From the historical point of view the Ni-base welding alloy type 70/15 (SG-NiCr15FeMn, alloy 182) was developed in North America and was standardized in AWS A5.14.

However it came out, that this welding alloy was susceptible to hot cracks and to subsequent stress corrosion cracking (SCC), when welding thick plates, especially in reactor constructions.

Hence the Ni-base welding alloy type 70/20 (SG-NiCr19Nb, alloy 82) was developed in Europe, which has considerable lower hot cracking susceptibility (HCS) and susceptibility to SCC than type 70/15. From there it appeared, that no solid wires type 70/15 are available, as the higher alloyed type 70/20 meets its requirements. Consequently the similar base material alloy 600 is usually welded with "overalloyed" welding rod or solid wire, however welding with MMA-Electrode of type 70/15 is also permitted [1].

Table 2 shows the standard All-weld metal composition of Ni-base 70/20, Ni-base 70/15 and base metal alloy 600 as well.

The function of Nb in Ni-base weld metals and stainless weld metals as well, is to stabilize the C by forming NbC. By that the formation of Cr-Carbides (sensitization) is avoided and the susceptibility for intergranular corrosion should be lowered.

Additionally the very stable NbC should also improve the creep strength of the weld metal for high temperature applications. For the high cooling speeds in welding the Nb-content has to be distinctly higher than the stoichiometric content 8 x % C. The weld metal has to be so-called "over-stabilized". Unlike the base metal a stabilization using Ti is not advisable, because of the flashing of Ti in the welding arc, due to the high affinity to O [2].

In the similar base metal alloy 600 the formation of NbC seemed to correlate with increased protection from intergranular corrosion IGC, where a suitable addition of Nb/C>30 had a positive influence [3]. However, increasing the Nb content for alloy 690 welds, resulted in decreased IGC resistance within the interdendritic spaces [4].

The stated Nb-contents of Ni-base welding alloy types 70/20 (bare rods and weld metal) in European and AWS standards are 1.5–3 and 2.0–3.0 % respectively. The maximum carbon content is stated to 0.10 and 0.05 % respectively [5 – 9]. The objective of this investigation was to evaluate the influence of Nb within and out of this tolerances, see Table 2.

Table 1. Standard requirements for the mechanical values of Ni-base 70/20 and 70/15 All-weld metal

Standard	Abbreviation	$R_{p0.2}$ [MPa]	R_m [MPa]	A5 [%]
DIN 1736-2	NiCr20Nb	360	600	30
DIN 1736-2	NiCr15FeMn	360	600	30
AWS A5.11	ENiCrFe-3	–	550	30
AWS A5.14	ERNiCr-3	–	550	30

Table 2. Standard all-weld metal composition of Ni-base 70/20 and 70/15, bare electrodes and similar base metal.

	Kurzzeichen –DIN EN Norm/ Classification-AWS Spec.	W.-Nr./ UNS No.		C	Si	Mn	P	S	Cr	Ni	Nb	Fe	Ti	Mo	Cu	Others[a]
Base metal	NiCr15Fe, alloy 600	2.4816 N06600	min						14	72		6				
			max	0.08	0.5				17			10	0.3		0.5	0.3 Al
	LC-NiCr15Fe, alloy 600 L	2.4817 N06600	min						14	72						
			max	0.03	0.5				17				0.3			
Bare Electrodes/ All-weld metal	NiCr20Nb DIN 1736 Draht-, Bandelektroden	2.4806	min			2.5			18.0	65.0	1.5					
			max	0.05	0.5	3.5		0.015	22.0		3.0	3.0	0.8		0.5	
	NiCr15FeMn DIN 1736 Schweißgut Stabelektrode	2.4807	min			5.0			13.0	61.0	1.0	2.0				
			max	0.10	1.0	10.0		0.015	17.0		3.5	9.0	1.0		0.5	0.3 Ta
	ERNiCr-3 AWS A5.14 Bare Electrodes and Rods	N06082	min			2.5			18.0	67.0	2.0					
			max	0.10	0.5	3.5	0.03	0.015	22.0		3.0	3.0	0.75		0.5	0.5
	ENiCrFe-3 AWS A5.11 Weld Metal SMAW-Electrode	W86182	min			5.0			13.0	59.0	1.0					
			max	0.10	1.0	9.5	0.03	0.015	17.0		2.5	10.0	1.0		0.5	0.5
	NiCr20Mn3Nb prEN ISO 14172, Schweißgut Stabelektrode	6077	min			2.0			18.0	63.0	1.5					
			max	0.10	0.5	6.0			22.0		3.0	4.0	0.5	2.0	0.5	
	NiCr15Fe6Mn prEN ISO 14172, Schweißgut Stabelektrode	6182	min			5.0			13.0	60.0	1.0	2.0				
			max	0.10	1.0	10.0			17.0		3.5	10.0	1.0		0.5	0.3 Ta
	NiCr20Mn3Nb prENISO 18274, Draht-, Bandelektroden	6082	min			2.5			18.0	67.0	2.0					
			max	0.10	0.5	3.5			22.0		3.0	3.0	0.7		0.5	
	NiCr16Fe8Nb prEN ISO 18274, Draht-, Bandelektroden	6062	min						15.0	70.0	1.5	5.0				
			max	0.08	0.3	1.0			17.0		3.0	9.0			0.5	

■ Ni-base Type 70/20 □ Ni-base Type 70/15.

[a] Sum must not exceed 0.5 wt%.

Experimental Procedures

Self-Loaded Hot Crack Testing: Hook Cracks on Tensile Test Specimens

The evaluation of the so called Hook cracks on tensile specimens according to DIN 32525 was recommended as a self-loaded hot crack testing method in 1990 [10] (Fig. 1). Hot crack investigations with this method were already performed in earlier decades [11, 12]. As described, with this simple method only distinctions can be made between lower or higher HCS of weld metals, i.e. the preliminary selection of weld metals. For detailed investigation this method is not suitable, as solidification cracks and liquation cracks can not be kept apart [12].

Fig. 1. Hook cracks on the surface of a tensile test specimen (SC and LC widened by loading)

For the evaluation of the HCS of ductile weld metals Pohle [13] has suggested to use the reduction of area over the number of so called Hook cracks found on the surface of broken tensile specimens by using a magnifier (6–10 fold).

The Testing program I is shown in Table 3: The tensile test specimens were made out of All-weld metal (AWM), acc. EN 1597-1 and weld metal of a Plate Deposit (PD).

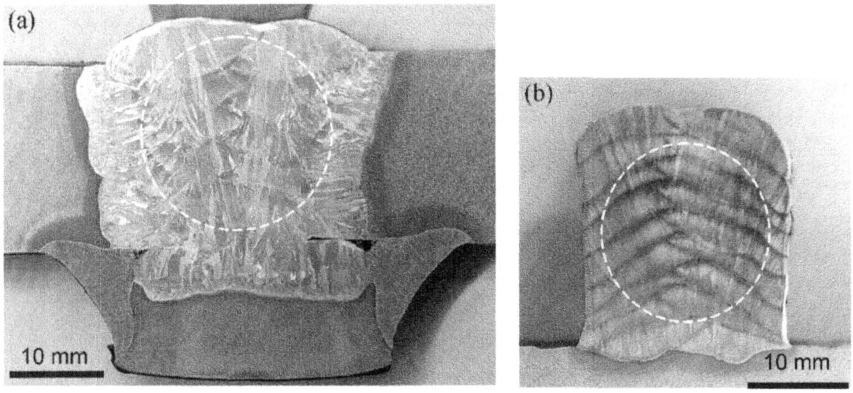

Fig. 2. **(a)** Cross section of All-weld metal AWM, 6 layers - 2 runs each and **(b)** weld metal of Plate Deposit PD, 8 layers - 2 runs each. Doted circles: Diameter of tensile specimen acc. to DIN 32525 (thread M16)

The weld metal of PD is partially hindered in shrinkage by the plate thickness and the underlying layers, and therefore a higher HCS can be expected than for AWM (Fig. 2).

The fabrication of the weld metals was performed using metal-cored wires (MCW) and FCW respectively. Cross sections of them are shown in Fig. 3. Additional weld metals were produced with solid wire (SW), MMA-Electrode and FCW on the market. The applied welding procedures were GMA-welding for FCW (shielding gas $82Ar/18CO_2$) and for MCW (50Ar/50He), with direct current electrode positive (DCEP). The SW Ni-base 70/20 was welded using pulsed current (50Ar/50He).

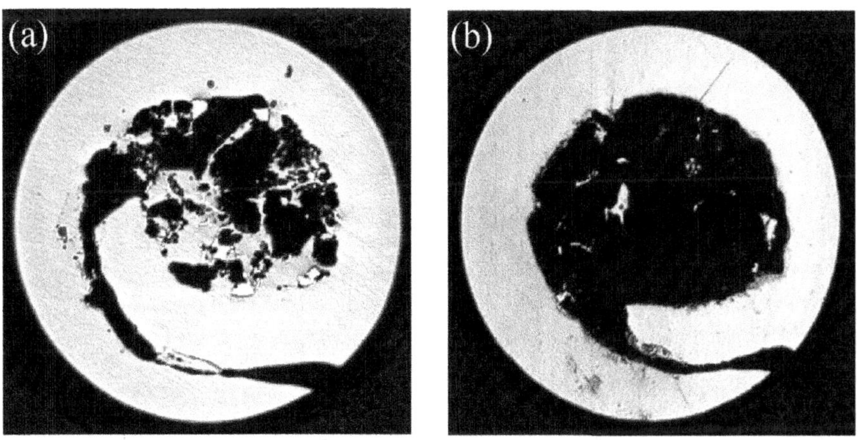

Fig. 3. Cross sections of Testing wires Ni-base type 70/20 ∅ 1.2 mm: **(a)** Metal-cored wire MCW No. 3–10 and **(b)** Flux-cored wire FCW No. 11–13 in Table 3

Table 3. Nb- and C-content of the investigated welding consumables and heat input for manufacturing weld metals Ni-base 70/20 and 70/15 (Testing program I)

Consumable type	Solid wire	MMA-Electrode		Metal-cored wires MCW								Flux-cored wires FCW								
	Market products			Testing wires								Testing wires			Market products					
Alloy type	Ni-base 70/20			Ni-base 70/20								Ni-base 70/20			Ni-base 70/20			Ni-base 70/15		
Alloy No.	1	2a	2b	3	4	5	6	7	8	9	10	11	12	13	14	15	16	17	18a	18b
Nb [%] in Filling	2.5	n.k.		0.0	1.0	2.0		3.0		3.5		1.0	2.0	3.0	n.k.	2.4	n.k.	n.k.	1.9	
C [%] in Filling	≤0.03	n.k.		0.1	0.1	0.1	0.1	0.1	0.1	0.1	0.1	0.1	0.1	0.1	n.k.					
Nb/C-ratio	83	n.k.		0	20	40	20	60	30	70	35	20	40	60	n.k.					
Heat input AWM [kJ/cm]	12.2	9.6–10.3		12	13	13	9.5	12.2	9.5	15	9.7	9.5	9.3	9.3	9.5	10.4	8.5	9.1	–	
Heat input PD [kJ/cm]	13.4	9.6–10.5	5.6–6.2	8.6	8.7	8.4	8.3	8.2	7.9	8.2	7.9	6.5	6.5	6.3	–	–	–	–	13.0	7.8

n.k. not known, – no weld metal manufactured.

External-Loaded Hot Crack Testing: PVR-Test

For the development and quantification of hot crack susceptible fully austenitic weld metals the controlled deformation crack test or controlled flat tension test, PVR-test for short, was developed and employed the first time more than two decades ago [14, 15, 16] (Fig. 4).

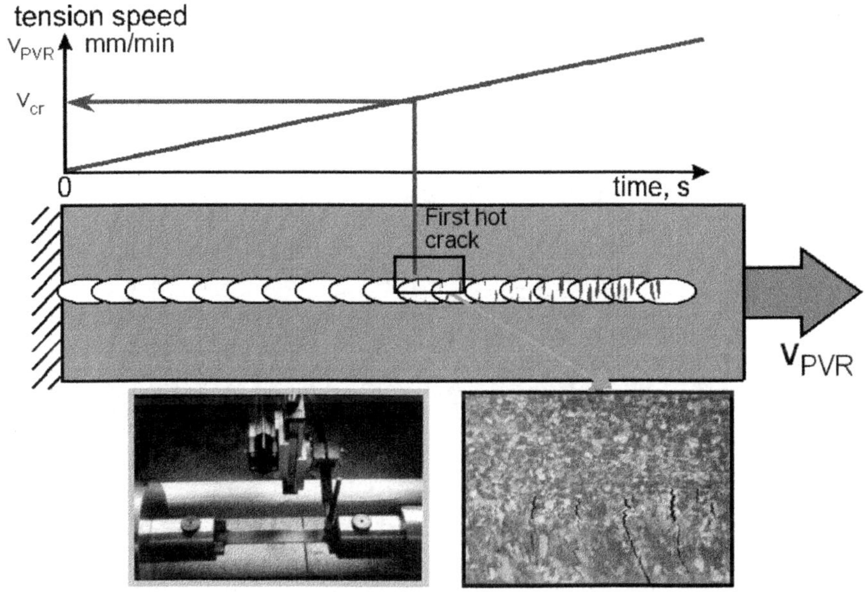

Fig. 4. PVR-test, scheme and test procedure to determine the critical tension speed v_{cr} initiating the 1st hot crack transvers to welding direction; U Magdeburg

The PVR-test belongs besides the Modified Varestraint-Transvarestraint-Test (MVT) and Hot tension test (using the Gleeble® physical simulator) and others, to the external loaded hot crack testing methods [17]. The PVR-test was created to evaluate the effects of the welding procedure on hot cracking [2]. Only a single test specimen is required to determine the HCS of a base metal and weld metal of the applied welding process. The ordinary test procedure of the PVR-test in Fig. 4 uses flat specimens with the dimension 40 x 10 x 300mm, clamped into a special tension fixture, lengthened in a horizontal servo-hydraulic system of the test equipment.

The welding process with a constant welding speed is superposed by a linearly increased tension speed v_{PVR} in welding direction. The standard test procedure of the PVR-test is carried out with tension rates linearly increasing from zero to 60 mm/min, during bead on plate TIG-welding with Argon shielding gas, applying two types of heat input per unit length about 7 and 10 kJ/cm (I_{weld}=180/220 A, U_{weld}=12/14 V, v_{weld}=19 cm/min) [18].

The critical tension speed v_{cr} or critical elongation speed (CES) is the test criterion for the PVR-test. CES corresponds to the first hot crack, detected visually at a magnification of 40. It can be determined for each of the hot crack types: Solidification cracks SC, liquation cracks LC and ductility dip cracks DDC (Fig. 5).

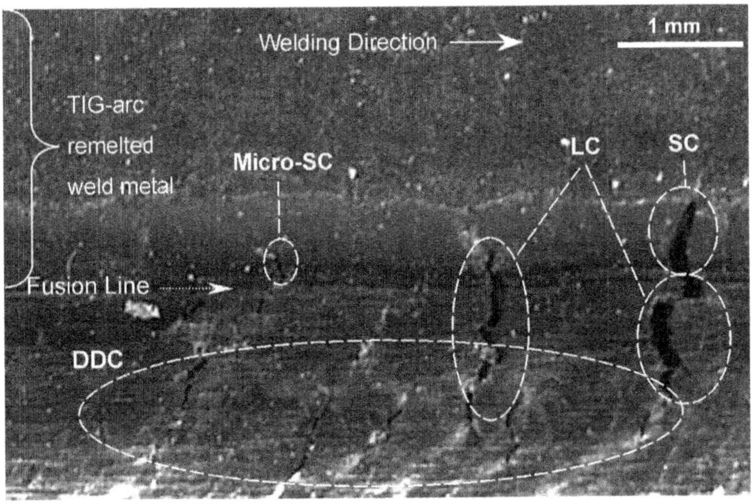

Fig. 5. Solidification cracks SC, liquation cracks LC and Ductility dip cracks DDC on a PVR-specimen

The success to assess all hot cracking types depends on the superposition of the local thermal cycle beside the bead with the tension rate. In addition, different indices for the critical tension speed v_{cr} are possible, dependent on the size and number of visible cracks. For example: v_{cr1} is the critical tension speed for the first microscopically visible hot cracking feature. v_{cr3} describes the first three hot cracks per 10mm of weld bead and v_{cr9} determines the first nine hot cracks per 10mm of weld bead. The modelling of the PVR-test has shown the correlation between test criterion, hot cracking theory by Prokhorov, and its applicability [19].

The PVR-test I was for to quantify the HCS of MMA-Electrode Ni-base type 70/20 ∅ 4 mm (No. 2) and the basic FCW Ni-base type 70/15 (No. 18), using two different heat inputs from the TIG-arc for remelting the PVR-weld metal (6.8 and 9.7 kJ/cm). To evaluate the HCS of the optimized FCW (No. 12b) in comparison to FCW market products (No. 14,

15, 16) the PVR-test II was performed. The final PVR-test III was to compare it's HCS to MMA-Electrode Ø5mm (No.2'), Solid Wire (No.1), and similar base metal alloy 600, see Table 4.

Table 4. Parameter for manufacture of PVR-specimens with different welding consumables (PVR-test I, II, III)

	PVR-Tes I				PVR-Test II				PVR-Test III		
Consumable type	MMA-Electrode, Ø 4 mm		Flux-cored wire, Ø 1.2 mm		Flux-cored wire, Ø 1.2 mm				MMA-Electrode Ø 5 mm	Solid wire, Ø 1.2 mm	FCW, Ø 4 mm
Alloy type	Ni-base 70/20		Ni-base 70/15		Ni-base 70/20				Ni-base 70/20		
Alloy No.	2		18		16	14	12b	15	2'	1	12b
PVR-specimen No.	N 30	N 31	N 32	N 33	1	2	3	4	P 29	P 30	P31
Heat Input [kJ/cm]	6.4	6.4	10.5	9.0	8.4	8.5	8.9	9.3	5.5	7.6	7.4
Number of layers	4, á 5 runs		4, á 3 runs	4, á 4 runs	4, á 4 runs				4, á 4 runs		

The standard heat input in the PVR-test was 6.8 kJ/cm, except PVR-specimen N31 and N33 having 9.7 kJ/cm.

Evaluation of the Slag Basicity

The basicity of the solidified slag was determined by EDX-analysis of slag micro-sections, using the modified basicity index B.I. by Bauné et al. [20] in Eq. 1 below.

$$B.I. = \frac{CaF_2 + CaO + MgO + Na_2O + K_2O + 0.5\times(MnO + Fe_2O_3 + Cr_2O_3)\,[at.\%]}{SiO_2 + 0.5\times(Al_2O_3 + TiO_2 + ZrO_2 + Nb_2O_5)\,[at.\%]} \tag{1}$$

This basicity approach is a modification of Tuliani's expression [21], using mole fractions (atomic %) for the oxides-concentration in the solidified slag. Their concentration was estimated by stoichiometric calculations. However two adaptions were carried out: For Ni-base slags Cr_2O_3 und Nb_2O_5 had to be added in Eq. 1, this was accomplished using the optical basicity Λ of these oxides: As Cr_2O_3 has the same recommended optical basicity like Fe_2O_3 (Λ=0.69), it was added to the transition oxides Fe_2O_3 and MnO (Λ=0.95) in the dividend of Eq. 1. Nb_2O_5 (Λ=0.61) was added to the amphoteric oxides Al_2O_3, TiO_2 (Λ=0.65) and ZrO_2 in the denominator [22, 23, 24].

Table 5. Analysis of All-weld metals AWM manufactured with different welding consumables.

	Ni	C	Nb	Fe	Cr	Mo	Si	Mn	Ti	P	S	O	N
1 SW 70/20[a]	73.0	0.006	2.72	0.37	20.2	0.00	0.11	3.15	0.33	0.005	0.003	0.010	0.019
2 MMA 70/20	68.5	0.027	2.19	3.17	18.52	1.25	0.39	5.33	0.06	0.008	0.006	0.055	0.023
3 MCW 70/20	73.3	0.015	0.04	2.42	20.48	0.09	0.41	3.20	0.01	–	0.005	0.093	0.015
4 MCW 70/20	73.0	0.017	1.04	2.45	19.60	0.09	0.38	3.38	0.02	–	0.005	0.100	0.016
5 MCW 70/20	72.0	0.020	2.01	2.04	19.90	0.10	0.38	3.45	0.03	–	0.005	0.098	0.017
6 MCW 70/20	72.4	0.092	2.24	2.62	19.40	–	0.36	3.04	–	<0.005	0.005	0.033	0.019
7 MCW 70/20	71.5	0.021	3.03	2.03	19.55	0.10	0.36	3.27	0.04	–	0.005	0.086	0.018
8 MCW 70/20	70.9	0.093	3.32	2.70	19.61	–	0.37	3.01	–	–	0.007	0.032	0.020
9 MCW 70/20	71.2	0.020	3.43	1.88	19.33	0.12	0.36	3.39	0.04	–	0.006	0.110	0.019
10 MCW 70/20	69.9	0.092	3.66	2.77	19.61	–	0.37	3.10	–	–	0.007	0.028	0.021
11 FCW 70/20	73.6	0.022	1.00	1.97	19.17	0.08	0.34	2.57	0.08	–	0.003	0.072	0.018
12 FCW 70/20	72.7	0.024	2.00	1.65	19.29	0.08	0.35	2.79	0.09	–	0.003	0.068	0.019
13 FCW 70/20	71.2	0.027	3.14	1.66	19.66	0.09	0.34	2.84	0.10	–	0.004	0.086	0.018
14FCW 70/20	73.3	0.030	2.36	1.68	19.27	–	0.50	3.00	–	<0.005	0.004	0.054	0.011
15 FCW 70/20	–	0.063	2.63	2.13	21.52	–	0.22	3.18	0.36	0.003	0.005	0.071	–
16 FCW 70/20	72.3	0.031	2.28	1.25	20.70	–	0.36	2.83	–	<0.005	0.002	0.078	0.011
17 FCW 70/15	66.6	0.024	1.72	8.22	16.22	–	0.54	6.80	–	–	–	0.064	–
18 FCW 70/15	76.5	0.023	1.47	1.04	13.74	–	0.36	6.53	0.08	<0.005	0.004	0.049	–

– not analyzed.
[a] Analysis Solid wire.

Results and Discussion

Analysis of All-Weld metal

Except the Nb-contents of the consumables No. 3 and 4 MCW 70/20 and No. 11 FCW 70/20, all of them are within the tolerances of Ni-base weld metal 70/20 from SMAW-Electrode (prEN ISO 14172) and bare electrodes and rods (DIN 1736, AWS A5.14) respectively. Only the Fe-content of weld metal No. 18 FCW 70/15 of approx. 1 % is below the min. standard content of 2%Nb in Ni-base 70/15 from SMAW-Electrode (DIN 1736), see Table 5, compare Table 2.

For the optimization of the manufactured Flux-cored wires the alloy No. 12 FCW 70/20 was chosen as a basis (2 % Nb in filling). By increasing the Nb-content to about 2.5 % and making some variations mainly in Fe, Mn and Si-content, the required value for the fracture strain of 30% could be reached.

Taking into consideration the requirements for the arc stability and slag viscosity of the FCW, the final product was found to be 12b FCW 70/20, see Table 6.

Table 6. Analysis of All-weld metal AWM manufactured with optimized Flux-cored wires

FCW 70/20	Ni	C	Nb	Fe	Cr	Si	Mn	P	S	O	N
12	72.7	0.024	2.00	1.65	19.29	0.35	2.79	–	0.003	0.068	0.019
12a	73.2	0.027	2.37	2.01	19.57	0.37	2.45	<0.005	0.004	0.086	0.018
12b	72.7	0.042	2.46	1.71	19.83	0.29	3.01	<0.005	0.004	0.063	0.013
12c	71.8	0.077	2.58	3.53	19.50	0.33	2.54	<0.005	0.004	0.068	0.015
12d	72.4	0.030	2.44	2.09	18.43	0.34	2.80	<0.005	0.006	0.060	–

– not analyzed

Mechanical Values of Weld Metals

With increasing Nb-content up to 3 % in the weld metal, higher fracture strain and tensile strength was measured for FCW and MCW as well. At 3.5 % Nb the fracture strain for weld metal of MCW becomes lower again, see Table 7 and Fig. 6. With increasing Nb-content from 0 to 3 % in AWM

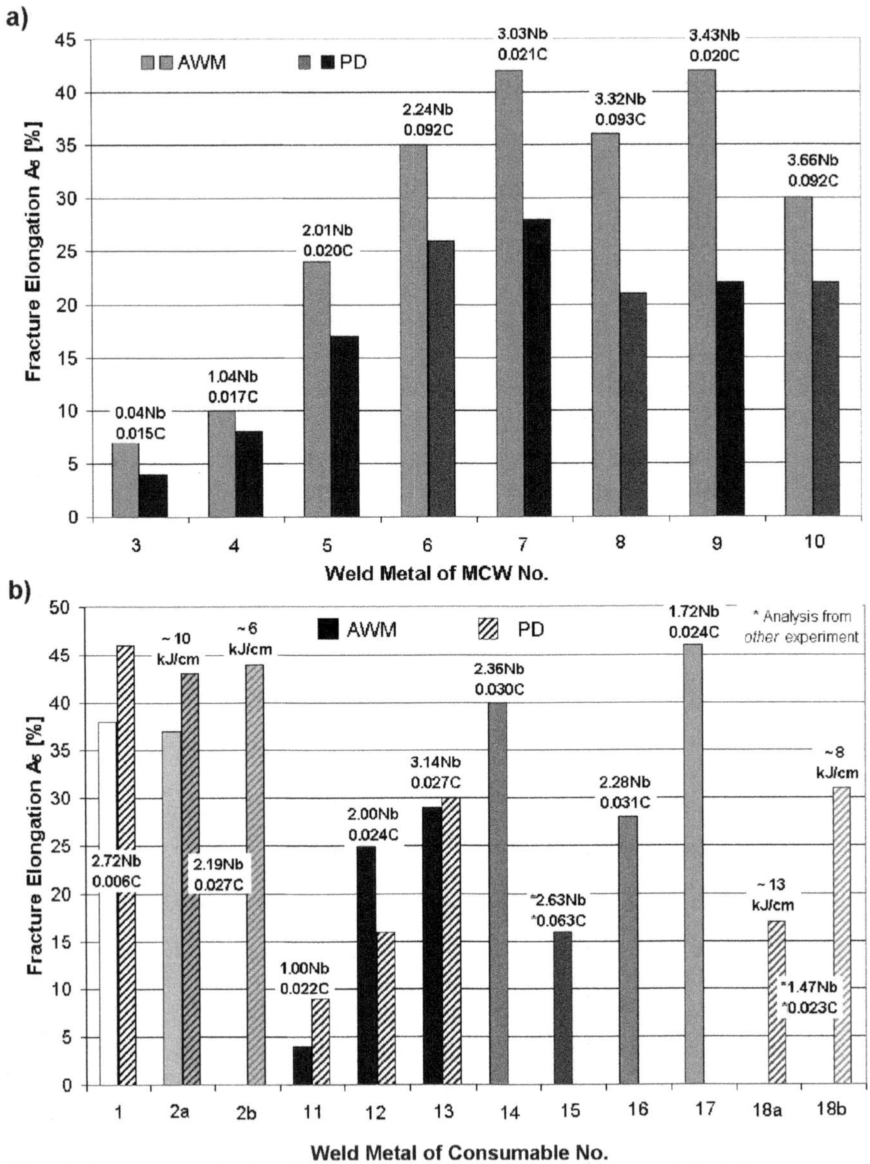

Fig. 6. Fracture strain A_5 of All-weld metal AWM and weld metal of Plate Deposit PD, manufactured with Metal-cored wire **(a)** and different welding consumables **(b)**

and PD of low carbon MCW a strong improvement of the fracture strain is found (7/4 to 42/28 %).

For weld metals with high C-content between 2.2 % and 3.6 % Nb no significant change in the fracture strain can be observed (Fig. 6a).

With increasing Nb-content from 1 to 3 % in AWM and PD manufactured with No. 11, 12 and 13 FCW a strong improvement of fracture strain can be found, too. Two of five FCW from market are below the required 30 % fracture strain, whereas No. 18 FCW 70/15 seems to be susceptible to higher welding heat input. The Solid wire and the MMA-Electrode, for lower and higher heat input as well, give very good A_5-values (Fig. 6b).

With increasing Nb-content from 1 to 3 % in the AWM and PD produced with No. 11, 12 and 13 FCW a strong improvement of the fracture strain can be found, too. Two of five commercial FCW are below the required 30 % fracture strain, whereas No. 18 FCW 70/15 seems to be susceptible to higher welding heat input.

The Solid wire and the MMA-Electrode (for lower and higher heat input as well) give very good A_5-values (Fig. 6b). With some variations in the metallurgical system of No. 12 FCW(Table 6), the fracture strain could be improved to about 40 % (12b–12d), so the required standard value of $A_5 \geq 30$ %, could easily be fulfilled (Table 8).

Reduction of Area and Hook Cracks

The differences in the mechanical values (Table 7, Fig. 6) was due to the appearance of Hook cracks, which have been evaluated over the reduction of area (magnification 8 x), as suggested by Pohle [13] (Fig. 7).

The AWM and PD of MCW with high-carbon content (HC) No. 6 (6A, 6P) with 2.2 % Nb and 0.09 % C reached the highest RoA (47/39 %) and the fewest Hook cracks (6/31) as well. HC MCW No. 8 and 10 with 3.3/3.7 % Nb and 0.09 % C give comparable good results too.

Among weld metals with low-carbon content (LC) MCW No. 9 and 7 with 3.4/3.0 % Nb and 0.02 % C give the best values for AWM (9A, 7A). The Plate Deposit PD has the highest RoA and lowest no. of cracks for No. 7 and 5 (7P, 5P) (Fig. 7a).

The weld metals of No. 1 Solid wire type 70/20 (1A, 1P) and No. 17 FCW type 70/15 (17A) show superior ductility, i.e. RoA more that 50 %, whereby only a few Hook cracks appeared. Astonishingly the Plate Deposit DP of the basic No. 18 FCW (18bP) produced with ~8 kJ/cm heat input show a RoA of 51 %, having 102 small Hook cracks.

The same PD 18aP welded with 13 kJ/cm show a distinctly lower RoA (28 %), but lesser cracks too (33). From this it can be expected that the weld metal is sensitive to overheating, compare Table 3.

Table 7. Mechanical values of All-weld metal AWM and weld metal of Plate Deposit PD (Testing program I)

Consumable type		Solid wire	MMA-Electrode		Metal-cored wires MCW								Flux-cored wires FCW								
		Market products			Testing wires								Testing wires			Market products					
Alloy type		Ni-base 70/20			Ni-base 70/20								Ni-base 70/20			Ni-base 70/20			Ni-base 70/15		
Alloy No.		1	2a	2b	3	4	5	6	7	8	9	10	11	12	13	14	15	16	17	18a	18b
$R_{p0.2}$ [MPa]	AWM	407	419		335	365	383	431	400	437	409	451	396	426	430	414	428	423	387	–	
	PD	358	386	417	318	362	390	406	411	420	386	431	355	376	414	–	–	–	–	325	345
R_m [MPa]	AWM	660	676		381	446	588	686	666	707	684	712	416	616	674	655	587	635	633	–	
	PD	651	668	679	337	424	558	676	664	676	633	687	407	525	644	–	–	–	–	458	560
A_5 [%]	AWM	38	37		7	10	24	35	42	36	42	30	4	25	29	40	16	28	46	–	
	PD	46	43	44	4	8	17	26	28	21	22	22	9	16	30	–	–	–	–	17	31
RoA [%]	AWM	51	43		23	26	36	47	41	42	45	36	21	33	38	46	32	37	55	–	
	PD	58	49	48	17	23	29	39	32	32	28	35	18	29	39	–	–	–	–	28	51
Av[J]	RT	178	129		105	86	111	94	125	78	106	71	137	146	113	142	99	146	154	–	
	-196 °C	179	124		90	104	97	83	117	69	94	60	142	134	101	134	84	140	138	–	

$R_{p0.2}$ Yield Strength, R_m Tensile Strength, A_5 Fracture strain, *RoA* Reduction of Area, *Av* Charpy-Impact Value.

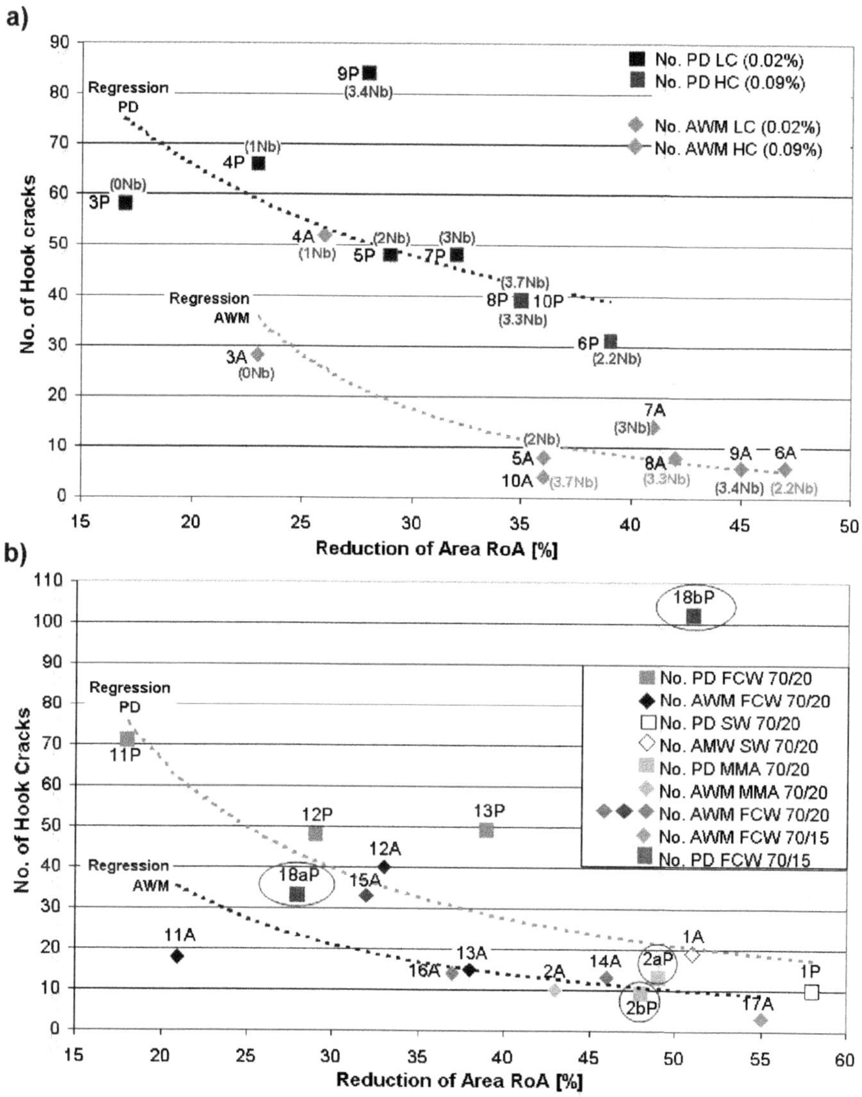

Fig. 7. Number of Hook cracks against RoA [%] of All-weld metal AWM (A) and weld metal of Plate Deposit PD (P) manufactured with Metal-cored wire **(a)** and different welding consumables **(b)**

For FCW testing wires the RoA increases significantly with the Nb-content (1.0, 2.0, 3.1 %), like the fracture strain (Fig. 6). The lowest amount of Hook cracks appeared for No. 13 All-weld metal (13A).

The No. 2 MMA-Electrode shows also very good ductility and just a few Hook cracks. Besides it no deterioration of the values using ~10 kJ/cm welding heat input (2aP) compared to ~6 kJ/cm (2bP) for Plate Deposit PD can be observed (Fig. 7b).

Table 8. Heat Input for the manufacture of AWM and PD, mechanical properties and no. of Hook cracks of optimized FCW (Testing program II)

Consumable type		Flux-cored wires FCW				
		Testing wire	Optimized testing wires			
Alloy type		Ni-base 70/20				
No.		12	12a	12b	12c	12d
Heat Input [kJ/cm]	AWM	9.3	10.0	9.6	8.5	9.8
	PD	6.5	9.5			
Yield Strength $R_{p0.2}$ [MPa]	AWM	426	418	416	403	423
	PD	376	358	380	362	363
Tensile Strength R_m [MPa]	AWM	616	521	669	654	669
	PD	525	436	641	628	648
Fracture strain A_5 [%]	AWM	25	12	43	41	39
	PD	16	8	32	37	37
Reduction of Area [%]	AWM	33	10	22	41	28
	PD	29	11	31	36	40
No. of Hook Cracks	AWM	40	34	4	2	1
	PD	48	50	92	40	49

A satisfying optimization of the manufactured No. 12 FCW could be reached by setting the Nb-content to about 2.5 % and the C-content in AWM to about 0.05 %. Other important factors are the ratios of Nb/Si and Ni/Cr/Fe, as well as the Mn- and Fe-content [25, 26].

According to the investigation of Dupont [27] a higher C-content lowers significantly the solidification interval. Herewith the no. of Hook cracks could be decreased strongly from 40 of No. 12 FCW to 4 cracks of No. 12b FCW at high RoA of almost 50 % (Table 8).

The AWM of No. 12a FCW was somehow out the tolerances of this metallurgical system, i.e. slightly lower/higher C-/Si-content, compare

Table 6, what can be seen at the lower Basicity Index (BI=0.59) which is shown in Fig. 10.

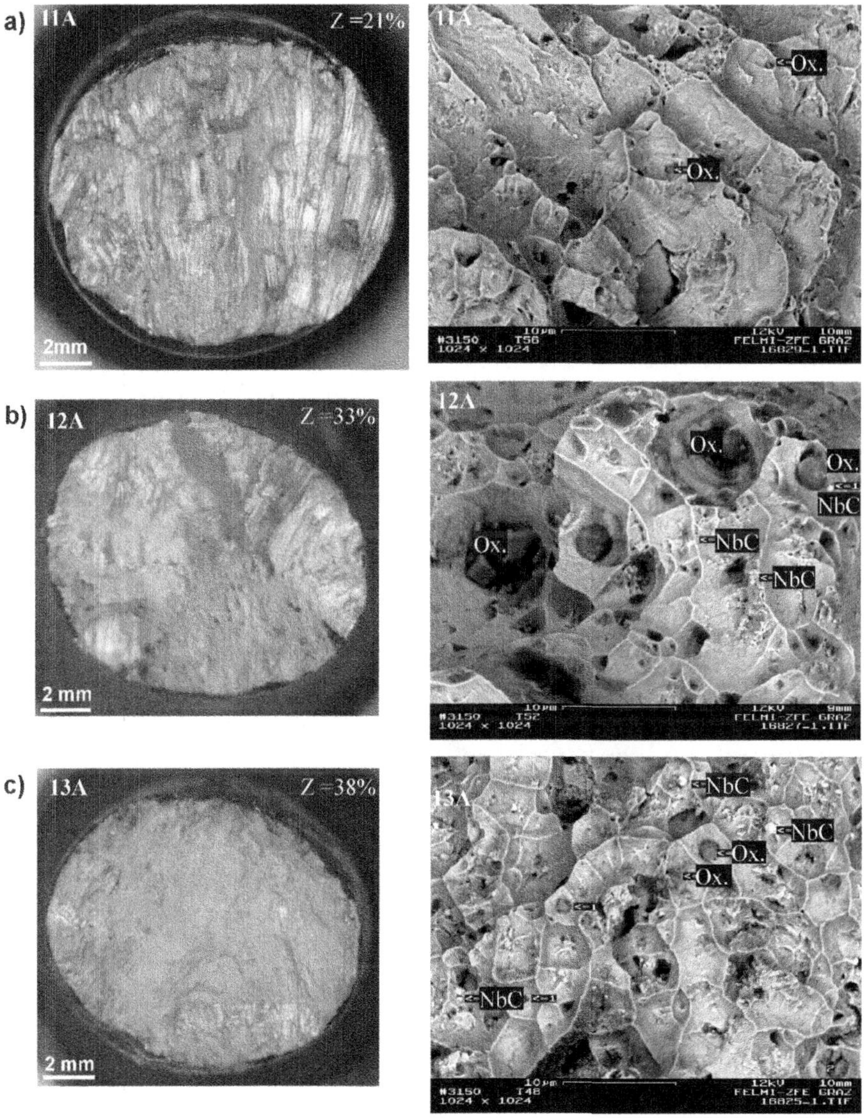

Fig. 8. Fracture surface of tensile test specimens from AWM of No. 11 **(a)**, 12 **(b)** and 13 **(c)** FCW from Stereo microscope (left) and SEM-fractographs in BSE mode (right)

Fractography of All-Weld Metal from FCW

For the necessary development and optimization SEM-investigations of the fracture surface from the AWM tensile test specimens No. 11, 12 and 13 FCW were performed. As the ductility and tensile strength was improved as well (Table 7), it could be expected to find a finer subgrain structure, i.e. smaller dendrite spacings, what could be confirmed.

The All-weld metal of No. 11 FCW (11A) shows brittle fracture behavior, i.e. due to the intercrystalline cracks leading to fracture, a typical fibrous structure can be seen in the stereo microscope (Fig. 8a left). In the SEM-fractograph plane structures with partially melted zones appear (Fig. 8a right). All-weld metals of No. 12 and 13 FCW (12A, 13A) show a microstructural ductile fracture and the bright precipitations found in the SEM were analyzed to be NbC (Fig. 8b and 8c).

All three weld metals show oxidic inclusions of different size and shape, marked with "Ox." The All-weld metal of the optimized No. 12a and 12b FCW 70/20 show totally different fractographs, despite the similar chemical analysis in Table 6: 12a shows partially melted zones and a crack surface topography as well. On the contrary 12b shows a distinctive dimple fracture (Fig. 9a and 9b).

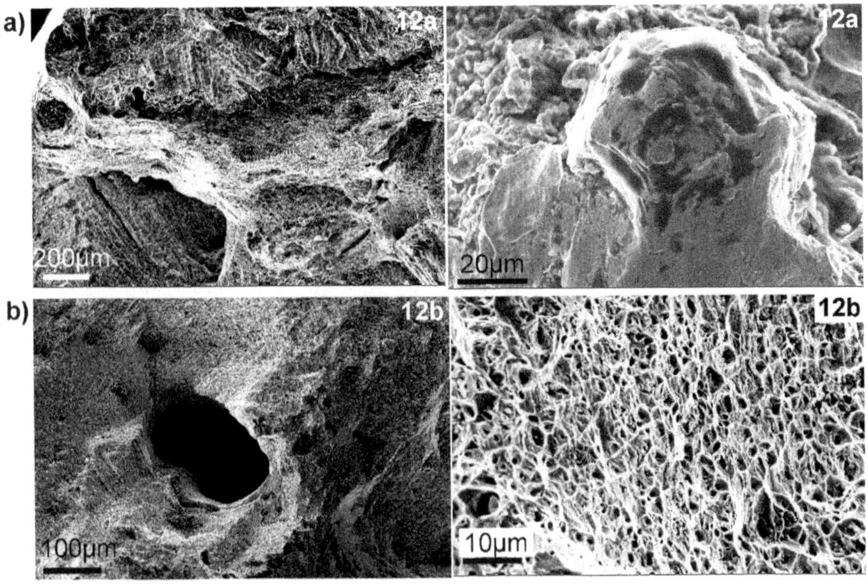

Fig. 9. SEM-fractographs of tensile test specimens AWM FCW 12a **(a)** and 12b **(b)** in SE mode

Table 9. Basicity Index BI of the slag, O-content and Charpy-Impact Values Av of All-weld metal AWM (Testing program I)

Consumable type	Solid wire	MMA-Electr.	Metal-cored wires MCW								Flux-cored wires FCW							
	Market products		Testing wires								Testing wires			Market products				
Alloy type	Ni-base 70/20		MC Ni-base 70/20								FC Ni-base 70/20			FC Ni-base 70/20			FC Ni-base 70/15	
Alloy No.	1	2	3	4	5	6	7	8	9	10	11	12	13	14	15	16	17	18a
BI slag [at.%/at%]	–	n.a.	–	–	–	–	–	–	–	–	0.72	0.76	0.71	n.a.	0.91	0.81	n.a.	2.02
O-content [ppm]	100	550	930	1000	980	330	860	320	1100	280	720	680	860	540	710[a]	780	640	490[a]
Charpy-Impact Value Av [J] RT	178	129	105	86	111	94	125	78	106	71	137	146	113	142	99	146	154	n.a.
Charpy-Impact Value Av [J] -196°C	179	124	90	104	97	83	117	69	94	60	142	134	101	134	84	140	138	n.a.

n.a. not analyzed, – no slag.
[a] Analysis from other experiment.

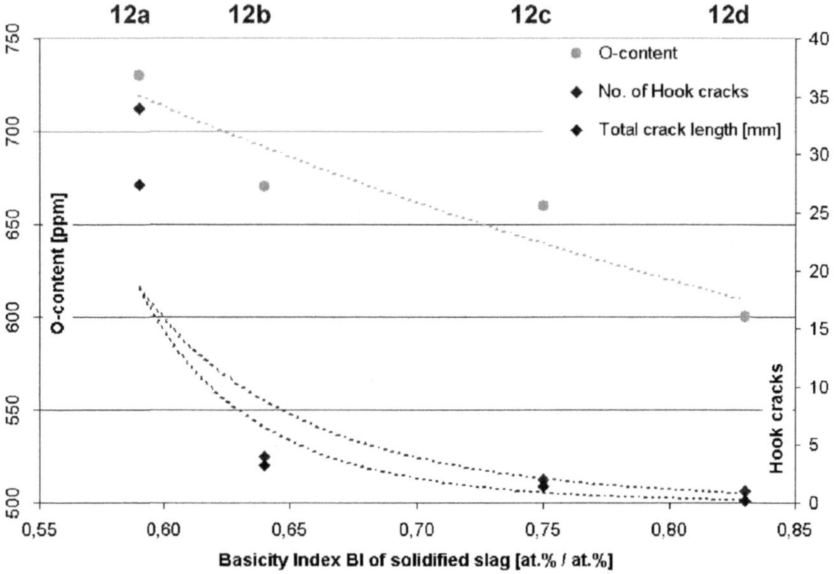

Fig. 10. Correlation of Basicity Index BI to the O-content in All-weld metal and the Hook cracks for optimized No. 12a - 12d FCW

Slag Basicity and O-Content

The basicity of the solidified slag was determined by the modified basicity index by Bauné et al. [20, 28] in Eq. 1. Herewith a good correlation could be achieved with the O-content and the impact values for the Testing wires FCW No. 11, 12, 13 (Table 9). By this means the relatively small variations of metals (1, 2, 3 Nb) in the filling could be investigated.

For the optimized FCW No. 12a-12d also a good correlation was found: Decreasing O-content in All-weld metal with higher BI of the solidified slag and decreasing Hook cracks (Fig. 10).

PVR-Test Results

The results of PVR-test I in Fig. 11 show a reduction in the critical elongation speed (CES), i.e. higher HCS for Solidification cracks SC and Liquation cracks LC as well, when using higher heat input of the TIG-arc. That applies to No. 2 MMA-Electrode Ni-base type 70/20 ∅ 4 mm and the

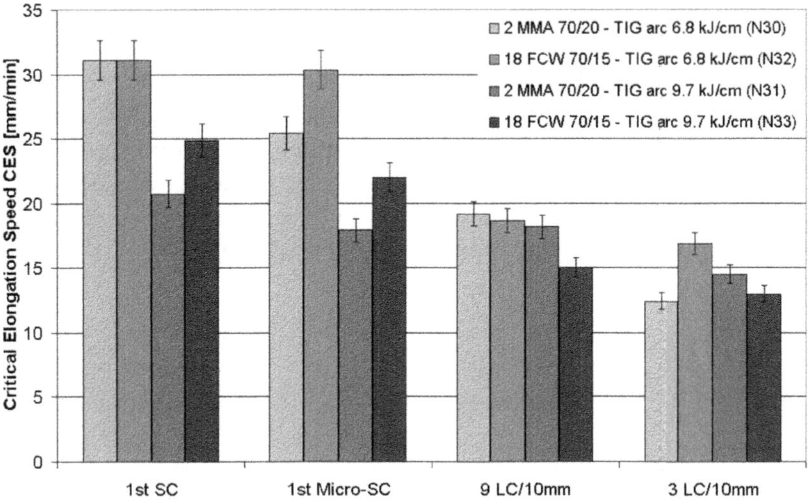

Fig. 11. Evaluation of the PVR-test I: MMA-Electrode ∅4mm Ni-base 70/20 and basic FCW Ni-base 70/15 show comparable low Hot crack susceptibility HCS to Solidification and Liquation cracks SC/LC

basic No. 18 FCW Ni-base type 70/15, too. The reason for this is the bigger size of the mushy zone at 9.7 kJ/cm [19]. Apart from that the weld metals have comparable low HCS, like it was estimated in the RoA-Hook-

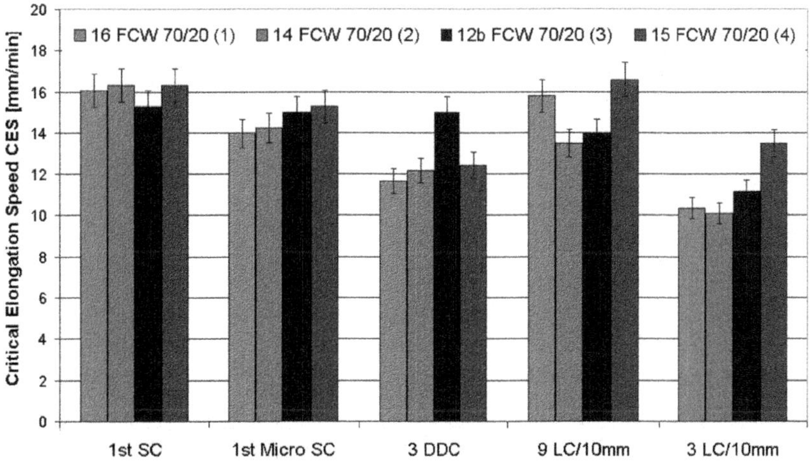

Fig. 12. Evaluation of the PVR-test II: The optimized 12b FCW 70/20 (3) show comparable HCS to Solidification, Liquation and Ductility Dip Cracks SC / LC / DDC (TIG arc 6.8 kJ/cm) to FCW on the market

cracks diagram, Fig. 7, 2aP/2bP/18bP. Ductility Dip cracks DDC could not be found in these weld metals.

The optimized weld metal of FCW No. 12b show comparable CES to the FCW 70/20 market products for each of the hot crack types (Fig. 12), but it is quite lower than the CES of No. 2 MMA-Electrode ∅ 4 mm and the basic No. 18 FCW 70/15 in Fig. 11. A further comparison shows, that the CES of 12b is far away from that of the No. 1 Solid wire. It is just comparable to No. 2 MMA-Electrode ∅5mm, concerning the v_{SC} / $v_{Micro\text{-}SC}$ criterion. Still these values are quite below the similar base metal alloy 600 (Fig. 13).

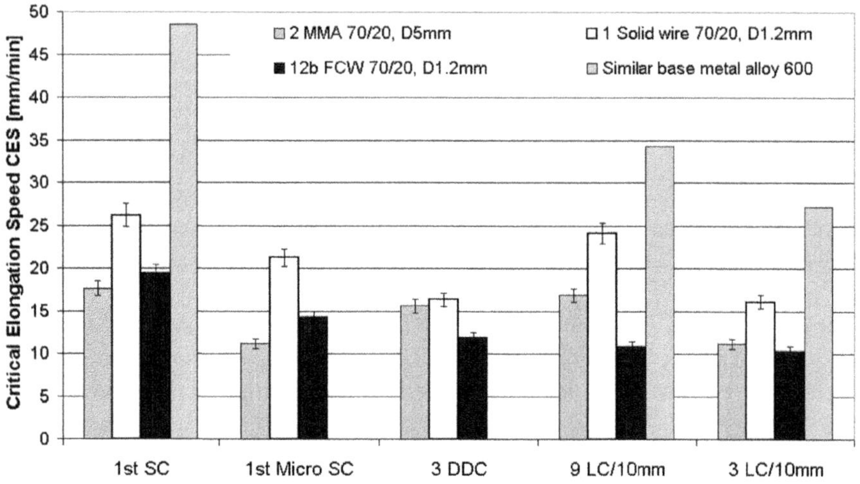

Fig. 13. Evaluation of the PVR-test III: MMA, SW, FCW and Base metal (TIG arc 6.8 kJ/cm)

Conclusions

Nb-additions to weld metals Nibas-70/20 can reduce the hot crack susceptibility HCS strongly, because of the formation of NbC, what makes finer subgrains in the weld metal. By that also the fracture strain and the tensile strength is improved.

The tested weld metals of Plate Deposit PD show mostly more Hook cracks than All-weld metal AWM. Thus it should be a useful method for pre-selecting a weld metal concerning its HCS. For the optimization of slag bearing consumables the determination of the Basicity Index of the solidified slag can be a support to find the right metallurgical system.

Higher heat input in the manufacture of the weld metals as well as for the remelting in the PVR-test is increasing the HCS strongly. The HCS of weld metals from FCW 70/20 in the PVR-test is distinctly higher than for weld metals of MMA-Electrode and Solid wire. The HCS of weld metals Ni-base 70/15 for the investigated products was lower than for 70/20.

In general the HCS of similar base metal alloy 600 is still much lower than of the weld metals.

Acknowledgments

The author would like to thank Prof. Herold, Mrs. Streitenberger and Mr. Pchennikov from the Institute of Joining and Beam Technology at TU Magdeburg for the performing and evaluation of the PVR-tests. Likewise Mr. Tösch and Mr. Klagges, Böhler Schweißtechnik Austria and Mr. Heinemann, UTP-Schweißmaterial for their expert advices, as well as Prof. Cerjak, head of the institute, for the possibility to carry out this investigation.

References

1. Heinemann J, Kilian R Personal communication
2. Folkhard E (1984) Welding Metallurgy of Stainless Steels. Springer Wien New York
3. Yamanaka K et al. (1993) Development of High Corrosion Resistant Alloy 600 with Niobium Addition. Sumitomo Search 51: 1–15
4. Lee HT, Kuo TY (1999) Effects of niobium on microstructure, mechanical properties, and corrosion behavior in weldments of alloy 690. Science and Technology of Welding and Joining, Vol 4, No 4: 246–256
5. DIN 1736 (Aug 1985) Schweißzusätze für Nickel- und Nickellegierungen
6. DIN EN ISO 18274 (Aug 2000) Draht- und Bandelektroden, Drähte und Stäbe zum Lichtbogenschweißen für Nickel und Nickellegierungen
7. AWS A5.14 Specification for Nickel and Nickel-Alloy Bare Welding Electrodes and Rods
8. DIN EN ISO 14172 (Mai 2001) Umhüllte Stabelektroden zum Lichtbogenschweißen von Nickel und Nickellegierungen
9. AWS A5.11/A5.11M-97 Specification for Nickel and Nickel-Alloy Welding Electrodes for Shielded Metal Arc Welding
10. Deutscher Verband für Schweißtechnik e.V. (Sept 1990) Heißrissprüfverfahren mit selbstbeanspruchenden Proben. Merkblatt DVS 1004 Teil 3. DVS Verlag Düsseldorf

11. Kauhausen E, Kaesmacher P (1955) Bestimmung der Warmrißbeständigkeit von Schweißelektroden. Schweißen und Schneiden 7, H 2: 400–404
12. Düchting W (1978) Einige Betrachtungen zur Heißrißprüfung von Schweiß-zu-satzwerkstoffen. Schweißen und Schneiden 30, H 12: 498–502
13. Pohle C (1985) Möglichkeiten für die Beurteilung der Heißrissanfälligkeit von austenitischem Schweißgut. Schweißen und Schneiden 37, H 2: 55–60
14. Folkhard E et al (Nov 1977) Der PVR-Test, ein neues Verfahren zur Ermittlung der Rißsicherheit von Schweißwerkstoffen mit hoher quantitativer Aussagekraft. Sonderdruck aus Jubiläumsschrift "50 Jahre BÖHLER-Schweißtechnik"
15. Rabensteiner G, Tösch J, Schabereiter H (Jan 1983) Hot Cracking Problems in Different Fully Austenitic Weld Metals. Welding Journal: 21-s–27-s
16. Kobayashi T et al (1986) Hot Cracking Susceptibility Evaluation of Austenitic Stainless Steel Weld Metals. The Japan Welding Engineering Society, IIW Doc II-1068-86/IX-1395-86
17. Deutscher Verband für Schweißtechnik eV (Nov 1996) Heißrissprüfverfahren mit fremdbeanspruchten Proben. Merkblatt DVS 1004-2. DVS Verlag Düsseldorf
18. Herold H, Pchennikov A, Streitenberger M (2003) Current problems in hot cracking research described on the example of PVR test. Materials Science Forum Vols 426–423: 4093–4098
19. Herold H, Streitenberger M, Pchennikov A (2001) Hot Cracking Theory by Prokhorov and Modelling of the PVR-test. Welding in the World 45, 3/4: 17–22
20. Baune E, Bonnet C, Liu S (March 2000) Reconsidering the Basicity of a FCAW Consumable-Part 1: Solidified Slag Composition of a FCAW Consumable as a Basicity Indicator. Welding Journal: 57-s–65-s
21. Tuliani SS, Boniszewski T, Eaton NF (1969) Notch Toughness of commercial submerged arc weld metal. Welding and Metal Fabrication 37, 8: 327–339
22. Mills KC (1995) Slag Atlas. VDEh 2nd Edition, Verlag Stahleisen GmbH, Düsseldorf
23. Sommerville ID, Yang Y (2001) Basicity of metallurgical slags. The AusIMM Proc 306, 1: 71–77
24. Lebouteiller A, Courtine P (1998) Improvement of a Bulk Optical Basicity Table for Oxidic Systems. Journal of Solid State Chemistry 137: 94–103
25. Pease G R (1957) Metallkundliche Vorgänge beim Schweissen von Nickel und Nickellegierungen. International Nickel, Zürich 1957
26. Anik S, Dorn L (1983) Metallphysikalische Vorgänge beim Schweißen von Nickelwerkstoffen - Einfluss der Werkstoffzusammensetzung. Schweißen und Schneiden 35: 445–450
27. DuPont N, Robino CV, Marder AR (1998) Solidification and Weldability of Nb-Bearing Superalloys. Welding Journal 10 (1998)
28. Baune E, Bonnet C, Liu S (March 2000) Reconsidering the Basicity of a FCAW Consumable – Part 2: Verification of the Flux/Slag Analysis Methodology for Weld Metal Oxygen Control. Welding Journal: 66-s–71-s

Hot Cracks as Stress Corrosion Cracking Initiation Sites in Laser Welded Corrosion Resistant Alloys

K. Stelling, Th. Böllinghaus, M. Wolf, A. Schöler, A. and A. Burkert, B. Isecke

Federal Institute for Materials Research and Testing, Berlin, Germany

Abstract

Although hot cracks at weld surfaces of corrosion resistant alloys (CRAs) might provide access to respective aggressive media and thus, might provide the conditions for a local potential and pH drop, the interaction between hot cracking and corrosion phenomena has not been investigated up to the present. This particularly concerns the initiation of stress corrosion cracking inside such crevices.

As a first approach to clarify the influence of hot cracks on corrosion resistance, considering particularly stress corrosion cracking, hot crack afflicted laser welds of three different CRAs have been subjected to a series of different corrosion tests.

Besides the fact that the Drop Evaporation (DE) Test represents a more realistic procedure than standard immersion tests it turned out that hot cracks can obviously represent an initiation site of stress corrosion cracking in the austenitic stainless steel AISI 309 (German No. 1.4828). The susceptibility of such hot crack afflicted laser welds to stress corrosion cracking (SCC) significantly depends on temperature and chloride concentration of the test environment In comparison, the high nitrogen austenitic stainless steel AISI S 34565 (German No. 1.4565) did not exhibit any stress corrosion cracking associated with hot cracking. This material exhibited pitting corrosion and stress corrosion only in the base material. The Alloy 50 (German No. 2.4850), a Nickel based material, was susceptible to pitting corrosion, but showed no tendency towards stress corrosion cracking at all.

Introduction

Laser welding is increasingly applied to fabrication joining of components made of corrosion resistant alloys (CRAs), for instance in pipe manufacturing for offshore or chemical industry. As a particular advantage of the low heat input associated with laser welding, the risk of heat-induced precipitation and respective depletion of alloying elements is reduced at considerably lower residual stresses as compared to arc welding processes. On the other hand, most CRAs exhibit an austenitic microstructure providing large solidification intervals with respective phase segregation and, consequently, the conditions for the appearance of hot cracking during welding.

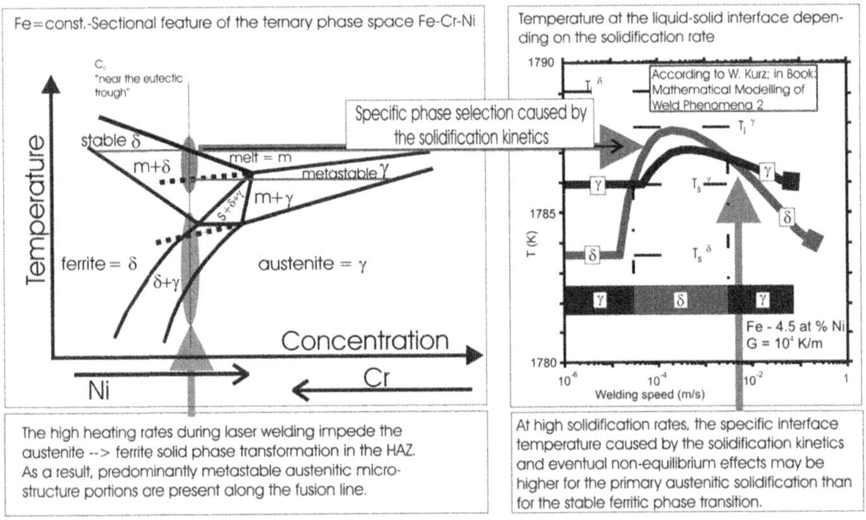

a) b)

Fig. 1. Change of the primary solidification mode in austenitic stainless steels near the eutectic rim

As a special effect it has to be considered that a group of widely applied stainless steels with a chemical composition located at the primary δ-ferrite side of the respective constitutional diagram (Fig. 1a) and which are considered as non-hot cracking sensitive, might undergo a primary solidification mode change to austenite caused by kinetic undercooling and also caused by epitaxial effects during rapid welding procedures (Fig. 1b). It has widely been proven in previous studies that laser welding of these materials increases the risk of hot cracking [1, 2].

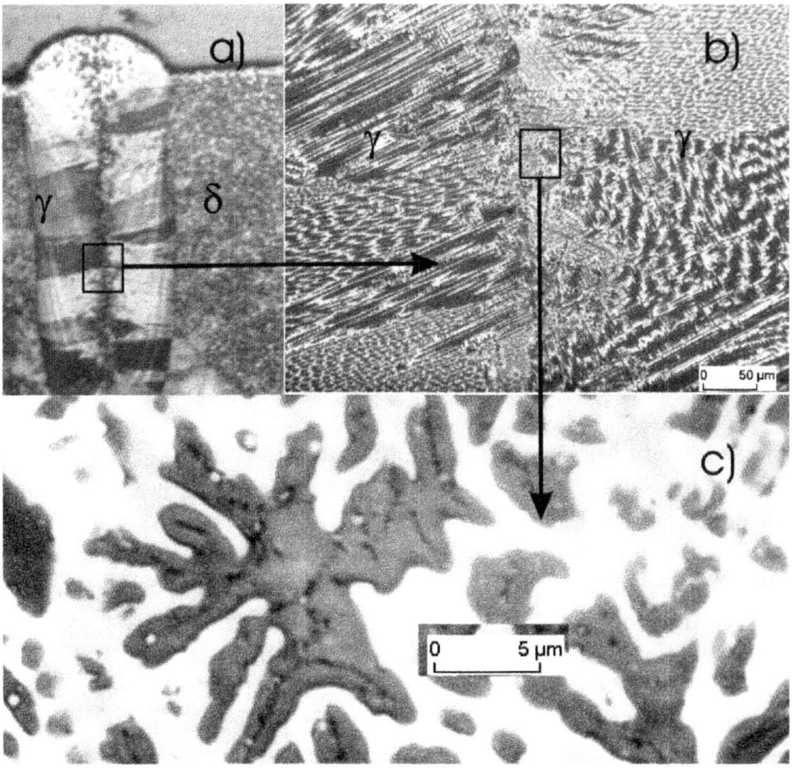

Fig. 2. Metastable ferritic crystallization of equiaxed dendritic morphologies in an austenitic stainless steel [2] produced by the following laser welding sequence: 1 –Welding with high nitrogen additions in the laser process gas to shift the solidification mode from primary ferrite to primary austenite, 2 – Remelting the primary austenitic solidified weld at high welding speeds to shift the primary solidification mode back to the ferrite side by kinetic undercooling

As a particular item, by respective nitrogen additions evidence has been produced for the first time that rapid welding processes might not only lead to a primary solidification
mode change from ferrite to austenite, but also vice versa from austenite to ferrite (Fig. 2), up to the present only known from non-chambered undercooling and quenching experiments, respectively [2, 3, 4].

Regarding the service behavior of components made of CRAs it has to be made sure that welds do not represent preferential sites for corrosion attack. Weld defects like undercut notches have thus to be avoided. Hot cracks as particular weld imperfections in austenitic CRAs cannot always be detected during non-destructive testing before a welded component is applied to service operation.

There are only a few studies on the service behavior of hot crack afflicted welded components and, in particular, no investigations have been undertaken as to whether hot cracks might represent preferential sites for corrosion attack in welded CRAs, except a brief study about intergranular corrosion starting at hot cracks in nickel-base alloys [5].

However, since austenitic CRA components are most frequently subjected to chloride environments, stress corrosion cracking represents the most safety-relevant failure type. Hot cracks opened to the weld surface might particularly provide the conditions for local pH and potential drops and thus, might represent respective stress corrosion cracking initiation sites.

The present study has been performed, in order to gain some insight into such phenomena. Hot cracks have been introduced into laser welds of three different types of austenitic CRAs using a special designed test facility. After loading by a screw bolt the specimens have subsequently been subjected magnesium chloride immersion tests at temperatures of up to 140 °C. In order to simulate high salt concentrations as they prevail in the splash water zone in marine environments, the more realistic Drop Evaporation (DE) Test has also been carried out at medium temperatures of up to 80 °C. Additionally, potentiodynamic polarization experiments were performed in order to investigate the critical corrosion potentials and repassivation characteristics.

Materials and Experimental

As an austenitic stainless steel known to have a tendency to change the solidification mode during rapid laser welding, an alloy AISI 309 (1.4828) was selected for the present study. Additionally, a high nitrogen austenitic stainless steel AISI S 34565 (1.4565) and a nickel-base material trade named Alloy 50 (2.4850) were tested, which are known to be susceptible to hot cracking for any fusion welding technique.

All experiments were carried out using sheet material of 2 mm and 3 mm thickness, respectively. The chemical compositions of the materials were determined by spectroscopic analysis. They are summarized in the Tables 1 to 3.

Welding was carried out using a CO_2-laser with a maximum power of 6 kW. The hot cracks were produced by the Laser Specific Hot Cracking (LSHC) Test, a BAM-designed reference procedure (Fig. 3).

Table 1. Chemical composition of the steel AISI 309 in wt%

Cr	Ni	Mo	N	C	S	P	Mn	Nb	Si
19.92	11.08	0.273	0.033	0.052	0.003	0.025	0.693	0.011	1.97

Table 2. Chemical composition of the steel AISI S 34565 in wt% (Fe bal.)

Cr	Ni	Mo	N	C	S	P	Mn	Nb	Si	Ti
23.4	17.1	4.2	0.46	0.029	0.003	0.021	5.6	0.041	0.148	0.017

Table 3. Chemical composition of the Alloy 50 in wt% (Ni bal. to ca. 50 wt%)

Cr	Fe	Mo	C	S	P	Mn	Nb	Si	Ti
19.2	14.71	10.33	0.007	0.002	0.008	0.341	0.294	0.070	0.014

By wrapping a thin plate of the respective material around a 72 mm diameter mandrel during laser welding, hot cracks were generated by mechanical straining of the specimen during the solidification process.

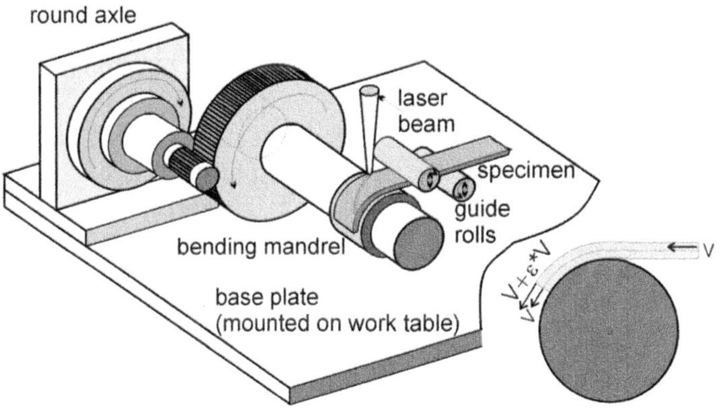

Fig. 3. Illustration of the Laser Specific Hot Cracking (LSHC) Test

The hot cracking resistance of different materials during laser welding is ranked by plots of the applied strain versus laser power, for instance [4].

Hot cracks can reproducibly be introduced into the laser welds by the LSHC test, if the welding parameters, i. e. welding speed v_s, laser power P_{Laser}, and focal point position are adjusted to the test parameters, i.e. deformation speed v_d, total strain, and distance of the weld pool from the bending line.

Four different specimen conditions were differentiated and are summarized for the various materials in Table 4.

Table 4. Tested materials and respective specimens

Specimen	AISI 309	AISI S 34565	Alloy 50
State 0 – not welded, plane sheet material	X	X	X
State 1 – welded, without any hot cracks	X	X	X
State 2 – welded, with numerous transverse cracks in the weld metal	X		
State 3 – welded, with a longitudinal crack in the weld centerline	X	X	X

Hot cracks produced in the AISI 309 steel by the primary solidification mode change are shown in Fig. 4. The different cracking characteristics occurred as a consequence of variations in the deformation speed v_d during the hot cracking test. Table 5 summarizes the different parameters applied to generate reproducibly longitudinal hot cracks.

Transverse hot cracks could only be introduced in the AISI 309 steel and only a few specimens of this type (State 2) have been investigated for comparison.

After the hot cracking test procedure the specimens were pickled and ultrasonically cleaned. For the corrosion tests, the pre-bent LSHC Test specimens were loaded with the aid of a screw bolt. Thus, the surface was exposed to tensile stress (in addition to the residual stresses caused by the welding process).

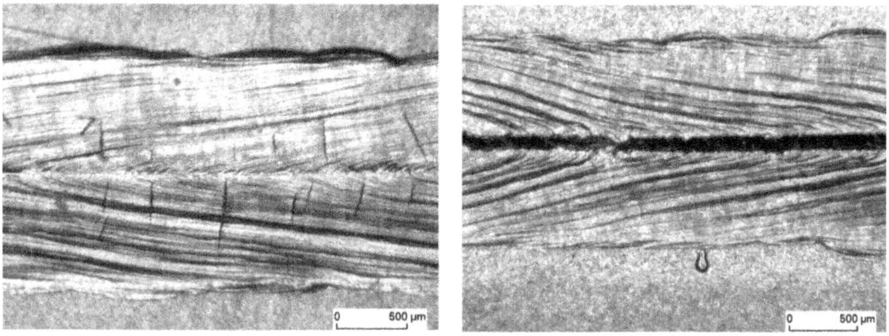

a) b)

Fig. 4. Hot cracks in AISI 309 laser welds **(a)** State 2, top view (v_d = 9 mm/min, v_s = 5000 mm/min, P_{Laser} = 2.6 kW), **(b)** State 3, top view (v_d = 22.5 mm/min, v_s = 5000 mm/min, P_{Laser} = 2.6 kW)

Table 5. Test parameters for introducing hot cracks of State 3

Material	Welding speed v_s [mm/min]	Laser power P_{Laser} [kW]	Deformation speed v_d [mm/min]	Total strain [%]	Distance laser beam – bending line [mm]
AISI 309	5000	3.0	15	4.0	0.98
AISI S 34565	5000	2.5	40	2.7	1.20
Alloy 50	2000	2.5	52	4.0	1.55

Fig. 5a illustrates the dimensions of the U-bend specimens while Fig. 5b shows their exposition to the Drop Evaporation Test. To ensure complete passivation of the materials, the specimens were exposed to air for one week before starting the corrosion tests.

Ahead of the stress corrosion cracking tests, potentiodynamic polarization experiments in artificial seawater were performed using a scan rate of 0.5 mV/s, in order to investigate whether the hot crack afflicted welds exhibit different pitting and repassivation characteristics in comparison to the non-cracked welds and the base materials. To determine the respective critical pitting and crevice corrosion potentials, the experiments were carried out three times for each specimen condition.

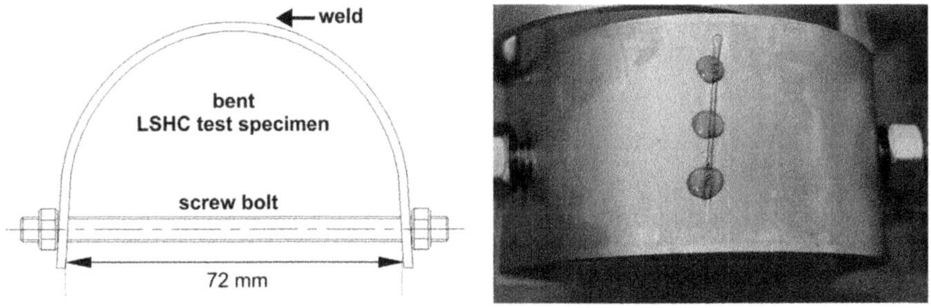

a) b)

Fig. 5. Stress corrosion cracking test specimens, **(a)** Specimen dimensions, **(b)** Exposure to the Drop Evaporation Test

For the Immersion Tests, the bend specimens were exposed to a $MgCl_2$-solution by covering the welded area with 10 mm of the electrolyte. The concentrations of the $MgCl_2$-solutions varied from 10 % to 40 %, the applied testing temperatures ranged from 80 °C to 140 °C. The test period was also varied from 720 h up to 1700 h depending on the tested material.

For the Drop Evaporation Test , the U-bend specimens have been mounted on a support inside a closed thermostatic box above a saturated

$MgCl_2$-solution. Thus, the desired temperature and constant air humidity of approximately 35 % was kept to ensure a constant atmosphere. Three drops of a saturated $MgCl_2$-solution were applied to every respective U-bend specimen (Fig. 5b).

Due to the expected different corrosion resistance, the three materials were tested at different temperatures, i. e. the steels AISI 309 and AISI S 34565 were exposed at 30 °C and 50 °C, respectively, while the Alloy 50 was tested at 70 °C. After every 1000 h period, one U-bend specimen of each alloy was taken out of the test environment. The longest test period was 5000 h in each case.

After the Immersion Test and the DE Test, the surface of the specimens was inspected using a stereo microscope. Subsequent, cracked samples were also subjected to metallographic and SEM investigations.

Results and Discussion

Potentiodynamic Tests

For a first insight into the corrosion behavior of the base materials, potentiodynamic reference measurements were carried out with the steel AISI 309 in comparison to the Alloy 50 at different test temperatures in artificial seawater.

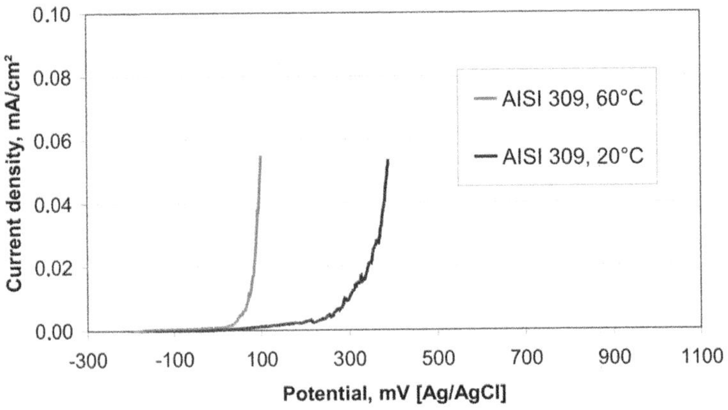

Fig. 6. Current density-potential plots with dU/dt = 0.5 mV/s, Steel AISI 309, base material, artificial seawater

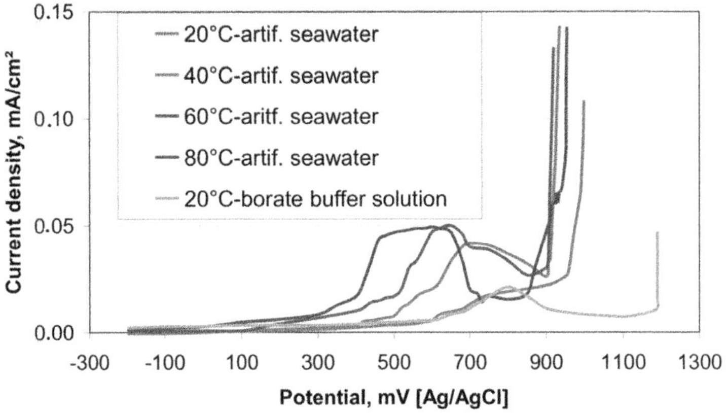

Fig. 7. Current density-potential plots with dU/dt = 0.5 mV/s, Alloy 50, base material, artificial seawater and borate buffer solution

As shown by the two characteristic current density potential plots in Fig. 6, the critical potential of the steel AISI 309 is reduced from values in the range between 290 mV and 320 mV to values of about 70 mV by increasing the temperature from 20 °C to 60 °C.

A comparison of Fig. 7 with Fig. 6 clearly shows that the critical potential of Alloy 50 is not as temperature dependent as that of the austenitic stainless steel. At 20 °C, the critical potential ranged at about 1000 mV, whereas from 40 °C to 80 °C the value leveled at about 900 mV. At 20 °C in a borate buffer solution (pH 6.4), the Alloy 50 showed an even higher corrosion resistance. Some temporary increase of the current density might be attributed to growth and reshape mechanisms inside the passive layer.

Additional potentiodynamic experiments were performed with various specimens of the same materials including reversed polarization in order to evaluate the repassivation behavior of hot crack afflicted welds as compared to the crack-free welds and the base materials.

The current density-potential plots in Fig. 8 exhibit no significant differences regarding the critical potential of the steel AISI 309 in all four states. The critical values still ranged between 270 mV and 320 mV. However, a significant difference was found with the reversed polarization.

Whereas the specimens of State 0 and State 1 showed a similar repassivation behavior with maximum current densities of about 100 µA/cm², repassivation was retarded at the specimen of State 2 where the maximum current density was up to 170 µA/cm². An even higher current density of 320 µA/cm² can be observed for the specimen with a longitudinal solidification crack (State 3).

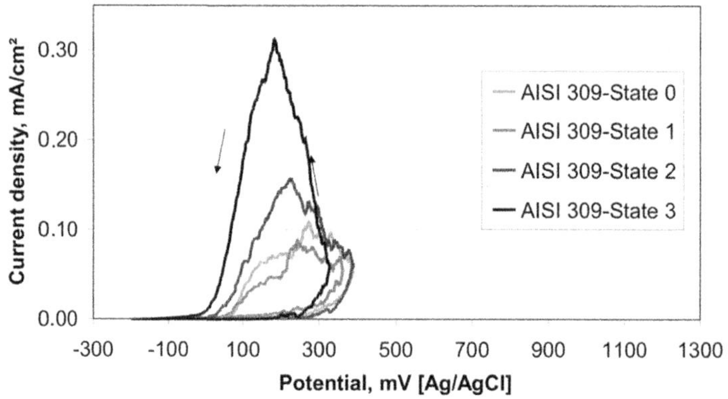

Fig. 8. Current density-potential plots of different specimen types with dU/dt = 0.5 mV/s, Steel AISI 309, base material, artificial seawater

This indicates that an increasing internal surface of hot cracks might reduce the repassivation capacity of the laser welded steel AISI 309 in aqueous environments.

Fig. 9 shows the current density-potential curves of similar experiments carried out with Alloy 50 at 60 °C, due to its already observed higher pitting corrosion resistance. With reversed polarization, a negative hysteresis was measured in every case which generally indicates a remarkably good repassivation behavior of this material.

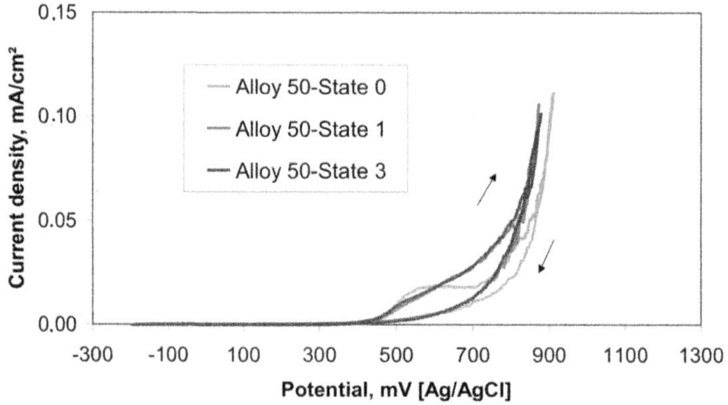

Fig. 9. Current density-potential plots of different specimen types with dU/dt = 0.5 mV/s, Alloy 50, base material, artificial seawater

A slight difference in the repassivation behavior might be deduced from the curves between the base material and the welded specimens. But no difference of the current density-potential plot of the State 3 specimen could be observed as compared to the crack free weld. Thus, hot cracking does not seem to have an influence on the local corrosion and repassivation behavior of this alloy.

Immersion Tests

The results from the Immersion Tests are summarized in Table 6. The high nitrogen steel AISI S 34565 showed no corrosion attack in the 20 % $MgCl_2$ solution at temperatures up to 110 °C. Even under aggravated conditions (1700 h in 40 % $MgCl_2$-solution at 140 °C test temperature), no indications of stress corrosion cracking were found in the Alloy 50 specimens. Only sporadically, pitting occurred in the vicinity of the weld. However, the steel AISI 309 definitely exhibited stress corrosion cracking in the Immersion Tests.

Table 6. Results of the Immersion Tests

Material	Temperature range [°C]	$MgCl_2$ concentration [%]	Test duration [h]	Results
AISI 309	80–120	10–40	720	Various SCC
AISI S 34565	80–110	20	1300	No corrosion attack
Alloy 50	90–140	10–40	720 and 1700	No SCC but pitting

As shown by the cracking tendency plot for the welded specimens of this material in Fig. 10, stress corrosion cracking generally appeared at temperatures above 80 °C and at $MgCl_2$ concentration above 25 %. It also turned out that longitudinal hot cracks in the laser welded alloy AISI 309 represent a preferential site for stress corrosion cracking. At 80 °C in the $MgCl_2$ solution, stress corrosion cracking started at hot cracks and increasing the temperature above 100 °C in the 30 % $MgCl_2$ solution caused complete cracking of the specimens (Fig. 10).

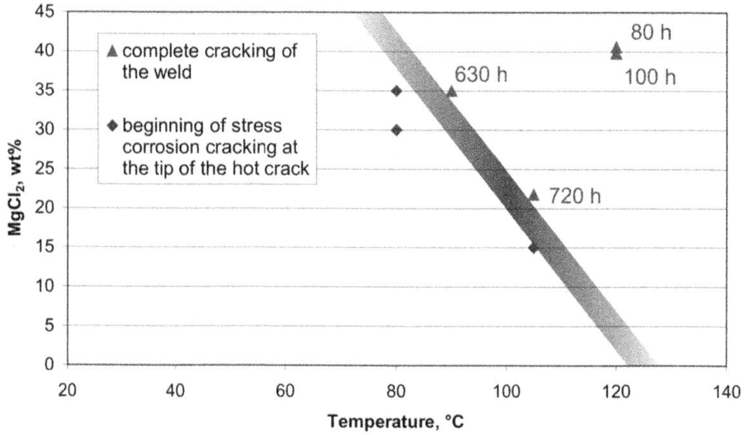

Fig. 10. Boundary between completely cracked specimens with the respective failure time and specimens with beginning stress corrosion cracking in the weld, AISI 309, State 3, maximum period of the Immersion Test: 720 h

At test temperatures of 120 °C and in 40 %-$MgCl_2$-solution, failure was observed already after 80 and 90 h, respectively.

The micrographs in Fig. 11 show two cross sections of two AISI 309 specimens, State 3, which exhibit stress corrosion cracking without complete fracture of the specimen. This confirms that the stress corrosion cracking process is initiated at the tip of respective hot cracks.

It also seems that the hot cracks must have certain dimensions in width and in depth to provide access of the electrolyte and to establish aggressive

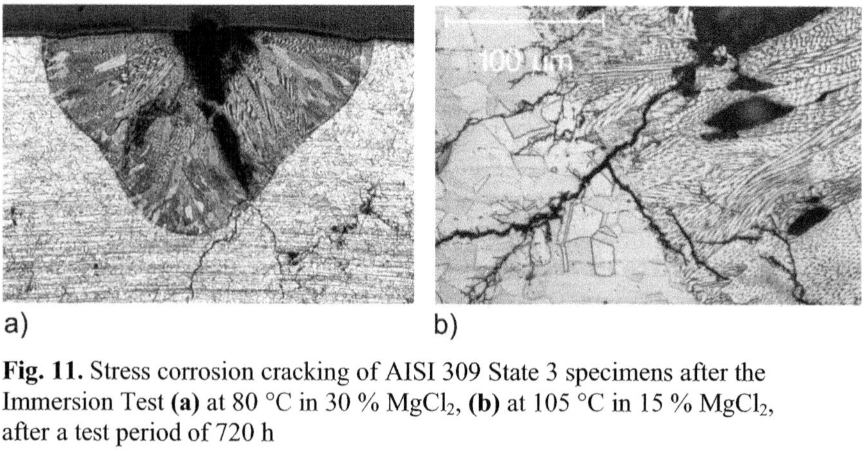

a) b)

Fig. 11. Stress corrosion cracking of AISI 309 State 3 specimens after the Immersion Test **(a)** at 80 °C in 30 % $MgCl_2$, **(b)** at 105 °C in 15 % $MgCl_2$, after a test period of 720 h

crevice like conditions in terms of a pH and potential reduction.

Also the direction of the hot cracks towards the tensile stress orientation seems to exert an influence on the stress corrosion behavior. This assumption is substantiated by the fact that State 2 specimens did not exhibit stress corrosion cracking in the weld, but in the base material at the same exposure conditions as applied to the respective State 3 specimens.

Stress corrosion cracking initiated at longitudinal hot cracks predominantly appears in a transgranular mode. This has been confirmed by respective SEM investigations of completely cracked specimens (Fig. 12). The fracture topography of the hot crack tip exhibits smooth molten dendrite surfaces and can clearly be distinguished from the transgranular cleavage mode of stress corrosion cracking.

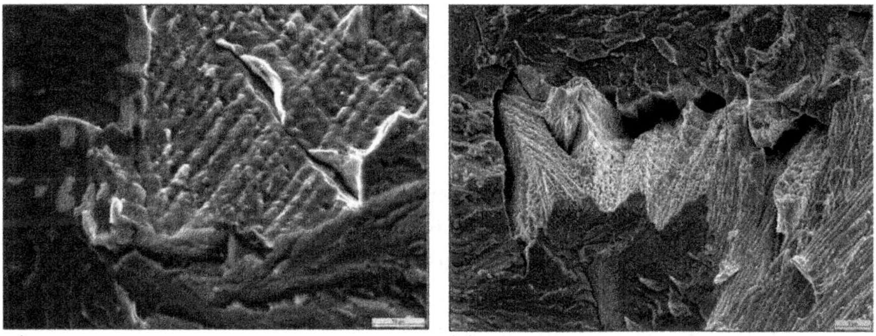

a) b)

Fig. 12. Fracture topography at the hot crack tip of completely cracked U-bend specimens, AISI 309, State 3, **(a)** in 41 % $MgCl_2$ at 120 °C, time to failure 80 h **(b)** in 40 % $MgCl_2$ at 120 °C, time to failure 100 h

Drop Evaporation Test

As confirmed by the results listed in Table 7, also in the Drop Evaporation Test stress corrosion cracking of the steel AISI 309 was found which had been initiated at hot cracks in State 3 specimens. Pitting was observed on all specimens. Stress corrosion cracking also occurred in the hot crack-free welded specimens. But in contrast to the State 3 specimens it was localized in the basc material (Fig. 13a).

Thus, the welds themselves do not represent preferential sites for stress corrosion cracking. In other words, hot cracks have to be present in such welds to initiate stress corrosion cracking. In this context, it can only be emphasized that although pitting occurred in the vicinity of the hot crack

Table 7. Results of the Drop Evaporation Tests with AISI 309 specimens

Exposure time at 30 °C	AISI 309, State 0	AISI 309, State 1	AISI 309, State 3
1000 h	No corrosion attack	Pitting in the vicinity of the weld	Pitting in the vicinity of the weld
2000 h	Pitting	Pitting in the vicinity of the weld	No corrosion attack
3000 h	No corrosion attack	Pitting in the vicinity of the weld	SCC initiated at hot cracks, pitting in the vicinity of the weld
4000 h	Pitting	No corrosion attack	Pitting in the weld and in the vicinity of the weld
5000 h	Pitting	SCC in the base material	SCC initiated at hot cracks, pitting in the vicinity of the weld

afflicted laser welds, stress corrosion cracking was initiated inside the longitudinal hot cracks, as shown in Fig. 13b. Regarding this material, the electrochemical conditions inside hot cracks are more critical than inside the pits which gives further evidence of the severity of hot cracks acting as stress corrosion cracking initiation sites.

In contrast to the Immersion Tests, the AISI S 34565 specimens (Table 8) exhibited pitting at 50 °C in the DE Test with nearly no exception. The pits occurred in the base material and also in the vicinity of the weld.

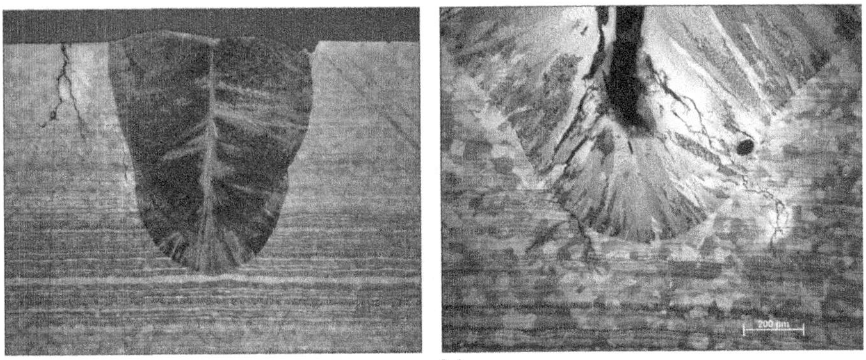

a) b)

Fig. 13. Stress corrosion cracking of AISI 309 in the Drop Evaporation Test **(a)** State 1, exposure time: 5000 h, **(b)** State 3, exposure time: 3000 h

Table 8. Results of the DE Test with AISI S 34565 specimens

Exposure time at 50 °C	AISI S 34565, State 0	AISI S 34565, State 1	AISI S 34565, State 3
1000 h	Pitting	Pitting in the vicinity of the weld	No corrosion attack
2000 h	Pitting	Pitting in the vicinity of the weld	Pitting in the vicinity of the weld
3000 h	Pitting	Pitting in the vicinity of the weld and in the weld	Pitting in the vicinity of the weld and in the weld
4000 h	SCC initiated at pits in the base material	SCC initiated at pits in the base material	Pitting in the vicinity of the weld and in the weld
5000 h	SCC initiated at pits in the base material	SCC initiated at pits in the base material	SCC initiated at pits in the base material

Fig. 14a shows exemplarily a strong localized corrosion attack in the weld and in the base material. Stress corrosion cracking initiated at such pits (Fig. 14b) was observed in the specimens of State 0 and State 1 after 3000 h and 4000 h, whereas the specimen of State 3 exhibited stress corrosion cracking only after a 5000 h period. This already indicates that, longitudinal hot cracks do not represent a preferential site for stress corrosion cracking in these welds as compared to such pits, in contrast to the AISI 309 steel. Stress corrosion cracking usually appeared at pits in the base material near the weld region.

a) b)

Fig. 14. Alloy AISI S 34565, State 3, subjected to the DE-Test **(a)** pitting attack in the weld and in the basc material, exposure time: 3000 h, **(b)** pitting induced stress corrosion cracking in the base material, exposure time: 4000 h

Table 9. Results of the DE Test with Alloy 50 specimens

Exposure time at 70 °C	Alloy 50, State 0	Alloy 50, State 1	Alloy 50, State 3
1000 h	No corrosion attack	No corrosion attack	No corrosion attack
2000 h	No corrosion attack	No corrosion attack	No corrosion attack
3000 h	Pitting	Strong pitting in the weld	Slight corrosion attack in the hot crack
4000 h	Pitting	Pitting in and in the vicinity of the weld	Slight corrosion attack in the hot crack
5000 h	Pitting	Strong pitting in the weld	Slight corrosion attack in and in the vicinity of the weld

The U-bend specimens of the Alloy 50 exposed to the DE Test at 70 °C exhibited no stress corrosion cracking in the base material and also not in the weld region (Table 9). A slight brown coloration of the drops on the welded specimens of State 1 and State 3 was observed after exposure times of 2000 h, and after 3000 h, an accumulation of corrosion products was detected even in the base material specimens.

The pitting attack appeared stronger in the crack-free welded specimens and indicates that the hot cracks might represent a preferential site for localized corrosion in the alloy.

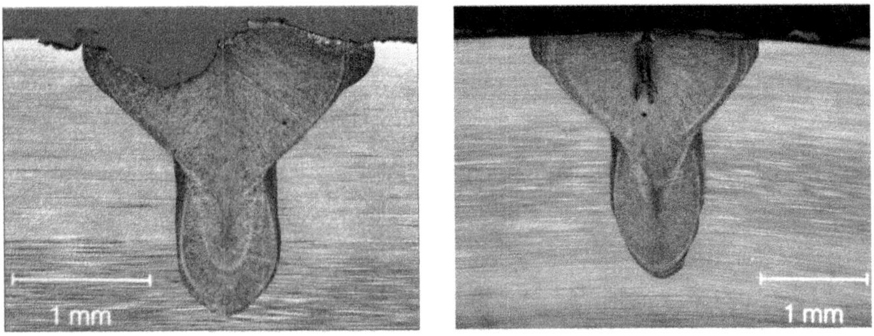

a) b)

Fig. 15. U-bend specimens, Alloy 50 **(a)** State 1, localized corrosion attack of the weld and the base material, exposure time: 3000 h,
(b) State 3, only slight corrosion attack, exposure time: 5000 h

Some evidence to this is given by the observation of a slight corrosion attack in the longitudinal hot cracks of specimens subjected for 3000 h and 4000 h to the DE Test. Fig. 15a shows a severe localized corrosion in the weld and in the base material of a State 1 specimen after 3000 h. In the specimens of State 3, only slight corrosion attack was found even after a 5000 h test period (Fig. 15b).

Conclusions and Perspectives

From the first approach to elucidate the interaction between hot cracking and localized corrosion, in particular, stress corrosion cracking, by using laser welded bend specimens in chloride environments, the following conclusions can be drawn:

1. In conjunction with respective lower repassivation capabilities, an enhanced susceptibility to stress corrosion cracking of hot crack afflicted AISI 309 laser welds was evidently shown by the Immersion Test in $MgCl_2$-solution and confirmed by the more realistic Drop Evaporation Test. Especially longitudinal hot cracks represent an initiation site for subsequent stress corrosion cracking, whereas numerous small hot cracks perpendicular to the welding direction do not seem to affect the susceptibility to stress corrosion cracking. The geometry as well as the orientation of the hot cracks to the stress state of the specimen thus significantly affect the initiation of stress corrosion cracking.
2. The influence of hot cracks on the corrosion resistance and, in particular, on the susceptibility to stress corrosion cracking strongly depends on the chemical composition, since the higher alloyed materials AISI S 34565 and Alloy 50 did not exhibited any stress corrosion cracking initiated at hot cracks. In contrast to the steel AISI 309, pits in the steel AISI 34565 represent a more favorable site for stress corrosion cracking than hot cracks. Also the Nickel base Alloy 50 partially showed severe pitting in the Drop Evaporation Test, but no stress corrosion cracking at all.

As a future perspective, also microstructural effects have to be investigated to clarify why hot cracks represent preferential stress corrosion cracking initiation sites in the metastable primary austenitic solidified steel AISI 309, but not in the two stable alloys. Further research work will also be performed to identify precisely the stresses and strains at the tip of hot cracks and how hot cracking might affect subsequent stress corrosion cracking in real restrained component welds.

References

1. Schobbert H, Böllinghaus Th, Wolf M (2003) Hot Cracking Resistance of Laser and Hybrid Welded Austenitic Stainless Steels. In: 6th International Trends in Welding Research Conference Proceedings, Pine Mountain, GA, ASM International, pp 76–81
2. Schobbert H, Wolf M, Böllinghaus Th (2004) Metastable Primary Solidification Modes in the Fe-Cr-Ni System During Laser Welding. In: Solidification and Crystallisation, Dieter M. Herlach, pp 216–226
3. Volkmann T, Löser W, Herlach DM (1997) Nucleation and Phase Selection in Undercooled Fe-Cr-Ni Melts: Part II. Containerless Solidification Experiments. Metall Mat Trans A 28A: 467–469
4. Böllinghaus Th, Schobbert H (2003) Laser Weld Metallurgy of Austenitic Stainless Steels. In: Conference Papers, Stainless Steel World 2003, Maastricht, pp 243–249
5. Herold H, Streitenberger M, Spieler S (1999) Verhalten von Heißrissen unter Korrosionsbeanspruchung. DVS-Bericht 204, pp 66–72

III Modeling and Simulation

Simulating and Predicting Weld Solidification Cracks

Y. Wei[1], Z. Dong[1], R. Liu[2], Z. Dong[2], Y. Pan[2]

[1] State Key Laboratory of Advanced Welding Technology Production, Harbin Institute of Technology, Harbin, China
[2] Harbin Welding Institute, Harbin, China

Abstract

Weld solidification cracks result from the competition between the material resistance and driving force to the cracks. The material resistance, namely the ductility curve in BTR, was obtained by in-situ observation and measurement of local strain at the trail of weld pool using a CCD-camera in the Trans-Varestraint Test. The driving force, i.e. mechanical strains against temperatures at the trail of a weld pool, was modeled with the finite element method (FEM) in three steps. Firstly, thermal distributions in both 3 mm and 10 mm thickness welds were modeled by two-dimensional and three-dimensional thermal models, respectively. Secondly, the strain/stress distributions arising during welding of 3 mm stainless steels were simulated by a two-dimensional model on the basis of the simulated thermal distributions. Thirdly, the driving force behind weld solidification cracking was determined from the simulated thermal cycle (temperature against time) and from the mechanical strain against time in the weld center line. Furthermore, a computer system was developed on the basis of two-dimensional simulation which provides a simple way of conducting complex work for simulating and predicting weld solidification cracking.

Introduction

The weld metal solidification cracking occurs when the mechanical driving force exceeds the material resistance. The fundamental parameter representing the material resistance is its ductility curve in the brittle temperature range (BTR) in the course of weld metal solidification, while the mechanical strain accumulated in the BTR is considered as the mechanical driving force. To obtain the ductility curve in the BTR, a variety of weldability tests have been advised [1–5]. Among them, the Varestraint-Test and the Trans-Varestraint-Test have been widely used to evaluate quantitatively the solidification cracking susceptibility of different materials by measuring their CTS (Critical Strain rate for Temperature drop), BTR, the plastic deformation ability – the solidification ductility curves in BTR and some other quantitative parameters such as crack length and crack number.

However, the strains obtained from these tests are the average bending strains on the surface of workpieces determined by the curvature radius of the die block, while a solidification crack is caused by the local strain at the trail of a weld pool. Consequently, it is necessary to measure the local strains at the trail of weld pool so as to get a more accurate ductility curve in BTR to predict weld solidification cracks.

Meanwhile, studies on the driving force, namely the mechanical strain in BTR, are less frequently undertaken owing to the complex solidification process involved in welding. With the rapid development in both computational and numerical methods, a great deal of progress has been accomplished in computational modeling of welding-related thermal and mechanical responses [6–10].

However, most of them focus on the residual stress and distortion of welded structures in which assumptions and simplifications on the material behavior at elevated temperatures are often made. Such treatments can provide results accurate enough for residual stress and distortion analyses, but obviously not sufficiently accurate for computing the transient thermal stress/strain fields in the vicinity of a weld pool.

To solve the problem, Z. L. Feng, W. P. Liu and S. Q. Guo had conducted related researches concerning aluminium [11–14] and D. Dye has also made profound research on superalloys [15]. Furthermore, weld solidification crack predicting and simulation not only deal with measuring ductility curves but also with simulating thermal distributions and with strain distributions as well which is usually difficult for welding engineers to perform.

This paper mainly deals with the methodology of simulating and predicting solidification cracking of stainless steels [16–19]. The work includes measurement of local strain around weld pool during the Trans-Varestraint-Test in combination with a CCD-camera to get a more accurate ductility curve, simulation of welding thermal/strain/stress fields to get the driving force, and development of a computer system for simulating and predicting weld solidification cracks.

Measurement of Material Resistance

Testing Device and its Basic Principle

In the early 1980ies, Japanese scientists had measured the local strain with the MISO method and indicated that there was strain concentration at the trail of the weld pool [20].

In order to obtain accurate material resistance, they measured the local strain through in-situ observation with a CCD-camera during conducting the Trans-Varestraint-Test. The test was carried out with a HHPL-1 Trans-Varestraint Testing Machine constructed by the Harbin Welding Institute. Fig. 1 shows the device used in the testing.

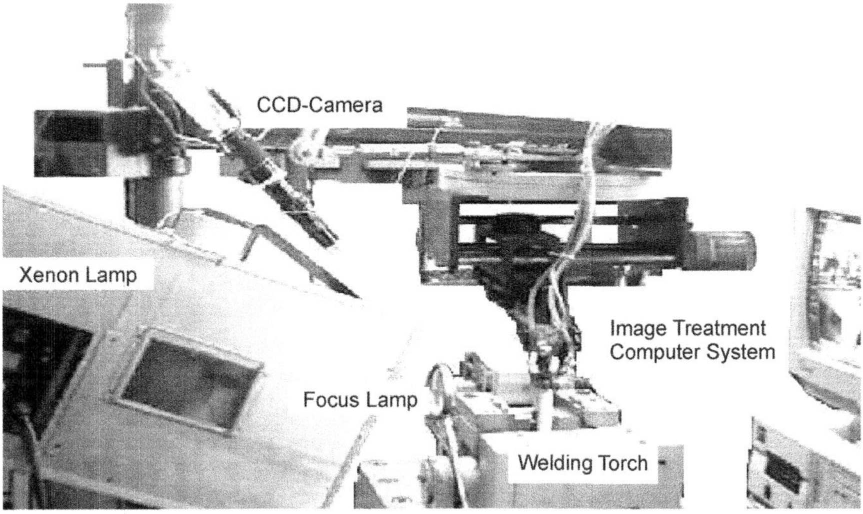

Fig. 1. Testing device used for measuring the material resistance against weld solidification cracking

As shown in Fig. 1, the device is composed of a CCD-camera, an image treatment card, a xenon lamp, a focus lamp, an image treatment computer system and the Trans-Varestraint Testing machine that is also illustrated in Fig. 2a.

The workpiece is shown in Fig. 2b and 2c.There are two identified lines along each side of the weld which are 5 and 6 mm away from the weld center line, respectively, as shown in Fig. 2c. During welding, hard particles were put inside the molten pool.

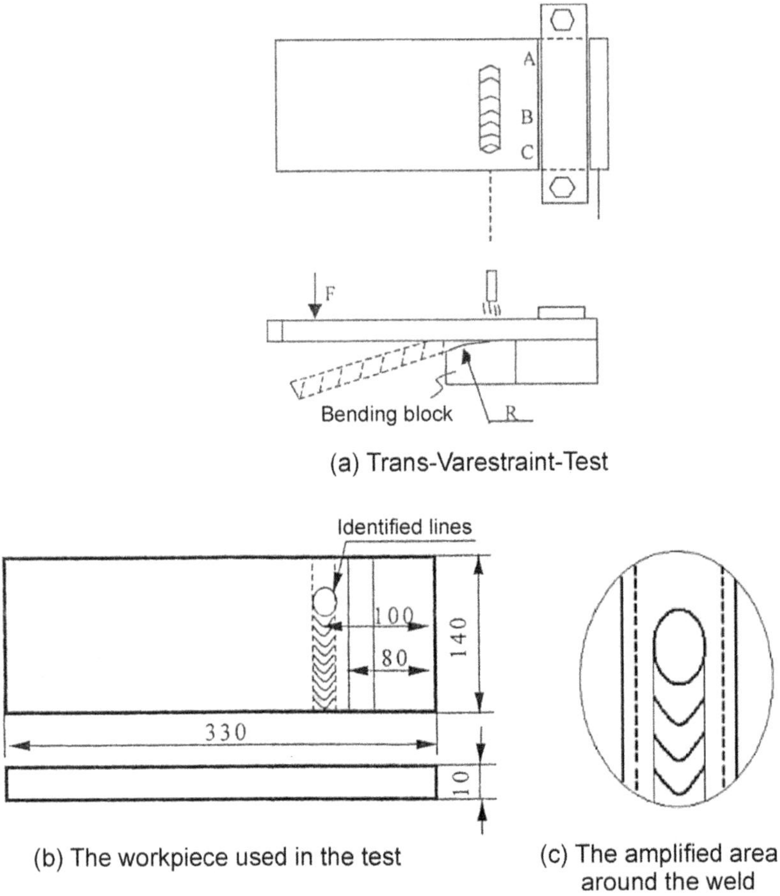

Fig. 2. Trans-Varestraint-Test and samples used in the testing

The changed distance of the identified lines and the hard particles will be used to calculate the parameters representing the susceptibility to weld solidification cracking as shown in Fig. 3a–c.

Fig. 4a and 4b show the weld appearances recorded with the CCD-camera before and after welding. According to Fig. 3a and Fig. 4a, the multiple of images can be calculated which is useful for calculating strains arising during testing.

Fig.5 shows the recorded process of crack initiation and development at the trail of weld pool during loading.

With the devices shown in Fig. 11, the following parameters are measured and calculated on the basis of Fig. 3 and Fig. 5.

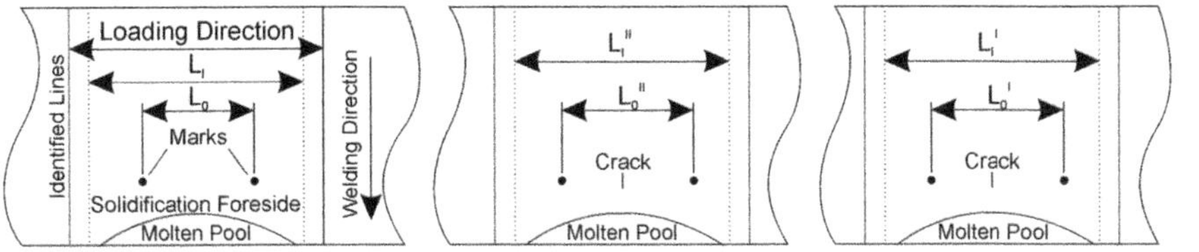

(a) Before loading (b) Crack initiation (c) After loading

Fig. 3. The principle for measuring parameters of the weld solidification cracking susceptibility

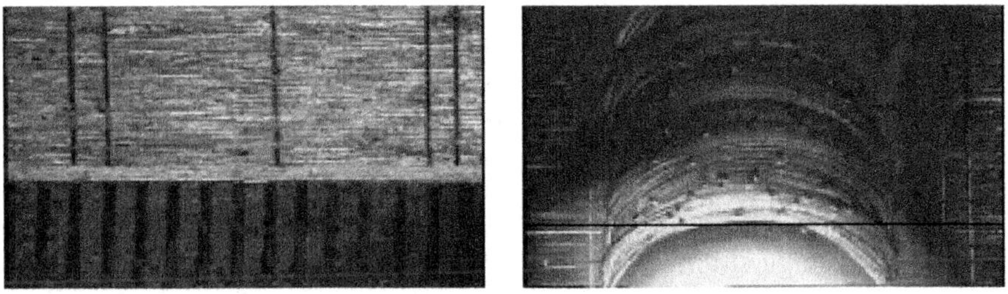

(a) The workpiece before welding (b) The weld after welding

Fig. 4. The workpiece before welding and the weld after welding recorded with a CCD-camera

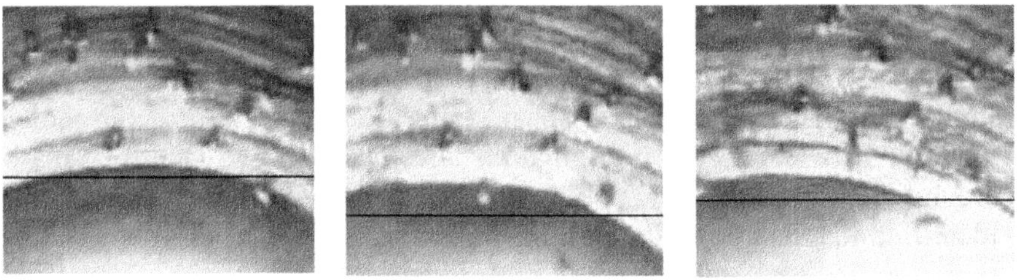

(a) Before loading (b) Crack initiation (c) After loading

Fig. 5. Crack initiation and development recorded with a CCD-camera

Nominal Strain. The strain level applied to the specimen is determined by the curvature radius of the die block as shown in Fig. 2a. As it is a pure bending process during which the nominal strain is calculated by Eq. 1.

$$\varepsilon_n = \frac{h}{2R} \times 100\% \tag{1}$$

In this formula, ε_n represents the nominal strain of specimen in %, h is the thickness of plate in mm and R the curvature radius of block in mm.

Average Strain across the Weld Pool. During welding, the strength and deformation ability of the weld pool differ from those of the cold metal around it so that its strains should be different, too. Average strain across the weld pool takes the form as expressed by Eq. 2.

$$\varepsilon_a = \frac{L_1^{'} - L_1}{L_1} \times 100\% \tag{2}$$

The parameter ε_a represents the average strain across the weld pool % and L_1 is original distance between two identified lines at each side of weld in mm where L_1' is the distance between two identified lines at each side of weld during or after bending in mm.

Local Strain around the Trail of Weld Pool. Solidification cracks usually occur at the trail of the weld pool at which the strain is also different from the average strain ε_a. In order to calculate the local strain around the trail of the molten metal, that is around the cracks, hard particles have been put inside the weld pool and two nearest points around the trail of weld pool were chosen to assess the local strain that takes the form as given in Eq. 3.

$$\varepsilon_l = \frac{L_0^{'} - L_0}{L_0} \times 100\% \tag{3}$$

In this formula, ε_l represents thelocal strain around the trail of the molten metal in %, while L_0 is the original distance between two identified marks around the trail of the weld pool in mm and L_0' is the distance between two identified marks around the trail of the weld pool during or after bending, mm.

Critical Strain at the Trail of Weld Pool. Critical strain is used to indicate the strain when a crack is initiated during loading which represents the minimal ductility of weld metal as formulated by Eq. 4.

$$\varepsilon_{l-c} = \frac{L_0^{''} - L_0}{L_0} \times 100\% \tag{4}$$

Where ε_{l-c} is the critical strain at the trail of the weld pool, in %, L_0 is the original distance between two identified points around the trail of the weld pool in mm and L_0'' is the distance between two identified points when cracks are initiated in mm.

Concentration of Strain at the Trail of Weld Pool. The ratio of ε_l and ε_n is used to illustrate the concentration of strain at the trail of weld pool which takes the form as expressed by Eq. 5.

$$\eta = \frac{\varepsilon_l}{\varepsilon_n} \tag{5}$$

Test Materials and Results

Two kinds of carbon steels were tested with the method described above and appropriate parameters representing the susceptibility to weld solidification cracking were measured and calculated. The chemical compositions of the carbon steels are shown in Table 1 and the test results are shown in Table 2.

Table 1. Chemical compositions (wt.%) of carbon steels used in the testing

	C	Mn	Si	S	P
B	0.19	0.54	0.27	0.039	0.041
D	0.16	0.20	0.46	0.015	0.011

Referring to Table 2, the following explanations can be given:

- $\varepsilon_{l\text{-}c}$ (critical local strain when cracks are initiated) is nearly constant for the same steel with similar welding conditions which demonstrates that the critical local strain is repeatable, and which further proves that the method is practical. The critical strain around the cracks for steel B is 1.88 % to 2.47 % and the average is 1.91 % while $\varepsilon_{l\text{-}c}$ of steel D is from 2.94 % to 3.98 % and the average is 3.46 %.
- The local strain around crack tips, ε_l,, is much higher than the average strain represented by block curvature ratios which illustrates that there is strain concentration around the trail of the weld pool. The average strain concentration is about 2.2 for steel B, that is, $\eta = 2.2$.
- As strain concentration cxists at thc trail of a weld pool, the ductility curves obtained with the common Trans-Varestraint-Test should be modified by a strain concentration coefficient, η. Fig. 6 shows both the original and modificd ductility curves of steel B.

Table 2. Tested materials and calculated results of in-situ Trans-Varestraint-Tests

work-piece number	Block number	Images /frame	ε_n	ε_a	ε_l	η	$\varepsilon_{a\text{-}c}$	$\varepsilon_{l\text{-}c}$
			[%]					
B-17	U	11	4.12	6.96	7.17	1.74	1.65	1.94
B-19	U	12	4.16	6.69	7.61	1.83	1.69	2.24
B-20	U	14	4.00	9.04	8.22	2.06	2.35	2.02
B-22	N	9	3.26	7.13	4.97	1.53	2.18	2.33
B-26	S	8	2.60	5.45	7.34	2.82	2.26	2.47
B-27	S		2.55				1.76	2.22
B-28	S	9	2.58	6.21	6.37	2.47	2.28	1.88
B-33	H	7	2.04	5.83	5.79	2.83	1.81	1.88
D-36	U	12	3.81	9.41			2.05	3.98
D-37	U	12	3.80	9.07			1.99	3.76
D-38	U	13	3.67	9.34			2.98	3.42
D-40	U	13	3.83	8.50			2.37	3.62
D-43	U	12	3.73	8.45			2.16	3.85
D-46	U	12	3.98	7.79			1.81	3.43
D-63	U	11	3.47	8.52			2.46	3.56
D-66	U	12	3.30	9.39			2.04	3.29
D-69	U	13	4.14	6.09			2.17	3.55
D-47	H	7	2.09	4.45			2.12	3.19
D-60	H	7	1.88	5.62			1.69	2.94

$\varepsilon_{a\text{-}c}$ critical ε_a when cracks are initiated.

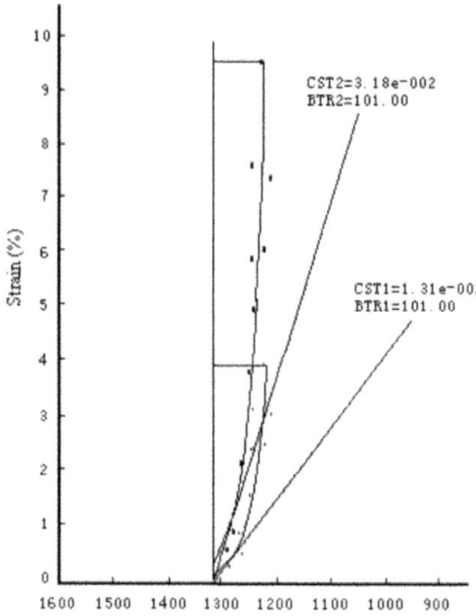

Fig. 6. Comparison between original and modified ductility curves of steel B

As shown in Fig. 6, the modified ductility curve of steel B moves up and its CTS becomes greater compared with the original one, which demonstrates that the ductility curve from the ordinary Trans-Varestraint-Test should be modified when used in predicting weld solidification cracks. The method described in the paper provides an effective way to get more accurate parameters representing the susceptibility to weld solidification cracking and founds basis for predicting them, too.

Numerical Models for Simulating Temperature Fields

Description of the Problem

As the mechanical strain is mainly caused by the transient non-uniform heating and cooling at the weld and near-weld position, it is first of all necessary to analyze the thermal distribution in the vicinity of the weld pool. The problem of interest is outlined as follows:

- Material: austenitic stainless steels, SUS310 and SUS316, are analyzed.
- Welding process: tungsten inert gas welding (TIG) is simulated since it has been used in most of the weldability tests related to solidification cracks, which makes the results comparable.
- Numerical analytical method: the finite element analysis is used for modeling both full penetration welding on 3 mm plate and bead-on-weld on 10 mm plate.
- Coordinate system: the coordinate system is fixed on the workpiece, the X-axis is along the centerline of the plate and the arc starts at point (X_0, 0, 0), the Y-axis is along the edge of the plate and the Z-axis is along the plate thickness, as shown in Fig. 7.

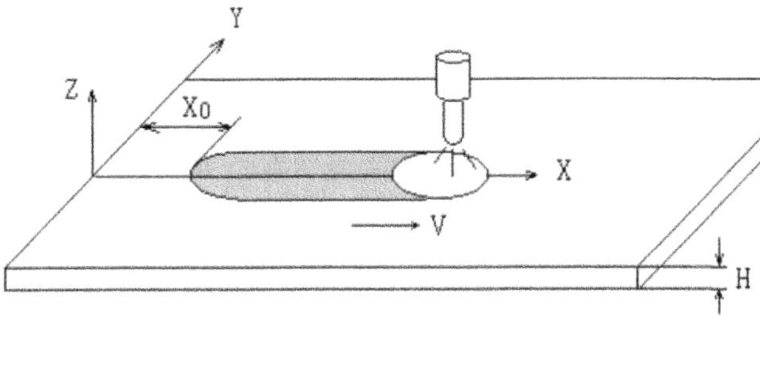

Fig. 7. The coordinate system

The Relevant Factors and their Treatments in Thermal Computational Modeling

Arc Heat Input

For TIG-welding of 3 mm plate, a two-dimensional heat flux disk with a radially symmetric welding torch is used.

Gaussian distribution is assumed to apply to the surface of the work-piece. If the electrode moves along the X-axis and in positive direction of the X-axis (Fig. 7), the Gaussian distribution of the arc energy input takes the following form:

$$q_{arc}(x,y,z) = q_0 e^{\frac{3r^2}{r_b^2}} = \frac{3\eta VI}{\pi r_b^2}\exp\left\{-\frac{3\left[(x-vt-x_0)^2+y^2\right]}{r_b^2}\right\}, \tag{6}$$

where V, I, v are the arc voltage, current and travel speed, respectively, η is the arc efficiency and r_b is the characteristic dimensional distribution parameter that defines the region in which 95 % of the heat flow is deposited.

For three-dimensional simulation of thermal fields for welding 10 mm plate, the double surface ellipsoid Gaussian heat model has been proven to be a more realistic representation of the heat source, of which the power density distribution is shown in Fig. 8 [21].

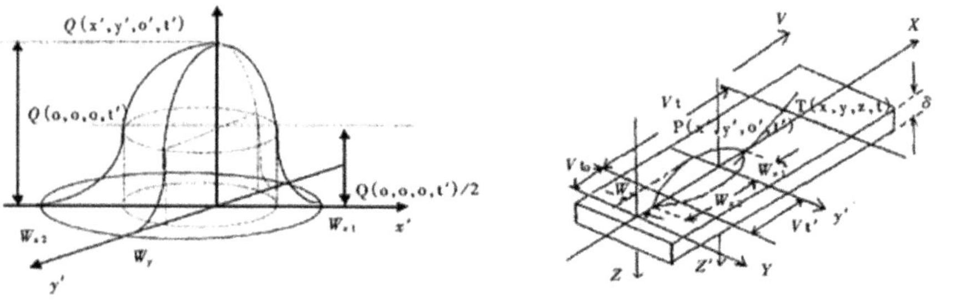

(a) Model of heat source (b) Moving coordinate for heat source

Fig. 8. Double ellipsoid Gaussian distribution model of heat source

With the energy input rate Q in W, the fractions of the heat deposited in the front and rear ellipse, f_f, f_r,, the parameters of double ellipsoid Gaussian distribution, w_{x1}, w_{x2}, w_y and the moving coordinates x', y', z' the power density distribution can be written by the following equation [21]:

$$Q(x',y',o',t') = \begin{cases} \dfrac{f_f Q}{2\pi w_{x1} w_y} \exp\left(-\dfrac{x'^2}{2w_{x1}^2} - \dfrac{y'^2}{2w_y^2}\right) & x' > 0 \\[2ex] \dfrac{f_r Q}{2\pi w_{x2} w_y} \exp\left(-\dfrac{x'^2}{2w_{x2}^2} - \dfrac{y'^2}{2w_y^2}\right) & x' < 0 \end{cases} \quad (7)$$

However, the above Gaussian surface heat flow model ignores the digging action of the arc that transports heat well below the surface. Another double ellipsoid source was proposed by Goldak et al. [22] as shown in Fig. 9. The power or heat flow distribution is Gaussian along the longitudinal axis. The front half of the source is the quadrant of one ellipsoidal source while the rear half is the quadrant of another ellipsoidal source. The power density distribution inside the front quadrant becomes:

$$q(x,y,z,t) = \frac{6\sqrt{3} f_f Q}{a_1 b c \pi \sqrt{\pi}} e^{-3x^2/a_1^2} e^{-3y^2/b^2} e^{-3z^2/c^2}\,. \quad (8)$$

Similarly, for the rear quadrant of the source the power density distribution inside the ellipsoid becomes:

$$q(x,y,z,t) = \frac{6\sqrt{3} f_r Q}{a_2 b c \pi \sqrt{\pi}} e^{-3x^2/a_2^2} e^{-3y^2/b^2} e^{-3z^2/c^2}\,. \quad (9)$$

In Eqs. 8 and 9, four characteristic lengths must be determined which physically correspond to the radial dimensions of the molten zone. The parameters *a*, *b*, *c* can have different values in the front and rear quadrants, since they are independent. In this study, two kinds of double ellipsoid heat source were used to simulate the three–dimensional thermal distributions, the former is called Gaussian model and the later is called double ellipsoid model in this paper.

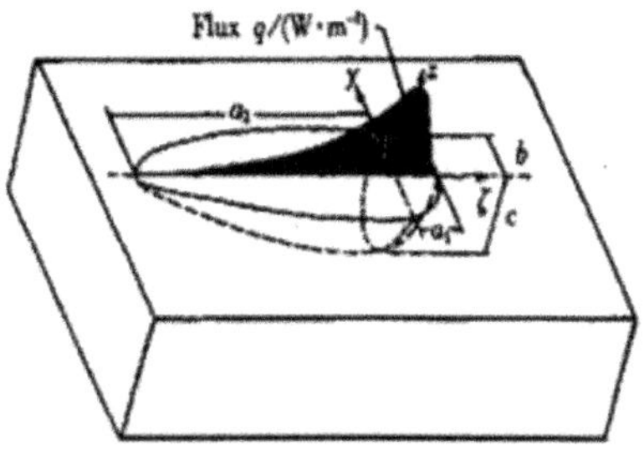

Fig. 9. Double ellipsoid heat source configuration

Fluid Flow in the Weld Pool

The fluid flow in the weld pool has a great effect on its temperature distributions, as the flow speeds the heat conductivity. Research in modeling heat and fluid flow in weld pools has been very active and much progress has been made in the past [23–26]. Instead of explicit simulation, the effect of the fluid flow in the weld pool on the overall temperature distribution of the workpiece will be considered in the heat transfer model by means of effective thermal properties such as the effective thermal conductivity. Such an approach has been used by many investigators in the past and the effectiveness of the approach has been well demonstrated [27–28]. Fig. 10 shows the effective thermal conductivity for different temperature ranges of the stainless steels SUS310 and SUS316.

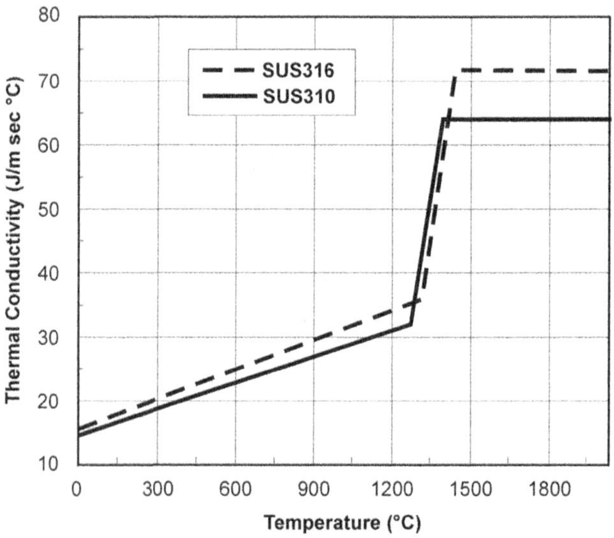

Fig. 10. The effective thermal conductivity for different temperature ranges

Latent Heat of Fusion

Previous studies have indicated that the latent heat of fusion highly influences the shape and size of the weld pool as well as the temperature distributions in the vicinity of the weld pool. Therefore, the latent heat of fusion should be considered in this study.

Two approaches have been considered in the past to include the effect of latent heat of fusion in the welding heat transfer analysis. The first one essentially assumes that solidification occurs at one constant temperature.

The release or absorption of the latent heat is modeled by application of a constant specific heat of the metal.

Such a treatment fails to recognize the fact that solidification of an alloy under most practical circumstances, such as welding and casting, takes place over a temperature range encompassing liquidus and solidus temperature of the alloy. In the second approach, a complex differential equation has been adopted to calculate the latent heat release for binary Al-Cu alloy [29]. Such computational approach is not suitable for ternary or multicomponent alloys such as SUS310 and SUS316 studied in this thesis.

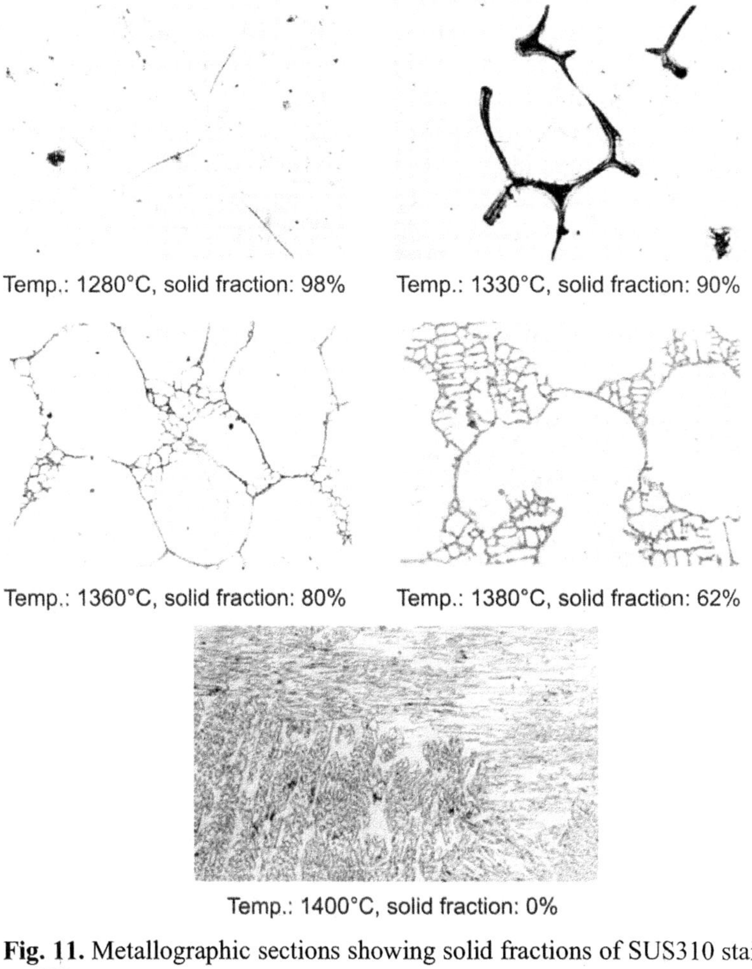

Fig. 11. Metallographic sections showing solid fractions of SUS310 stainless steel at different temperatures

It is known that the release rate of latent heat is in direct proportion to solid fraction, as described by the following equation:

$$q_i = \Delta H \frac{\partial f_s}{\partial t}, \tag{10}$$

where q_i is the release rate of the latent heat, f_s is the solid fraction, and ΔH is the latent heat of fusion per volume.

By means of liquid quenching tests, f_s, the solid fraction has been measured from which the release rate of the latent heat can be obtained.

The experiment has been carried out in the following procedures.

First of all, put a stainless steel cube, $10 \times 10 \times 10$ mm^3 in volume, in a crucible, and heat until it reaches a temperature a little higher than the liquidus temperature at which it is kept for 5 minutes.

Next, cool it to a given temperature in the range between liquidus and solidus and keep this temperature constant also for 5 minutes.

Then, take the sample out of the crucible and put it into the coolant where it is quenched. Last, corrode the quenched cube and measure the solid fraction with a metalloscope.

By this way, a series of solid fractions of the alloys SUS316 and SUS310 are measured at different temperatures between solidus and liquidus. The experimental results are demonstrated in Fig. 11.

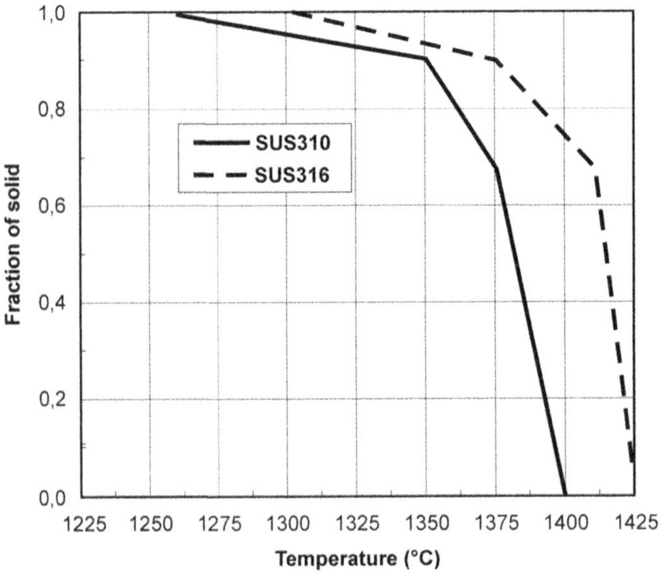

Fig. 12. The linearizing curves of solid fractions

Fig. 11 shows the solid fractions of SUS310 stainless steel at different temperatures in the range from solidus to liquidus. The measured solid fractions were linearized between the different temperatures as shown in Fig. 12.

Material Properties

The physical properties of SUS310 stainless steel are shown in Table 3. Both the thermal conductivity and the specific heat are dependent on temperatures. However, the mass density remains constant all the time.

Table 3. Relationship between physical properties of SUS310 stainless steel and temperatures

Temperature [°C]	Thermal conductivity [W/mK]	Specific heat [J/m^{3}K]	Latent heat release rate [J/m^{3}K]	Mass density [kg/m^3]
20–1250	15.013 $+1.363\times10^{-2}$T	4.109×10^6 $+1.138\times10^3$T		8000
1250–1340	56.8 $+1.98\times10^{-2}$T		2.58×10^6	
1340–1375			1.325×10^7	
1375–1400			6.496×10^7	
>1400	84.52	5.7026×10^6		

The Governing Equation

Based on the discussion above, the basic transfer models for 3 and 10 mm plate, respectively, are assumed to be two-dimensional and three-dimensional heat conduction models with prescribed heat flux moving along the weld, and simulated welding arc, convection and radiation heat loss from the top, bottom and edge (for 10 mm plate only) surfaces of the workpiece.

Because of the minor thickness of the 3 mm workpiece, the heat loss from its edges is ignored. Therefore, all boundary conditions, i.e. the surface heat loss and the heat flux from the welding arc, can be included in the governing heat diffusion equation in the terms of internal heat generation or loss.

In the heat transfer process, heat is lost from the surfaces in the form of convection and radiation, which occur on all surfaces of the plate except the plane of symmetry. The symmetry surface is defined as being subject to adiabatic boundary conditions.

According to Newton's law, the heat flux from a solid in contact with a flowing gas or liquid is proportional to the difference between the environment temperature T_∞ and the surface temperature T through a co-efficient of convective heat transfer h_c: $q_c = h_c\,(T - T_\infty)$.

The heat radiated per unit area and time, by a heated body, is according to Stefan-Boltzmann's law proportional to the fourth power of the difference between the environment temperature T_∞ and the surface temperature T through the radiation coefficient $\sigma\varepsilon$: $q_y = \sigma\varepsilon\,(T^4 - T_\infty^{\ 4})$.

In accordance with the coordinate system given in Fig. 7, the governing equation for two-dimensional simulation can be written as:

$$\rho c \frac{\partial T}{\partial t} = \frac{\partial}{\partial x}\left(k\frac{\partial T}{\partial x}\right) + \frac{\partial}{\partial y}\left(k\frac{\partial T}{\partial y}\right) + \overline{Q}\,. \tag{11}$$

For three-dimensional simulation, the governing equation takes the form as expressed by Eq. 12:

$$\rho c \frac{\partial T}{\partial t} = \frac{\partial}{\partial x}\left(k\frac{\partial T}{\partial x}\right) + \frac{\partial}{\partial y}\left(k\frac{\partial T}{\partial y}\right) + \frac{\partial}{\partial z}\left(k\frac{\partial T}{\partial z}\right) + \overline{Q} \tag{12}$$

$$Q = -\frac{q_c}{H} - \frac{q_r}{H} + \frac{q_{arc}}{H} + q_i = -\frac{2h(T - T_\infty)}{H} - \frac{2\sigma\varepsilon(T^4 - T_\infty^4)}{H} + \frac{q_{arc}}{H} + q_i\,, \tag{13}$$

where ρc is the voluminal specific heat, H is the thickness of the workpiece, and Q is the rate of total internal heat generation that consists of four terms: surface convection heat loss, q_c; surface radiation heat loss, q_r; heat flux input from the arc as described in Eqs. 6 or 7 or 8 and 9, and the release rate of latent heat, q_i, as described in Eq. 10.

Analysis of Simulating Thermal Distributions

The Finite Element Implementation

The finite element implementation of the mathematical models above can be divided in three parts: geometrical discretization of the workpiece, consideration of the thermophysical properties and imposition of boundary conditions.

To discretize the 3 mm workpiece, four-node elements are used, and smaller elements are adopted near the weld to account for possible severe temperature gradients in those regions.

Due to symmetry along the weld line, half of the 10 mm plate is discretized with eight-node isoparametric hexahedron elements.

The FEM mesh is shown in Fig.13. Smaller elements were used near the fusion zone and the adaptive mesh technology was used in and near the fusion zone.

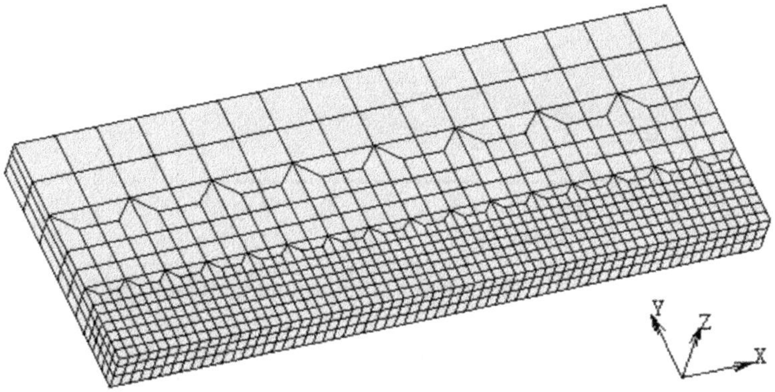

Fig. 13. Finite element mesh used for temperature calculations

In the study, the materials are assumed isotropic and homogeneous, but with temperature dependent properties. The treatment of boundary conditions is fairly simple.

The arc energy input and surface heat losses due to convection and radiation are converted into a single internal heat generation term in this work as described above.

Analysis of Simulating Results of Two-Dimensional Plate

The Influence of Arc Efficiency. As the arc efficiency highly affects the arc energy input in the workpiece, it will have an influence on temperature fields. Generally speaking, the arc efficiency is 0.65~0.70 for the TIG-welding process.

To compare the influences of the arc efficiency, temperature fields with different arc efficiencies, 0.65 and 0.70 respectively, were calculated as shown in Fig. 14.

Fig. 14 demonstrates that the arc efficiency has a great effect on the thermal distributions and isotherm shapes as well.

In this research work, 0.70 is finally used as arc efficiency in the calculation, as its corresponding temperature field coincides better with the experimental results.

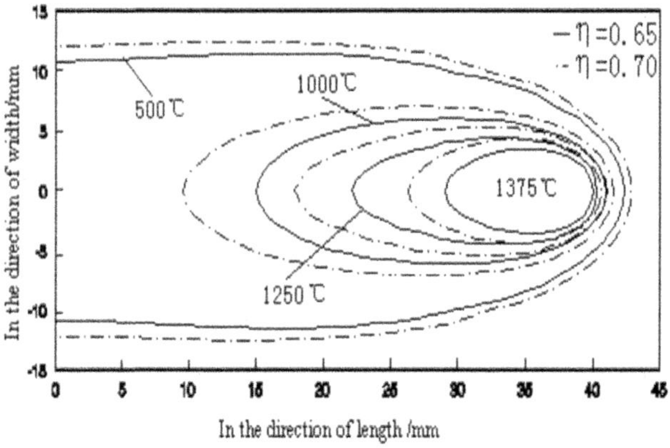

Fig. 14. The effects of the arc efficiencies on the temperature distributions

The Influence of the Characteristic Dimensional Distribution Parameter. The characteristic dimensional distribution parameter, r_b, defines the region in which 95 % of the heat flux is deposited. According to previous studies, r_b can be 2.5 mm or 3.0 mm for the TIG welding process when the electrode diameter is 2.5 mm. Thermal distribution simulation, as shown in Fig. 15, was performed for r_b equal to 2.5 mm and 3.0 mm, respectively.

As illustrated in Fig. 15, the value of r_b affects the thermal distributions. In this study, 3.0 mm is chosen for the calculation, of which the results are in better conformity with the experimental results [24].

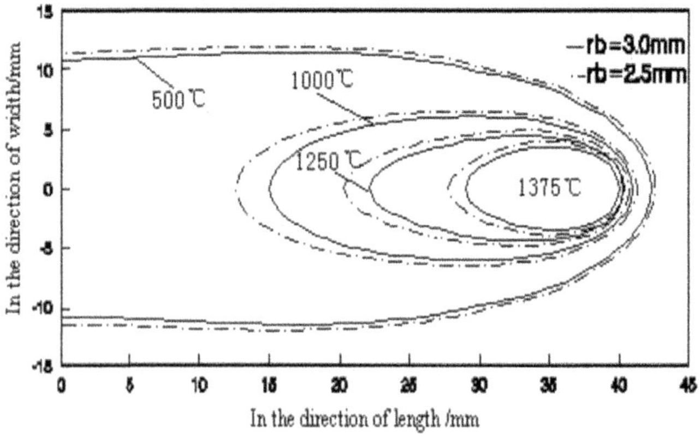

Fig. 15. The effects of the characteristic radial dimensional distribution parameter r_b on temperature fields

The Influence of Thermophysical Material Properties. Although both SUS310 and SUS316 are austenitic stainless steels, their chemical compositions and mechanical properties are much different. However, their thermophysical properties are a little different at elevated temperatures such as solidus, liquidus and thermal conductivity. The thermal distributions for the two materials are shown in Fig. 16.

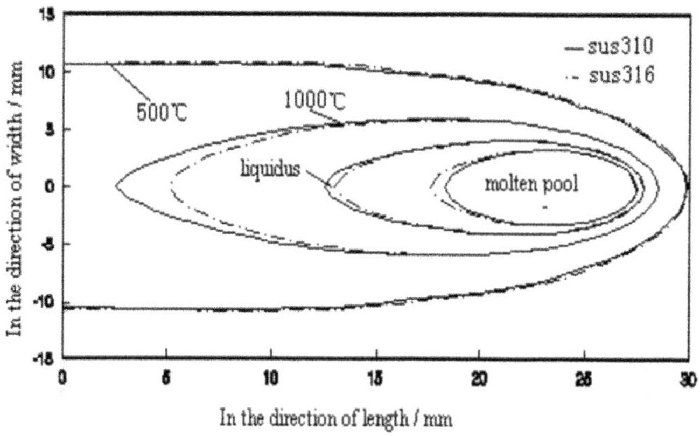

Fig. 16. The effects of the thermophysical material properties on temperature fields

As indicated in Fig. 16, the isotherms in longitudinal direction are a bit different while other parts are nearly the same. It may therefore be presumed that the stress/strain distributions will not have much difference too.

In other words, the reason why SUS310 is more susceptible to solidification cracking than SUS316 is mainly because the resistance to crack of SUS310, which is determined by metallurgical factors, is lower than that of SUS316.

Analysis of Simulating Results of Three-Dimensional Plate

Temperature Distributions of Two Heat Sources. Three-dimensional numerical simulation of the welding process is developed with the model above. The work is accomplished with commercial software MARC. Welding is assumed to be finished in 75 s and the cooling time is 200 s. The heat source is accomplished by a specially defined user subroutine. Fig. 17 illustrates the three-dimensional temperature distribution of a double ellipsoid model. The comparison between the temperature distributions of the double ellipsoid model and those of the common Gaussian model are shown in Fig. 18.

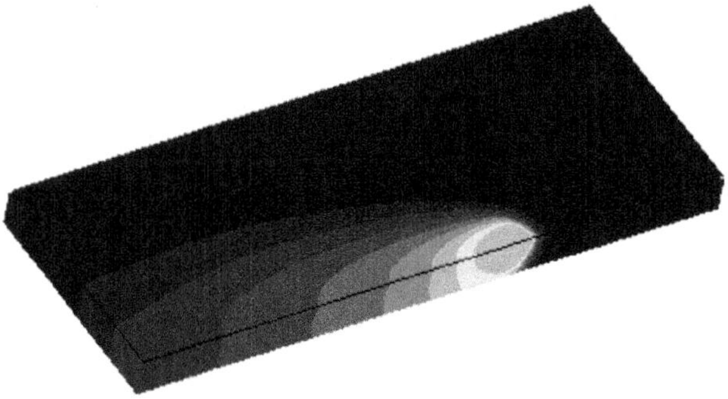

Fig. 17. The three-dimensional temperature distributions of two double ellipsoid model

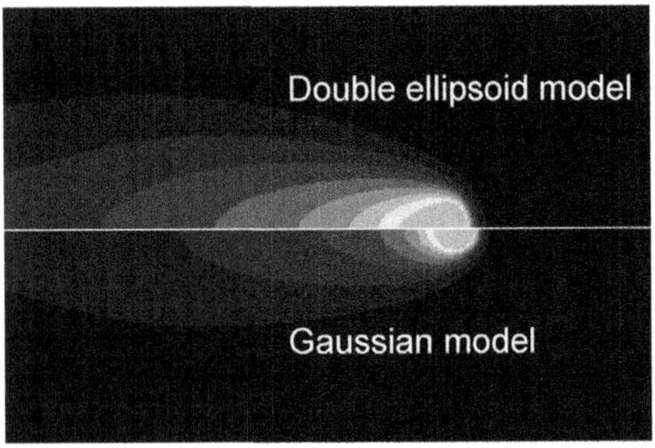

Fig. 18. The temperature distributions of different heat models

As indicated in Fig.18, the different heat source models have an important influence on the temperature distributions. The results have been modeled with the double ellipsoid model when no particular mention was made.

The Influence of Fluid Flow in the Weld Pool. The fluid flow in the weld pool greatly affects the temperature distributions in or near the fusion and heat-affected zones as it speeds the heat conductivity.

Fig. 19 shows the simulated temperature distributions. It can be seen that the weld pool is shorter when the influence of the fluid flow is considered, compared to the weld pool obtained without consideration of this influence. However, this influence gets weaker in the areas away from the weld pool.

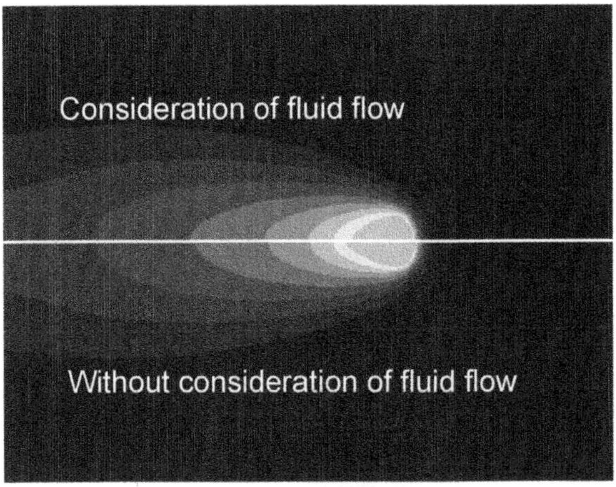

Fig. 19. The influence of the fluid flow on the temperature distributions

The Effects of Latent Heat of Fusion. The latent heat effects are more important for improving the accuracy of the simulated results. Fig. 20 illustrates the influence of different treatments of latent heat of fusion on temperature fields.

When latent heat is released linearly, the weld pool is longer than that simulated assuming that uniform latent heat is released from the solidus to liquidus. Near the weld pool and at lower temperatures, the isotherms are shorter than those for uniform latent heat.

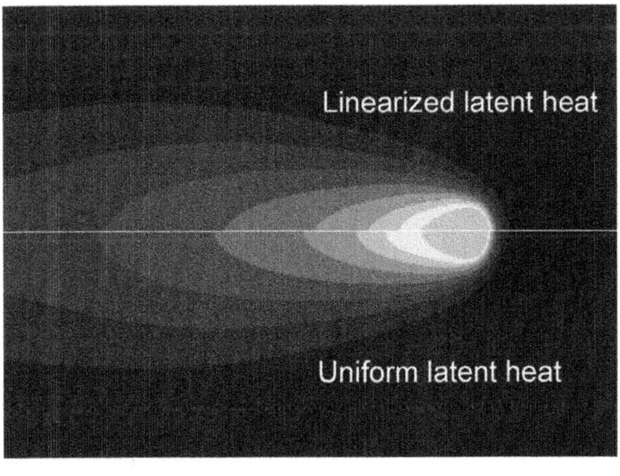

Fig. 20. The effect of different treatments of latent heat on the temperature distributions

The Influence of Welding Parameters. In this study, two different sets of welding parameters are adopted, the first: $I = 200$ A, $U = 13.5$ V, $\upsilon = 2$ mm/s and the second: $I = 325$ A, $U = 16.5$ V, $\upsilon = 4$ mm/s. Fig. 21 provides two kinds of temperature fields. Although with almost equal arc energy, the shapes and sizes of isotherms for the different welding conditions vary greatly. Larger welding current together with higher travel speed results in narrower and longer isotherms than at the other conditions.

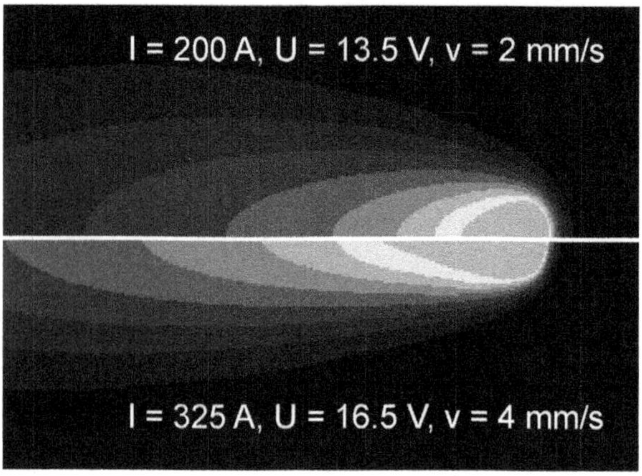

Fig. 21. The influence of the welding parameters on the temperature distributions

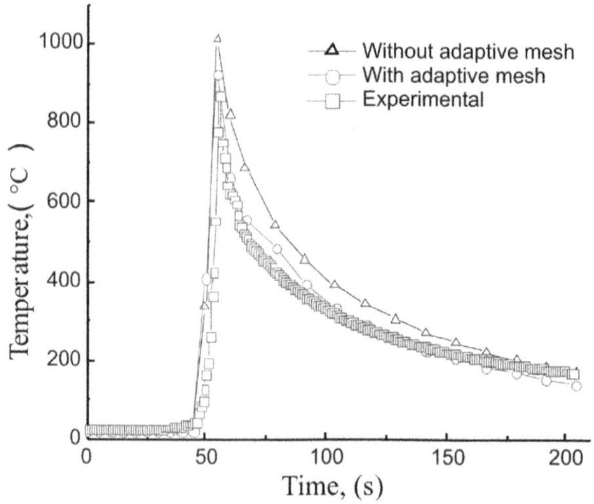

Fig. 22. Temperature change with the time-step

The Effects of Adaptive Mesh Technique. The temperature distributions in two different analyses with adaptive mesh and without adaptive mesh at different time-steps are shown in Fig. 22.

It can be observed that the temperature distributions of the two analyses are almost the same within the high temperature range, but there is big difference when the temperature decreases. The adaptive technique makes the FEM mesh much denser, so that the temperature distributions agree well with the experimental results.

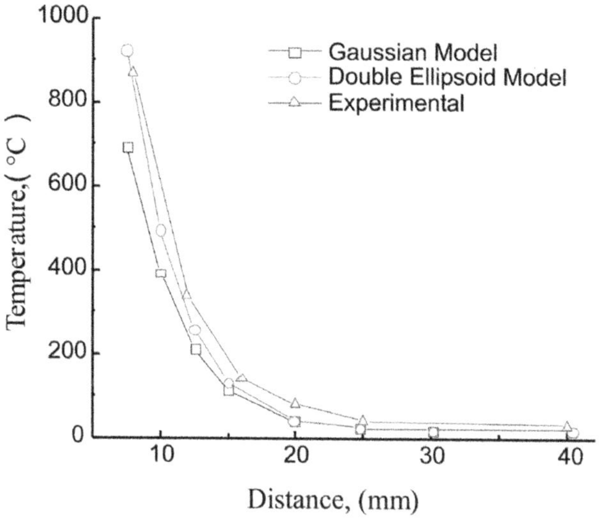

Fig. 23. The comparison between the simulated temperatures and the measured temperatures

Verification of the Simulated Temperature. To testify the computational results, the temperatures at different positions perpendicular to the weld centerline were measured using thermocouples.

The examined material is SUS310 stainless steel plate with 10 mm thickness. The welding current is 200 A. The welding speed is 2 mm/s. The fusion width of the weld is analyzed after welding. Fig. 23 shows the measured temperatures and simulated ones considering two kinds of heat sources. It can be observed that the temperature values obtained with the double ellipsoid model coincide well with those measured especially at high temperature.

The fusion width and shape of the weld are shown in Fig. 24. The results from the comparison between simulations and measurements are summarized in Table 4.

Table 4. The results of simulations and experiments

	Double ellipsoid model [mm]	Gaussian model [mm]	Experimental [mm]
Fusion Width	12.5	9.5	13.0

As shown in Fig. 24, the macrograph of weld pool geometry is similar to the calculated weld pool shape of double ellipsoid heat source.

From Table 4 it is apparent that the fusion width from the double ellipsoid model calculation shows excellent agreement with the experimental results.

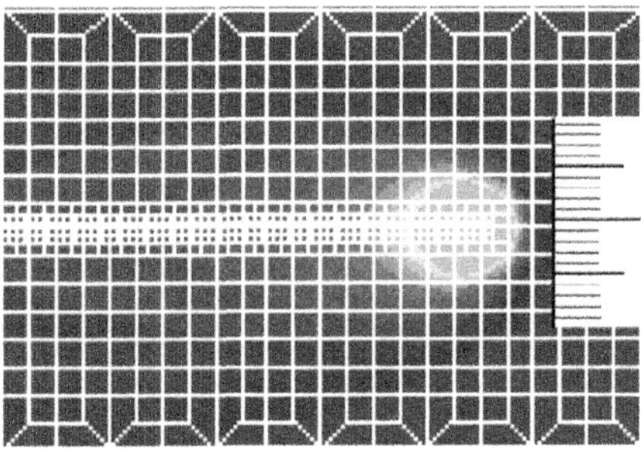

(a) The fusion width and shape of the Gaussian model

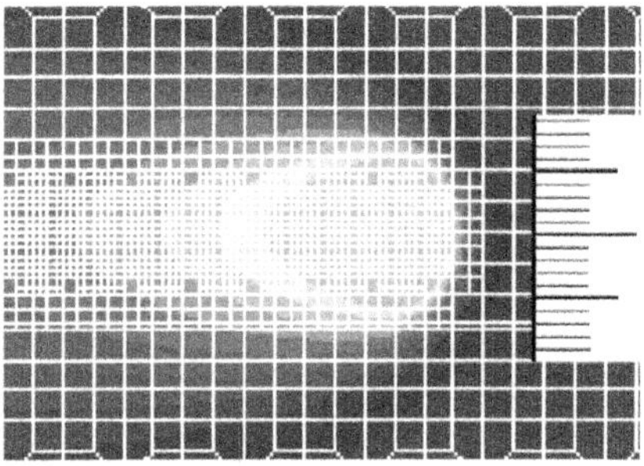

(b) The fusion width and shape of the double ellipsoid model

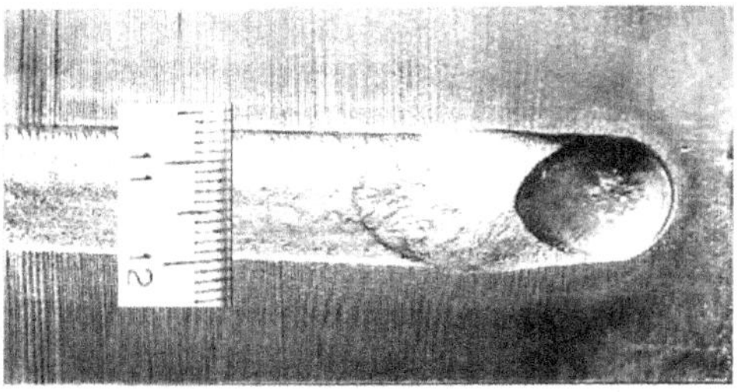

(c) The fusion width and shape measurement

Fig. 24. The fusion width and shape of the weld

Numerical Simulation of Stress/Strain Distributions

Basic Consideration during Development of Stress/Strain Models

The Deformation in the Weld Pool. It is known that the liquid metal of the weld pool has two characteristics. First, the liquid metal can only withstand minimum force which can also cause its flow and deformation. Second, the previous strain and deformation of resolidified weld metal will be "annealed" at the liquid-solid interface upon recrystallization because the weld metal ahead of arc melts and solidifies after the arc has moved forward.

Therefore, except for the effect of the fluid flow in the weld pool on the shape of the weld pool, which could be adequately dealt with in the heat transfer analysis, the fluid flow behavior in the weld pool can be omitted in finite element modeling of welding-specific stresses and distortions. A good way to deal with the weld pool in such analysis is to simply exclude it from the solution domain. Noting that the position of the weld pool continuously changes in order to keep up with the traveling electrode, the exclusion of the weld pool from the solution domain has to be done during dynamic and continuous fusion over the entire welding process.

The Elasto-Viscoplasticity in the Solid-Liquid Region. At present, there is no agreement on whether the elasto-viscoplasticity in the area of solid-liquid coexistence should be considered or not. In the late 1970's, researchers in the former USSR pointed out that the mechanical behavior of weld metal in solid-liquid state had characteristics of elasto-viscoplasticity and that its stress/strain was related not only to temperatures but also to the history and time of the load.

However, some researchers hold the viewpoints that the elasto-viscoplasticity of liquidus metal need not be included in the course of simulating stress/strain distributions. Among them, some believe that elasto-viscoplastic deformation is smaller than plastic deformation in the solid-liquid region and can be ignored. The others think that it is very difficult to measure elasto-viscoplastic properties at elevated temperatures and that it is also quite complicated to conduct such kinds of computations which will make inaccurate results.

To study the elasto-viscoplasticity in solid-liquid regions, a series of rheological tests of stainless steels SUS310 and SUS316 were conducted within the scope of this study with the elasto-viscoplasticity measuring device of the Beijing Institute of Science and Technology. Some of the measured results are shown in Fig. 25.

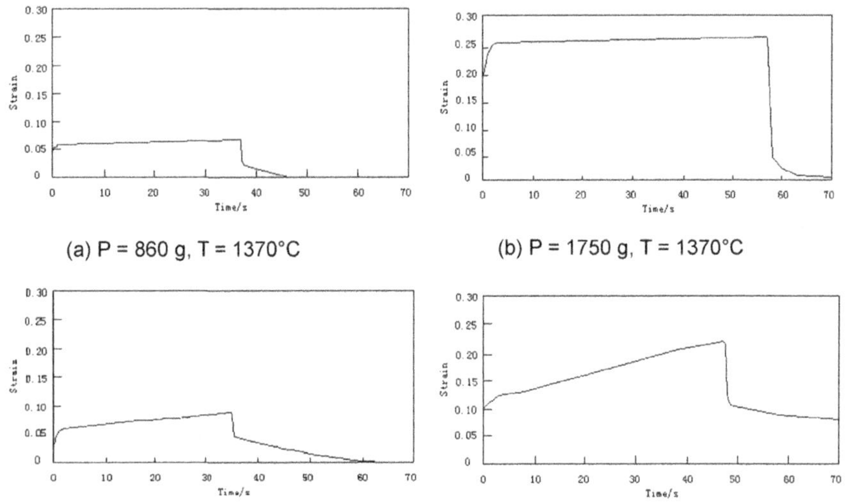

(a) P = 860 g, T = 1370°C

(b) P = 1750 g, T = 1370°C

(c) P = 900 g, T = 1380°C

(c) P = 1180 g, T = 1380°C

Fig. 25. The response curves of stainless steel SUS316 during loading and unloading

Fig. 25 demonstrates that when the temperature is below the coherent temperature as shown in (a) and (b), its behavior is mainly of the elastic deformation type, and that when the temperature of the metal is higher than the coherent temperature as shown in (c) and (d) its behavior is of the elastic, viscoelastic and viscoplastic deformation type.

Elasto-viscoplasticity is not obvious in welding owing to the high cooling rate and to the short exiting time of the solid-liquid region. When the temperature is higher than the coherent temperature the metal in the solid-liquid region exhibits liquid characteristics of molten metal because there is more liquid metal with interdentritic flow.

As a result, the region of solid-liquid coexistence can be divided into two areas: liquid/solid area and solid/liquid area. In the former case, the state of the metal is close to liquid metal of the weld pool and its mechanical behavior can be treated with the element rebirth method, while the state of the metal in the later case approximates solid metal and its stress/strain constitutional relation is in accord with the elasto-plastic heat theory.

In this study, the region of the SUS310 weld with temperatures from 1375 °C to 1400 °C was treated as liquid metal and the other region with temperatures from 1250 °C to 1375 °C was treated as solid metal.

The Solidification Shrinkage. It is well known that most metals and alloys contract during solidification and that solidification shrinkage has a significant impact on the stress and strain fields of a weld. As far as the thermal stress modeling of an alloy is concerned, the effect of solidification shrinkage is phenomenologically equivalent to thermal expansion/contraction due to temperature change. It causes mechanical deformations in the liquid region as well as in the other parts of the weld, owing to the compatibility requirement of the continuum solid mechanism.

In this study, the solidification shrinkage is thus assumed to be linearly distributed in the solidification temperature range. Therefore, it can be effectively treated as an additional thermal expansion/contraction term caused by temperature changes.

Based on the above discussion, thermal elastic-plastic theory was used to simulate the thermal stress/strain of interest. The computation was made on the following assumptions or conditions:

- Materials are isotopic and homogeneous;
- Mechanical properties are temperature-dependent;
- Material yields conform to Mises rules;
- Viscosity and creep of the materials are neglected;
- The characteristics in the plastic area are represented by flow rules of the plasticity and strain-hardening behavior.

Finite Element Implementation

As stress/strain analysis is based on the heat transfer analysis, the plate was discretized in the same way as in heat conduction analysis as shown in Fig. 26.

As discussed above, the best way to deal with the weld pool in such analysis is to exclude it from the solution domain. Therefore, the solution domain is composed only of continually changing solid parts of the welds.

The approach is the element rebirth technique by which the elements whose temperature is above the coherent temperature (so they are inside the traveling weld pool) are simply removed from the finite element model and added back into the model after their temperature drop to below the coherent temperature. When the elements are added back, they are assumed to be annealed and have zero deformation. Since the weld pool follows the moving welding arc, such element removal/inclusion is dynamic, namely, it follows the location of the weld pool.

In this study, a program is written to complete the above element rebirth scheme.

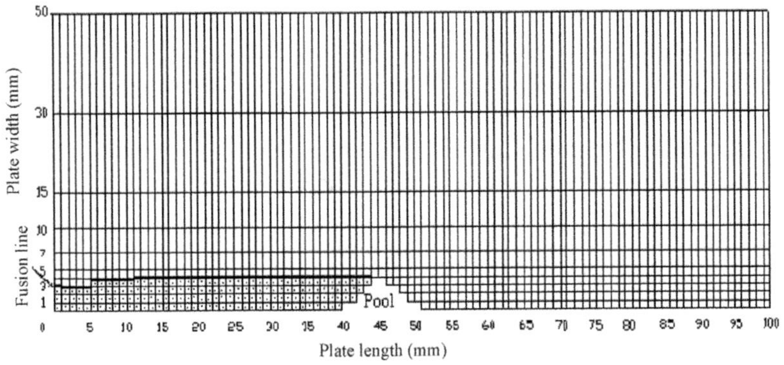

Fig. 26. The mesh used in calculating stress/strain distributions

Discussion of Numerical Analytical Results of the Stress/Strain Distributions

The Effects of Deformation in the Weld Pool. The effects of deformation in the weld pool are shown in Fig. 27 which indicates that larger compression existed when the deformation in the weld pool was not taken into account while the strain was completely annihilated when the element rebirth technique was used.

The Effects of Solidification Shrinkage. Fig. 28 shows different strains developed in the weld center with or without consideration of solidification shrinkage. As indicated in Fig. 28, the strains are much higher when solidification shrinkage is taken into account than those obtained disregarding solidification shrinkage.

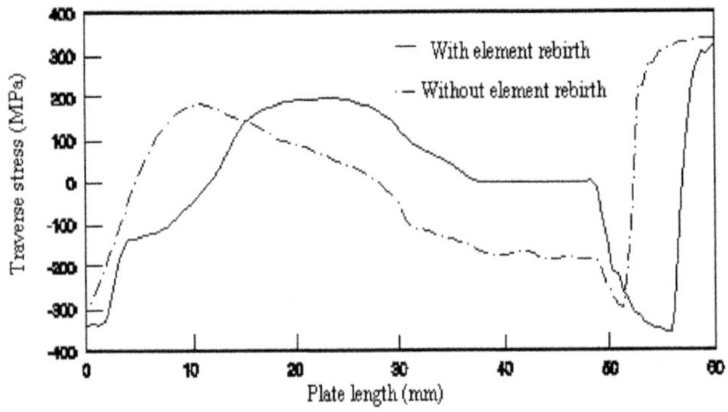

Fig. 27. The effects of deformation on stress shrinkage distributions

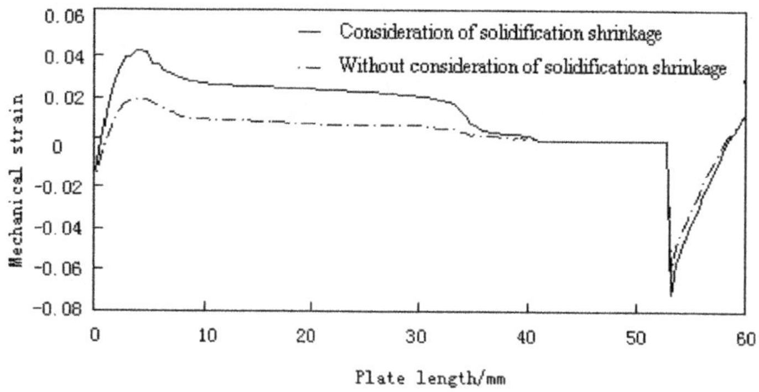

Fig. 28. The influences of molten metal on strain distributions

The Influences of Welding Conditions. As the welding conditions greatly affect the thermal distributions of stainless steels SUS310 and SUS316, especially when the thermal fields hold elevated temperatures, they will certainly influence the stress/strain development of the weld. On the condition of penetration, two kinds of welding conditions are used to simulate the stress/strain evolutions that are illustrated in Fig. 29.

From Fig. 29, we can find that the strains at the trail of the weld increase much earlier at lower welding speed than at higher welding speed which indicates that the driving force behind solidification cracking is stronger at higher welding speed. If the welding parameters did not affect the cracking resistance it could have been concluded that the steels are less susceptible to solidification cracking at higher welding speed.

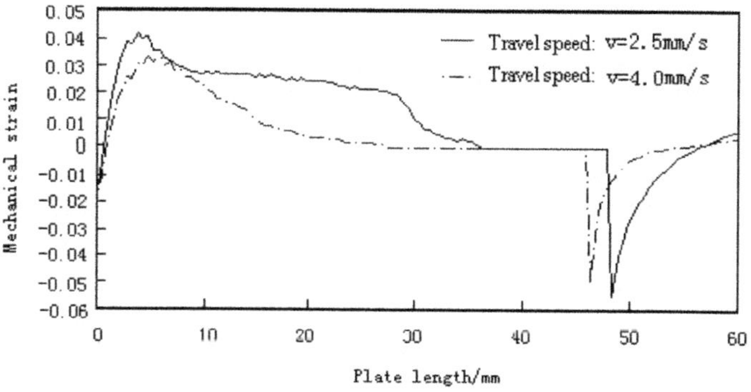

Fig. 29. The effect of welding conditions on strain distributions

In fact, this conclusion contradicts some phenomena in practical production because welding conditions do have effects on the cracking resistance of materials. For instance, dendrites will grow perpendicularly to the fusion line and meet at the center of weld when the welding speed is high, which results in the formation of the eutectic there and in the decrease of the weld metal cracking resistance.

The Influences of Restraint. To study the influences of restrained weld edge on the crack driving force, the strain development for two kinds of restraint was simulated and compared in Fig. 30.

From Fig. 30 we can find that the strain development curve for the free edge is lower than that of the restrained edge, which agrees well with the general concept of crack theory.

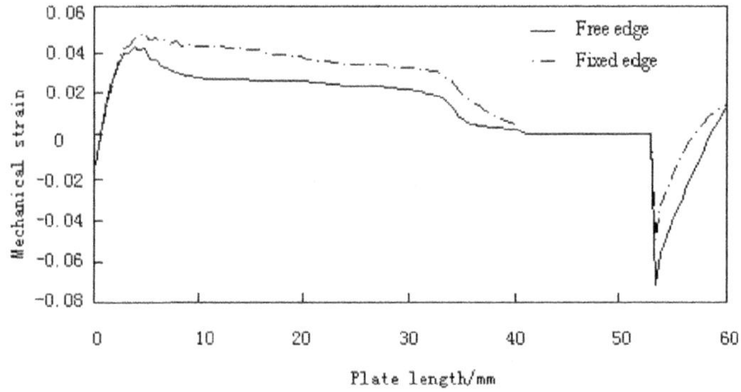

Fig. 30. The effects of restraint on strain distributions

Development of a Simulation and Prediction System for Solidification Cracks

From the work described above we can see that it is very complex to perform a complete simulation of solidification cracks with a common commercial software package, as various specific factors should be taken into account. To solve the problem, a simulation and prediction system for weld metal solidification cracks has been developed with which a welding engineer can easily carry out the task on the basis of two-dimensional simulation of weld cracks with Adina&T 8.4.

Generation of Input Cards

When Adina&T is used to execute simulation, there are two kinds of input data cards needed which require a strict data format, one is for simulating thermal distributions and the other is for computing stress/strain fields. To simplify the use of the software package, the system provides user-friendly interfaces to input relative parameters, performs corresponding calculation and data treatment, and generates input data files automatically. The interface for generating cards is shown in Fig. 31 from which we can find that there are three kinds of parameters for users to input, they are basic data about grid generation, material properties and welding conditions.

The system provides two ways to discretize the workpiece as shown in Fig. 31. One way asks users to enter the coordinate data and the other asks users to input the proportion of discretization. Four-node elements are used, and smaller elements are advised near the weld to account for the possible severe temperature gradient in those regions. After the data input, the workpiece is discretized, of which the results can be previewed and the corresponding data are then stored in input cards.

The thermophysical properties of alloys, including elastic modulus, Poisson's ratio, expansion coefficient, thermal conductivity and yield stress, are input easily and the curves of the thermophysical material properties versus temperatures can be viewed, too.

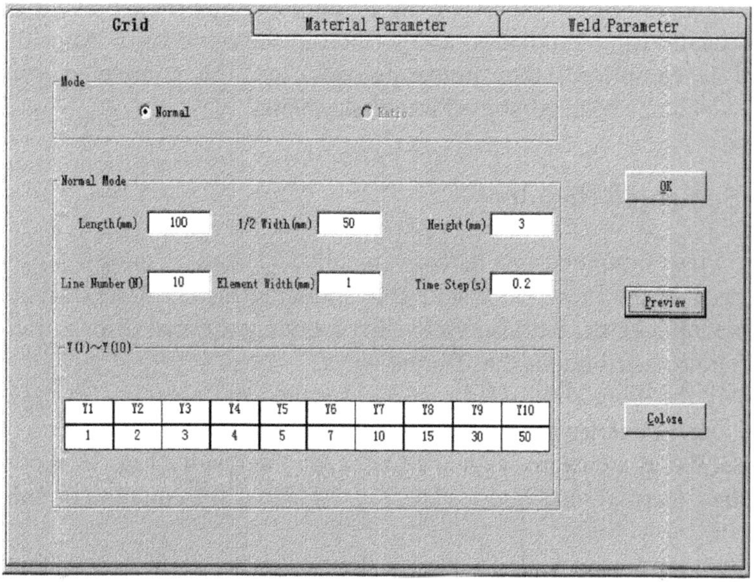

Fig. 31. Interface for generating input cards

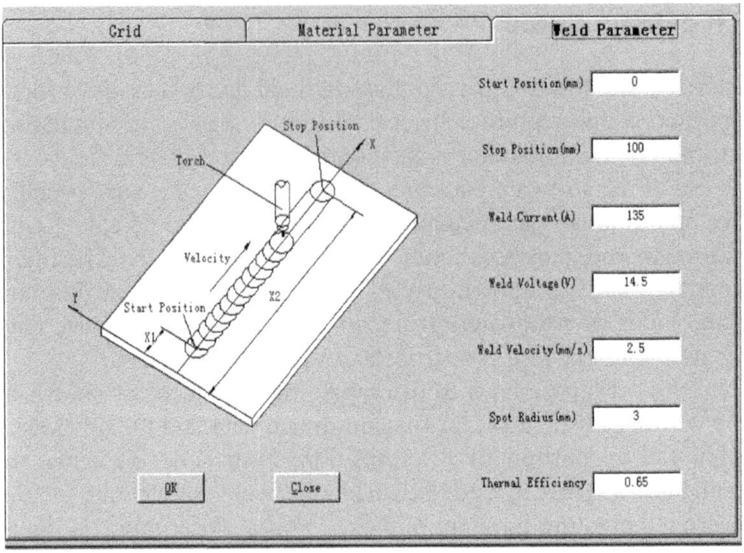

Fig. 32. Interface for inputting weld conditions

The welding conditions are input as shown in Fig. 32. After obtaining the welding parameters, the system calculates the heat generation according to the Gaussian distribution of the arc heat input.

The heat generation for each element is computed and treated as a part of internal heat during simulation and written in the input card. After the input card is completed, the simulating work can be completed with Adina&T which has been combined into the system.

Viewing Simulated Results

In order to view simulated results conveniently, the system adopts a program to organize the input cards and corresponding results. Users cannot only browse input cards, but also view corresponding thermal distributions and stress/strain distributions in different ways.

To display the simulation results in 3-D meshes and contours, the system makes use of a software package, Matlab5.3, as a server as well as of Visual Basic 6.0 as a custom language to display the results. Fig. 33 shows a 3-D mesh of thermal distributions and Fig. 34 shows the contours of temperatures.

The system uses MatrixVB to display thermal/stress/strain cycles, their distributions in transverse or vertical sections.

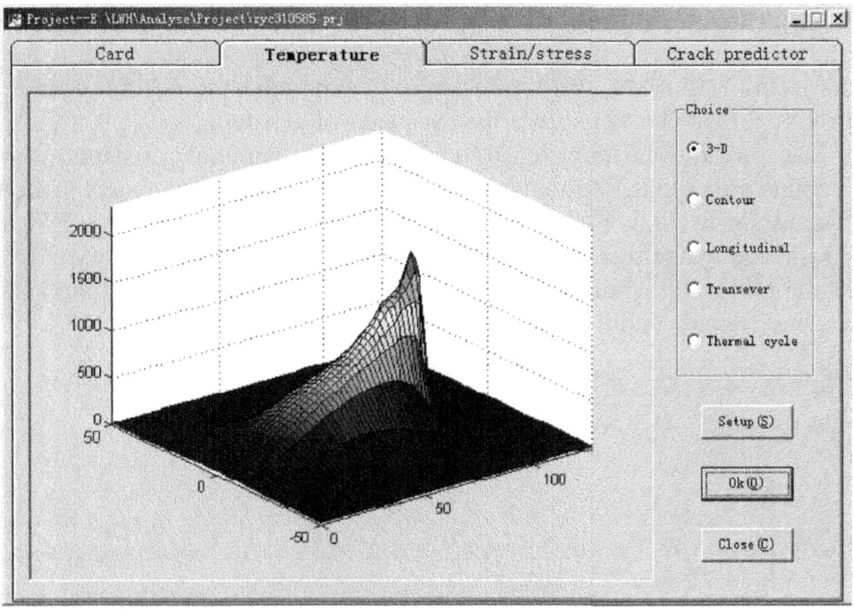

Fig. 33. 3-D mesh of thermal distributions

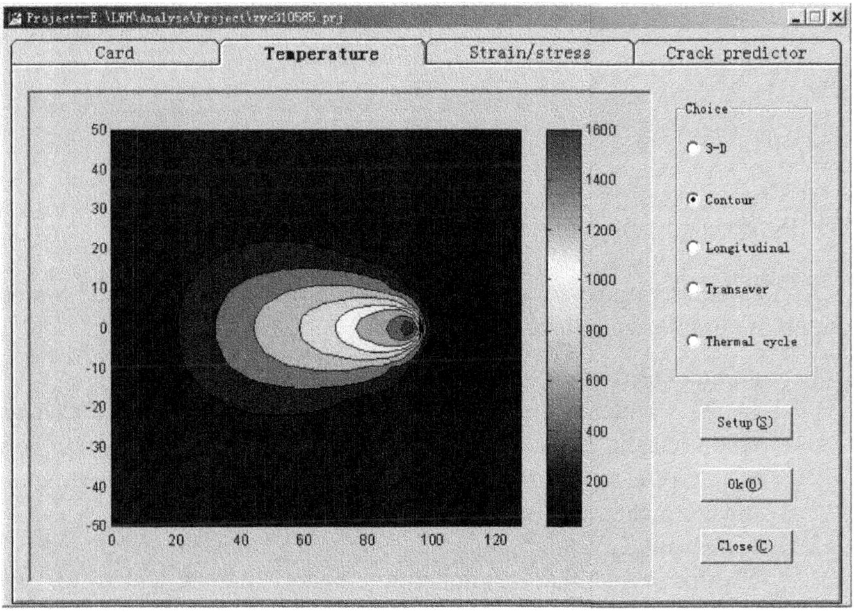

Fig. 34. Contours of temperatures

Predicting Solidification Cracks

As mentioned above, weld metal solidification cracking occurs when the driving force to crack exceeds the resistance of a material.

The fundamental parameter representing the material resistance is its ductility curve in the brittle temperature range (BTR) in the course of weld metal solidification, while the mechanical strain accumulated in the BTR is considered as the mechanical driving force. To obtain the ductility curves in the BTR, measurements were carried out in the Trans-Varestraint-Test under different conditions.

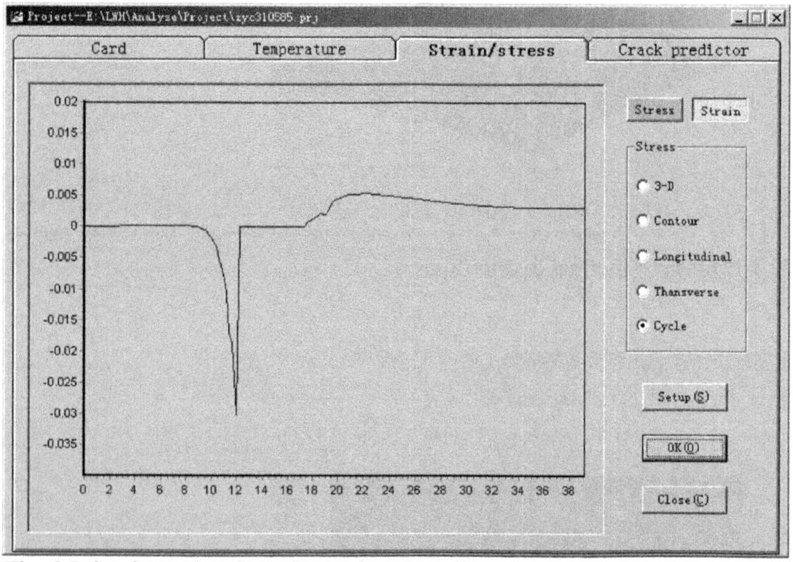

Fig. 35. Strain cycle of a point in the centerline

As for the driving force, a program module has been developed to deal with it in the following three steps:

1. *Obtain mechanical strain*
   In the simulated results, the total strain is composed of thermal strain, elastic strain and plastic strain. The mechanical strain consisting of elastic strain and plastic strain is obtained by subtracting the thermal strain from the total strain.
2. *Get thermal cycle*
   Thermal cycle, that is, temperature history, can be gotten from the simulated results through data treatment.

3. *Conversion of the strain-time curve into the strain-temperature curve*
   The mechanical stress obtained from the simulation is a function of time as shown in Fig. 35, while the driving force to solidification cracking is the function of transverse mechanical strain against temperature. By means of a line with a slope of -1 and the thermal cycle, the simulated strain can be converted into the driving force. As a result, weld solidification cracking can be predicted with the system as shown in Fig. 36.

According to the relative position of two curves, weld solidification cracking can be predicted. In Fig. 36, the driving force is lower than the material resistance, so there will be no solidification cracking.

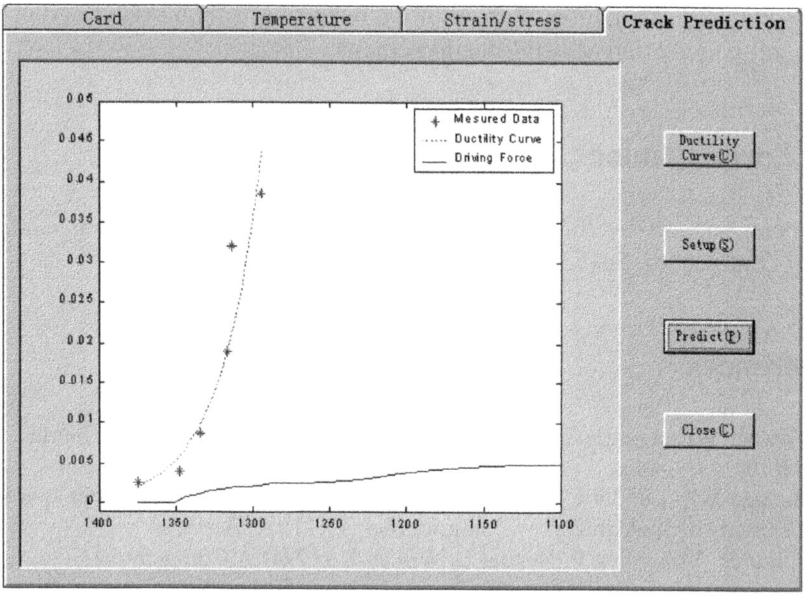

Fig. 36. Predicting weld metal solidification cracking

Conclusions

1. The strain concentration exists at the trail of the weld pool during Trans-Varestraint-Testing, so that the local strain at the trail of weld pool was measured and the ductility curve was modified by in-situ observation and recorded with CCD camera during Trans-Varestraint-Testing.

2. For three-dimensional simulation of thermal distributions, a double ellipsoid heat source model is more practical compared with Gaussian surface thermal models.
3. The latent heat of fusion, fluid flow in the weld pool and the welding parameters have influences on the thermal distributions in the vicinity of the weld pool, which indicates that it is necessary to consider these factors when weld metal solidification cracking is simulated.
4. The problem of deformation of the weld pool can be solved with the element rebirth scheme, and solidification shrinkage is taken into account by amplifying the thermal expansion/contraction coefficient.
5. The simulation and prediction system provides an easy way for welding engineers to perform simulation of weld solidification cracks, as the system needs minimal inputs that are related to each other and conducts other tasks automatically, including previous and post data treatment, carrying out calculation, display of the results in different ways and performing prediction of solidification cracks.

Acknowledgement

Financial support by the National Natural Science Foundation of China under contract No. 50175040 is gratefully acknowledged.

References

1. Savage WF, Lundin CD (1965) The Varestraint Test. Welding Journal, , 44(10): 433s–442s
2. Savage WF, Lundin CD (1966) Application of the Varestraint Technique to the Study of Weldability. Welding Journal, 45(11): 497s–503s
3. Zujue D, Yongming P, Renpei L, Wilken K (1994) A Comparative Study on the Criteria of Varestraint and Transvarestraint-Test of Solidification Cracking Sensitivity. IIW-Doc IX-1764-94
4. Sun Z (1992) Study of Solidification Crack Susceptibility Using the Solidification Cycle Hot-Tension Test. Materials Science & Engineering A: Structural Materials: Properties, Microstructure and Processing, A154(1): 85–92
5. Takeshi S, Katsuei H, Ryouichi Y (1990) Development of Solidification Cracking Test for MAG Narrow Gap Welding. Effect of Boron Contents on Solidification Cracking. Quarterly Journal of the Japan Welding Society, 8(1): 21–25
6. Brooks JA (1993) On Modeling Weld Solidification Cracking. In: Proceedings of Internal Conference on Modeling and Control of Joining Processes, Orlando, Fla., pp 174–185

7. Wu CS (1990) Numerical Analysis of Welding Heat Process. Publishing House of Harbin Institute of Technology
8. Yang YP (1995) A Study of Preventing Welding Hot Cracking of Thin High Strength Aluminum Alloy Plate with the Inverse Strain Method. Ph.D. Dissertation of Harbin Institute of Technology
9. Cao ZN (1993) Numerical Analysis of Fluid Flow and Thermal Distributions in Weld Pool for TIG/MIG Penetrations. Ph.D. Dissertation of Harbin Institute of Technology
10. Zheng W (1996) Numerical Simulation for Transient Behavior of Fluid Flow and Temperature Field in Pulsed Current TIG Weld Pool. Ph.D. Dissertation of Harbin Institute of Technology
11. Feng ZL (1994) Computational Analysis of Thermal and Mechanical Conditions for Weld Metal Solidification Cracking. Welding in the World, 33(5): 340–347
12. Feng ZL (1993) A Methodology for Quantifying the Thermal and Mechanical Conditions for Weld Metal Solidification Cracking. Ph.D. Dissertation of the Ohio State University, pp 15–24, 58–90, 280–300
13. Liu WP (1995) Computational Analysis and Prediction of Weld-Solidification Cracking. Computational Materials Science, 43(9): 211–219
14. Guo SQ (1995) A Study of Controlling Welding Hot Cracking and Distortion of Thin Aluminum Alloy Plate with High Susceptibility to Hot Cracking. Ph.D. Dissertation of Harbin Institute of Technology
15. Dye D, Hunziker O, Reed RC (2001) Numerical Analysis of the Weldabality of Superalloys. Acta mate, 49: 683–697
16. Shi JX (1997) The Numerical Simulation of the Driving Force of SUS310 Solidification Cracking. Ms. Thesis of Harbin Institute of Technology
17. Zhang HZ (1998) Numerical Simulation of Driving Force for Weld Metal Solidification Cracking of Stainless Steels. Master Thesis of Harbin Institute of Technology
18. Wei YH et al. (2000) Stress/strain Distributions for Weld Metal Solidification Crack in Stainless Steels. China Welding, 9(1): 1–6
19. Zhang YC (2000) A Simulation and Prediction System for Welding Solidification Crack. Master Thesis of Harbin institute of technology, 7
20. Matsuda, Nakagawa, Matsubara (1983) Quantitative evaluation of solidification brittleness of weld metal during solidification by in-situ observation and measurement (Report II) – Solidification ductility curves for steels with the MISO technique. Trans of JWRI, 12(1): 73–80
21. Shah AK et al. (1995) Weld heat-affected zone in Ti-6Al-4V Alloy, part I – Computer simulation of weld variables on the thermal cycles in the HAZ. Welding Journal, 76(9): 293s–305s
22. Goldak J (1984) A new finite element model for welding heat sources. Metallurgical Transactions B, 15B: 587–600
23. Varol I, Baeslack WA, Lippold JC (1989) Characterization of Weld Solidification Cracking in a Duplex Stainless Steel. Materials Characterization, 25(2-5): 555–573

24. Imoto I, Kim YC (1991) Mechanical Study on Quasi-solidification Cracks Under Pulsating Loads. Welding on Structures in Service Condition (report II). Quarterly Journal of the Japan Welding Society, 9 (1): 43–47
25. Ichikawa K, Bhadeshia HKDH, MacKay DJC (1996) Model for Solidification Cracking in Low-Alloy Steel Weld Metals. Science and Technology of Welding and Joining, 1(1): 43–50
26. Matsuda F, Nakagawa H, Tomita S (1989) Between Strain-Rate Dependence of Critical Strain Required for Solidification Crack Initiation and Grain Boundary Sliding - Investigation on Weld Solidification Cracking by MISO Technique (report IV). Quarterly Journal of the Japan Welding Society, 7(1), 105–110
27. Chan YW (1987) Computer Simulation of Heat Flow in Pulsed Current Arc Welding. Metal Construction, (10): 599–606
28. Zheng W (1996) Numerical Simulation for Transient Behavior of Fluid Flow and Temperature Field in Pulsed Current TIG Weld Pool. Ph.D. Dissertation of Harbin Institute of Technology
29. Liu W (1987) Numerical Simulation of Al-Cu Alloy Solid Mechanical Behavior, Stress-strain During Solidifying and Solidification Crack. Ph.D. Dissertation of Harbin Institute of Technology

Integrated Mechanical-Metallurgical Approach to Modeling of Solidification Cracking in Welds

V. Ploshikhin, A. Prikhodovsky, M. Makhutin, A. Ilin, H.-W. Zoch

Neue Materialien Bayreuth GmbH, Germany

Abstract

The presented approach is based on experimental observations which clearly show that the hot cracking phenomenon is a result of accumulation of macroscopic tensile strains in a microscopic intergranular film of liquid enriched with segregating elements at the final stage of solidification. The modeling takes into account the effects of strain accumulation as well as the influence of the microstructure and the thermo-dynamical properties of the welded material. The integrated approach provides a clear phenomenological interrelation between parameters of the welding process, properties of the welded material and the cracking susceptibility. It is successfully used for the development of technological means, such as a multi-beam welding, for eliminating solidification cracking.

Hot Cracking Problematic: Existing Approaches

The pioneering experimental research on solidification cracking in Al-alloy welds was carried out by Singer et al. [1–3] and later by Pumphrey et al. [4] at the end of the 1940ies. They have clearly shown the strong dependence of the cracking susceptibility upon the alloy composition.

At the beginning of the 1950ies, Palling introduced a clear explanation of the solidification cracking mechanism [5]. Based on the number of tensile tests of solidifying samples, Pellini suggested his "strain theory of hot cracking". According to this theory, the tensile strain developed during solidification is localized in the mechanically weak intergranular film formed at the end of solidification due to the segregation. This film limits the strain transmission and therefore it accumulates the major part of the whole strain (Fig. 1a).

The accumulated tensile strain exceeding some critical value leads to the tearing of the liquid film, i.e. to the formation of the hot crack. Pellini pointed out explicitly that, because of the strain localization, even relatively slight loading of the two-phase region leads to hot tearing. Providing a physically justified mechanical-metallurgical explanation of the mechanism of crack initiation, the theory of Pellini does not offer any quantitative criterion for predicting the cracking susceptibility.

The first quantitative description of solidification cracking was proposed by Prokhorov in the early 1950ies [6–8]. Prokhorov also considered the mechanical tensile strain as a reason for crack formation. However, the effect of the strain localization was not taken into account (Fig. 1b).

Prokhorov calculates the deformation of the liquid-solid region integrally and compares it with the so-called "deformability", a term introduced by Prokhorov in order to characterize the ability of a material to accumulate deformations without rupture at higher temperatures within the solidification range. It was considered as a material property, of which the value can be obtained experimentally. When the actual deformation of the two-phase region exceeds the deformability of the material, hot cracking occurs.

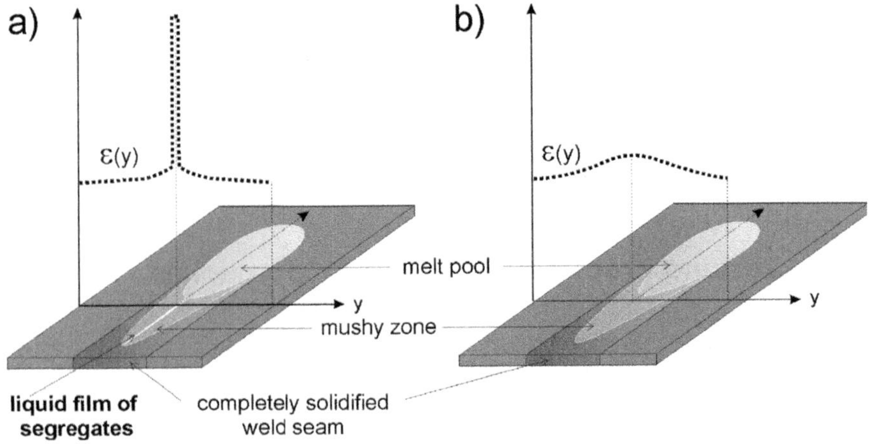

Fig. 1. Schematic illustration of the difference between the basic theories of solidification cracking: **(a)** Pellini [5] (considering the strain localization) and **(b)** Prokhorov [6] (without regard to the strain localization)

The indisputable advantage of Prokhorov's theory is the separation of the mechanical factors and the attempt to quantify them. However, the integral consideration without regard to the strain localization hinders the application of this approach to the solution of practical problems.

The methods developed for the experimental measurement of such an integral characteristics as the deformability are too sensitive to the investigational conditions. In these experiments, the controlled external loads of the welded specimen are usually used. In reality, the local deformation values of the two-phase region are not uniformly distributed over the welded joint. The larger this inhomogeneity, the bigger the experimental error. For example, experimentally measured ductility values reported in literature by Trans-Varestraint tests [9–11] and by MISO techniques [12, 13] differ by almost one order of magnitude.

The second critical point of Prokhorov's theory is the neglect of the metallurgical conditions in the mushy zone, i.e. the microstructure of the two-phase region formed during solidification is not taken into account. The experiments for determining the deformability are usually conducted only for certain process parameters.

However, the alteration of the welding conditions leads to changes in the microstructure and, correspondingly, to a different response to the external load. In turn, this results in different values of the deformability. Thus, the experimental determination of the ductility should be carried out for each possible set of process parameters. Because this procedure is very complicated, measurements are usually taken only for a certain set of process parameters, and then the obtained strength index is considered to be a material property independent of the welding conditions. This conclusion can be valid for some cases. For example, if a specific welding process for plates with a certain thickness is considered. However, the direct transfer of the measured deformability of one welding process to another (e.g. of MIG to laser welding), or to different plate thicknesses is always connected with some predefined errors arising from the changes in the state of the two-phase region.

Thus, while providing a possibility for a relative characterization of different materials with respect to their cracking susceptibility, the theory of Prokhorov does not offer a correct quantitative criterion for predicting the cracking susceptibility.

The majority of work on modeling of solidification cracking in welds [14–23] is based on Prokhorov's approach, i.e. neglecting the strain accumulation and the microstructure effects. Several existing criteria represent some mechanical conditions which are usually a critical value of strain or strain rate. A precise analysis has shown that none of the existing criteria can be considered as universal, which indicates the necessity for revision of existing approaches [16]. According to Feng [19], who made a considerable contribution to precise modeling of hot tearing, it is still impossible to predict the behavior of welded joints under real "field" conditions based only upon laboratory experiments.

There are only two modeling approaches appearing recently which are based upon the localization aspect of crack formation. The first one is proposed by Shibahara [24, 25] who uses the special numerical technique of the so-called "interface elements" for the simulation of crack initiation and crack propagation. The assumptions concerning the physical mechanism of crack formation are taken from the solid state fracture mechanics. The metallurgical conditions in the two-phase region are disregarded. The value of the stress in the liquid-solid region is taken as a criterion for the crack initiation. This stress is compared with the critical value predetermined through the correlation with the surface energy of the melt. The surface energy is considered to be a known quantity which has a unique value for the given temperature.

Despite the consideration of the localization aspect of crack formation, the approach of Shibahara has some critical points. First, the experimental determination of the surface energy at high temperatures is a very complicated task. The surface energy of the melt is very sensitive to any variation of the chemical composition. The addition of a negligibly small amount of a surface active element into the melt can lead to an excessive alteration of the surface energy over an order of magnitude. Second, the calculated value of stress in the two-phase region is very sensitive to the mechanical properties at high temperatures. The experimental measurements of such properties like the yield strength show large systematic errors. It can lead to significant errors during the calculation afterwards.

Another local modeling approach developed by Hilbinger [26–28] is based on the theory of Pellini. The simulation of hot cracking is carried out as in the previous approach using the finite element method. The localization of the tensile stress in the liquid film of the rest melt is taken into account through the introduction of the "liquid" elements in the middle of the weld seam. These elements have very low yield strengths in the liquidus-solidus temperature range. The criterion of crack initiation is the maximal allowable deformation of the "liquid" elements in the two-phase region. The critical deformation is the experimentally fitted parameter. Like the method of Shibahara, the method of Hilbinger also allows a visual presentation of crack initiation and propagation.

Experiments

The aim of the experiments was to determine the transition state to hot cracking in order to understand the mechanism of crack initiation. The strip-shaped samples of binary AlSi-alloys with different Si-concentrations were restrained on the edge on one side remaining free on the opposite side as shown in Fig. 2. The laser beam welds were produced at different distances from the free side. This experimental setup is based on the effect of the crack formation during welding close to the free edge of the plate. This effect was used by Hilbinger [26, 28] for similar investigations.

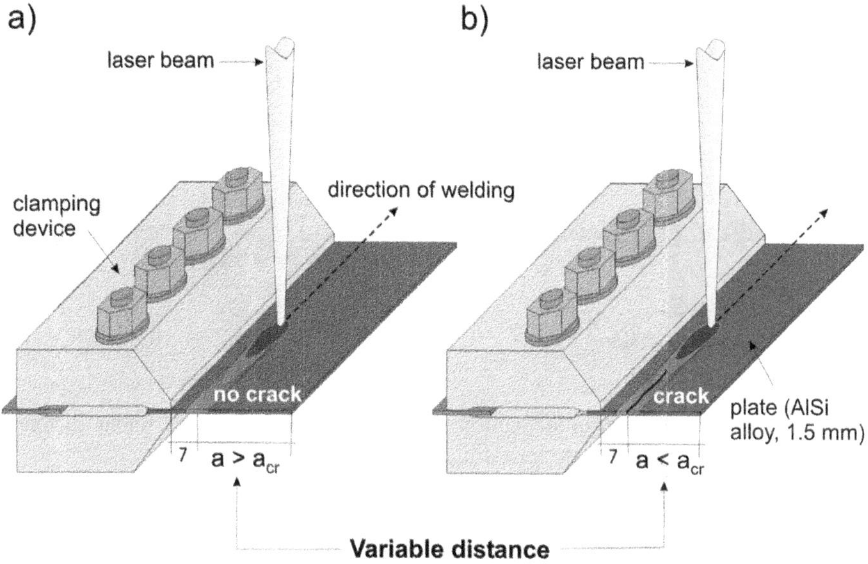

Fig. 2. Experiments on determination of the critical thermo-mechanical conditions needed for the crack initiation: **(a)** crack initiation by welding close to the free edge and **(b)** no crack initiation by welding at a distance bigger than the critical one

The parameters of welding made with Nd:YAG laser were constant for all samples. The laser power was 1.8 kW and the welding speed was 2.8 m/min. Due to variation of the distance between the weld and the free edge different thermo-mechanical conditions can be obtained. At small distances to the free edge, a centerline solidification crack is usually generated in these experiments [27, 28]. At a certain critical distance, the state of the threshold to solidification cracking can be achieved.

Fig. 3a shows typical top view photographs of the weld in the state of threshold to solidification cracking under different magnifications. That "crack" at lower magnifications (Figs. 3a–b) reveals a sequence of the holes separated by "bridges" of the eutectic, which are clearly seen at higher magnifications (Figs. 3c–d).

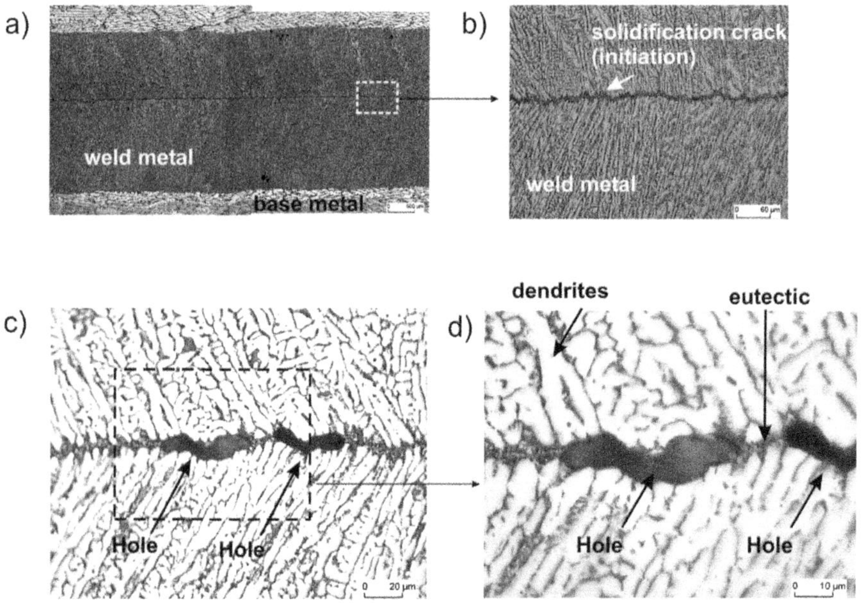

Fig. 3. Initiation of the solidification crack (uncompleted crack) under different magnifications (top-view-sections, binary alloy Al-3 % Si)

Such a threshold state is obviously a result of strain accumulation in the intergranular film which, because of segregation, remains liquid until the final stage of solidification is reached. An important result is that the value of the liquid film deformation needed for the crack initiation is relatively small. This value can be directly obtained from the metallographic view (Fig. 3d, Fig. 4c) and is approximately 10 µm. Considering the initial thickness of the liquid film, which is between 1 and 2 µm, the strain value can be estimated at approx. 1000 %. The accumulated tensile strain was high enough for crack initiation, but it was insufficient for complete tearing of the liquid film.

A very interesting transition state is shown in Fig. 5. The metallographic observation at low magnifications shows the existence of a centerline crack. However, precise examination at higher magnification reveals this seeming crack to be a thin strip of binary eutectic formed at last stages of solidification. Apparently, this weld seam was very close to form a solidification crack, but tearing of the liquid film did not happen. The value of the liquid film extension obtained from the metallography is approx. 3–5 µm. The tensile strain of this liquid film at a value of up to 500 % was relatively large. However, the accumulated strain was not high enough for crack initiation.

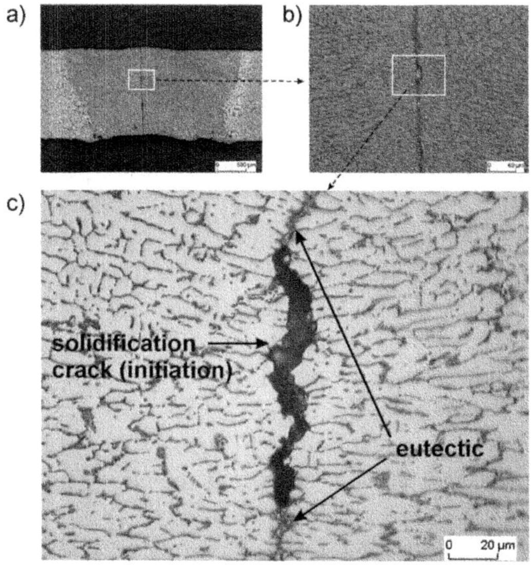

Fig. 4. Initiation of the solidification crack (uncompleted crack) under different magnifications (cross sections, binary alloy Al-3 % Si)

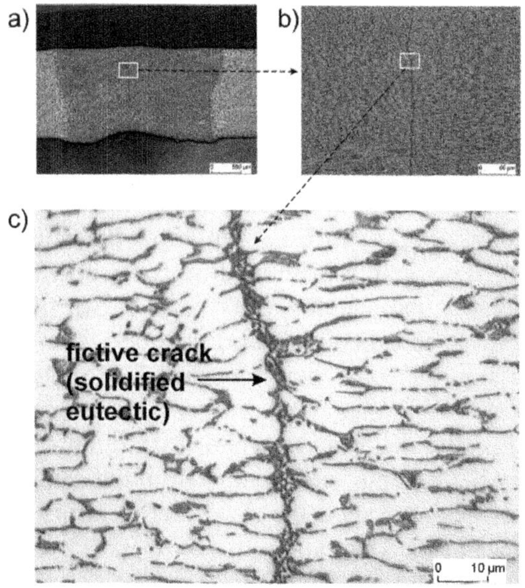

Fig. 5. Fictive solidification crack under different magnifications (top-view-sections, binary alloy Al-3 % Si)

These results support Pellini's theory and clearly show the importance of the local consideration of the deformation in the two-phase region. In accordance with conclusions of Pellini, the metallographic observations indicate that even minor deformation of the two-phase region can be sufficient for crack initiation. For the investigated alloys and welding parameters, the critical deformation had a value of approx. 10 μm.

From the photographs it can be seen that the value of the critical deformation is of the same order of magnitude as the dendrite arm spacing. This shows the importance of the microstructure state for hot tearing. For this reason, the neglect of the influence of the microstructure as practiced in many modeling approaches of hot cracking is inadequate. Evidently, the ability of the microstructure to accumulate the tensile strain without crack initiation depends on both the geometric parameters (primary and secondary dendrite arm spacing) and the rest amount of melt at the end of solidification.

Modeling

A mechanism of the solidification cracking, which can be derived from the experiments described in the previous section, is shown schematically in Fig. 6. The intergranular liquid film of the segregating elements at the weld centerline accumulates the tensile strain transmitted by the solidified dendrites from the completely solid regions of the base material. The displacement exceeding some critical value δ^{cr}_{acc} results in tearing of the liquid film of segregates, i.e. in the initiation of the solidification crack.

The strain accumulation strongly depends on the length of the mushy zone in the direction of welding. This length is defined by two isotherms: the real liquidus and the real solidus one denoted as T_L* and T_S*, respectively. These isotherms represent the so called real solidification range, i.e. the temperature interval of the simultaneous coexistence of the solid and liquid phases (the mushy state).

Neglecting the undercooling of the dendritic tips, the real liquidus can be assumed to be represented by the liquidus temperature from the equilibrium phase diagram. The real solidus represents the lowest temperature for the existence of the mushy state. For a certain alloy, this temperature must be calculated taking into account the effects of micro-segregation.

For example, for Al-Si binary alloys, the residual liquid usually solidifies as a binary eutectic. For this reason, the real solidus in this case can be reproduced by the temperature of the binary eutectic.

Correct modeling of the solidification cracking implies the solution of the following constituent problems:

- thermo-mechanical problem (evaluation of the maximal displacement of the liquid film of segregating elements taking into account the effect of strain localization);
- metallurgical problem (evaluation of the real solidification range taking into account the effects of segregation; calculation of the phase distribution within the mushy zone; evaluation of the geometrical microstructure parameters).

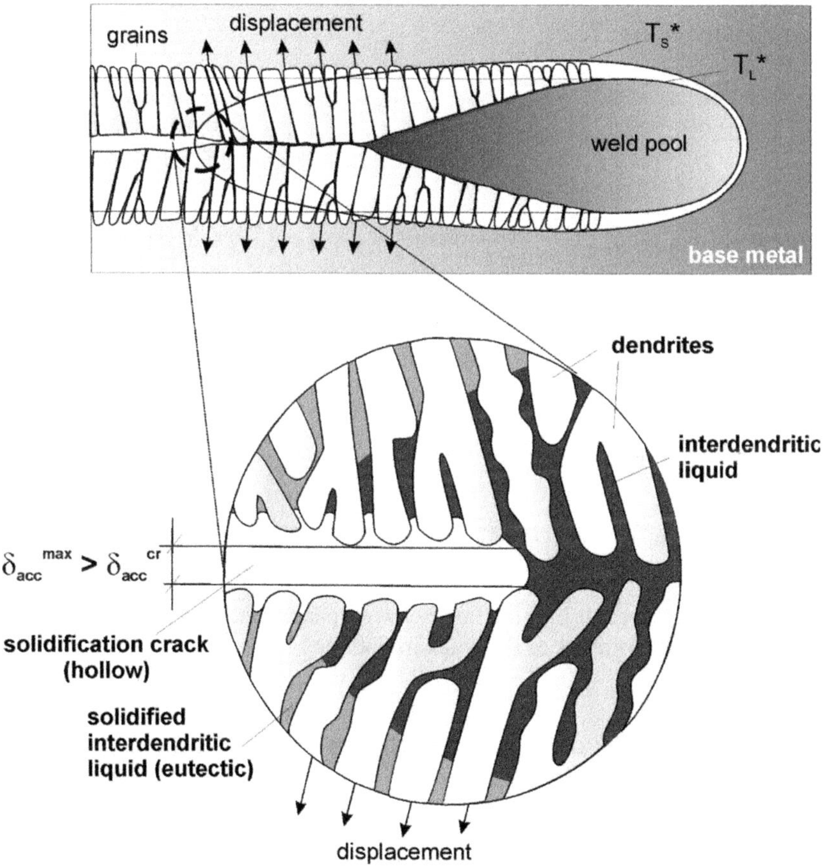

Fig. 6. Schematic representation of the solidification cracking mechanism

The general form of the criterion for crack initiation can be represented as follows:

$$\delta_{acc}^{max}(DS, WC, PBM, PFM) \geq \delta_{acc}^{cr}(PMS), \quad (1)$$

where δ_{acc}^{max} is the maximal displacement accumulated in the intergranular liquid film of the segregating elements, δ_{acc}^{cr} is the critical displacement which characterizes the ability of the liquid film to accumulate tensile strain without tearing, DS are the particularities of the design of the welded joint, WC are the welding conditions (for instance, velocity), PMB and PFM are the relevant properties of the base material and of the filler material, respectively; PMS are the parameters of the microstructure.

The phenomenological interrelation between the parameters included in the cracking criterion and the crack initiation mechanism is represented schematically in Fig. 7.

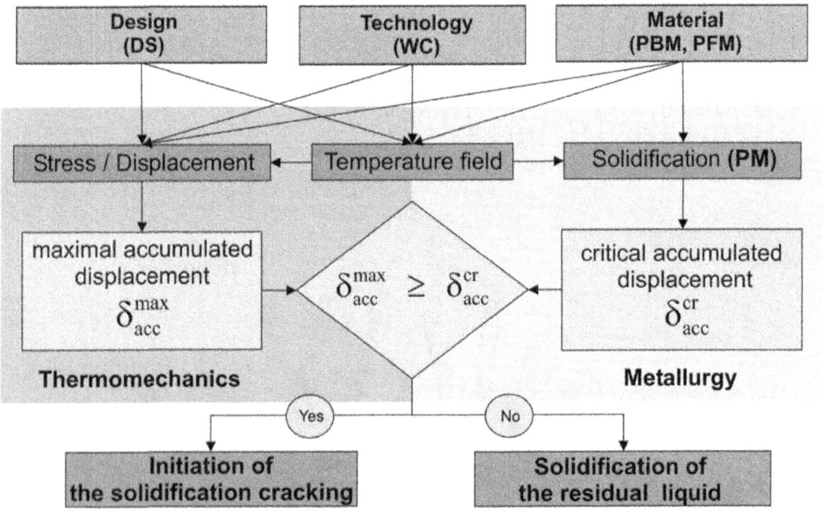

Fig. 7. Phenomenological interrelation between parameters included in the cracking criterion and the process of the crack initiation

A more detailed list of parameters relevant to the phenomenon of solidification cracking is given below:

- *Nature of Design (DS):* complete geometry of the welded joint (including the whole structure) and the position of the weld in the structure;
- *Welding Conditions (WC):* heat source (power, power distribution, velocity), shielding gas, filler material, clamping (thermal and mechanical influence), preheating (location and temperature), additional heating and/or cooling during welding (location, heating and/or cooling intensity);

- *Properties of the base material (PBM):* temperature-dependent thermal properties (thermal conductivity), thermo-dynamical properties (heat capacity, real solidification range), temperature-dependent thermo-mechanical properties (thermal expansion coefficient, yield strength, elasticity modulus);
- *Properties of the filler material (PFM):* the same properties as for the base material, especially the thermo-dynamic ones;
- *Parameters of the microstructure (PMS):* distribution of the liquid and the solid phase within the mushy zone, residual liquid phase fraction, primary and the secondary dendrite arm spacing.

The left part of the cracking criterion relates to the thermo-mechanical aspect of the total problem. It represents the mechanical loading of the liquid film of segregates. The value of δ_{acc}^{max} can be calculated only numerically because of the nonlinearity in material properties and the complex boundary conditions.

It should be emphasized that the solution of the thermo-mechanical problem directly depends on the solution of the metallurgical one. For this reason, special attention must be given to the correct evaluation of the real solidification range, since the real solidification range defines the length of the mushy zone and, therefore, the length of the liquid film. A proper preliminary study must be carried out also for the correct evaluation of the solid phases distribution in the mushy zone, since the last one governs the release of the latent heat and, therefore, the distributions of temperature, stress and accumulated strain.

The right part of the cracking criterion relates to the metallurgical aspect of the total problem. It represents the ability of the liquid film of segregates to accumulate tensile strain without tearing. The value of the δ_{acc}^{cr} depends on the thickness and on the morphology of the liquid film. The thickness is defined by the fraction of the residual liquid phase in the microstructure. The morphology is defined by the geometrical microstructure parameter, i.e. by the primary and secondary dendrite arm spacing. This dependence of the critical strain on the parameters of the microstructure can be expressed as follows:

$$\delta_{acc}^{cr} = \delta_{acc}^{cr}(f_L, \lambda_1, \lambda_2), \tag{2}$$

where f_L is the fraction of the residual liquid, λ_1 and λ_2 are the primary and secondary dendrite arm spacing, respectively.

On the basis of simple considerations of the accumulation mechanism it can be expected that the fraction of the residual liquid plays a primary role for the ability of the liquid film to accumulate strain without tearing. A higher liquid fraction in the microstructure leads to better strain accumulation and, therefore, to better resistance to solidification cracking.

Concerning the role of the geometric parameters of the microstructure it can be supposed that under the same distribution of the liquid fraction the coarser microstructures have a better ability of strain accumulation, i.e. a lower cracking susceptibility. More exact definition of the functional dependence of the critical displacement on the microstructure parameters is the subject of future development.

Direct calculation of the critical displacement will be possible on the basis of preliminary evaluation of the microstructure parameters. This can be done for the multi-component alloys using the modern numerical models [29]. The understanding of the interrelations between the microstructure parameters and the cracking resistance expressed in Eq. 2 can be effectively used for welding technology optimization by targeted development of filler materials. Such optimization can be made on the basis of thermodynamic calculations for multi-component systems [30], aimed at maximization of the residual liquid in the microstructure at the end of weld metal solidification.

Simulation

A simulation example of crack initiation and growth is shown in Fig. 8. The simulation is carried out for the case of laser beam welding close to the free edge described in the section "Experiments" (Fig. 2).

The thermo-mechanical problem is solved using the finite element method. For proper simulation of strain accumulation, the mechanical properties of the liquid film are assigned to the finite elements at the weld centreline (Fig. 9). For these elements, the value of the accumulated displacement is checked at each time step of the calculation. As soon as this value becomes critical, the material properties are changed to the properties of air (Fig. 9).

Particular attention is given to properly taking account of the latent heat release in accordance with the fraction of the liquid phase in the mushy zone. The real solidification range as well as the liquid phase fraction is calculated on the basis of the Scheil-Gulliver model [31].

Two isotherms, i.e. the real liquidus and the real solidus, outline the mushy zone. As seen, the crack appears in the inner part of the mushy zone. Once formed, the crack creates a thermal barrier which increases the asymmetry of the temperature field.

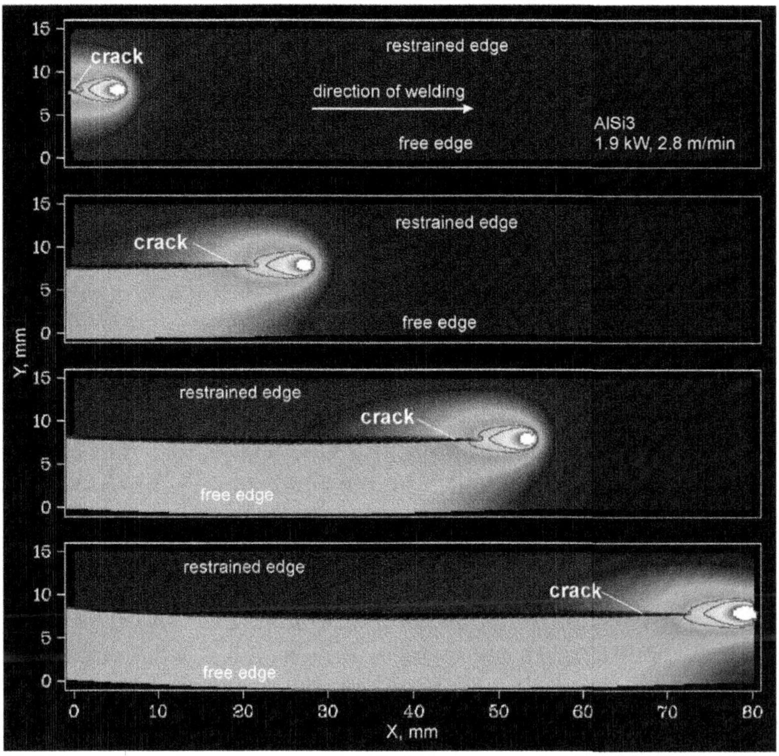

Fig. 8. Numerical simulation of crack initiation and growth: temperature fields at different steps of calculation

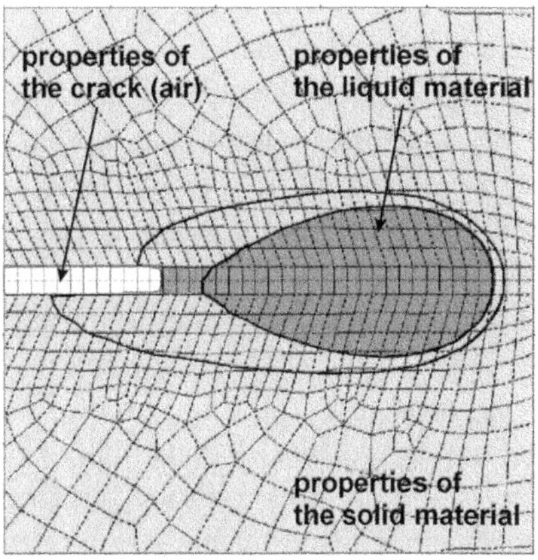

Fig. 9. Taking into account the effect of strain localization due to proper manipulation of the material properties during the numerical simulation

The position of the crack tip within the mushy zone is not constant; it changes in accordance with the local thermo-mechanical conditions. At the end of the weld seam, close to the right sample edge, the crack is forced out of the mushy zone, i.e. its growth is interrupted. The same behavior is also observed in all experiments. The reason for the growth interruption is the decrease of accumulated strain occurring due to overheating and the correspondent decrease of material strength in the near-edge region of the sample.

Concerning the Healing Effect

A controversial question in the theory of solidification cracking is the role of the healing effect. Here, we try to analyze the possibilities for healing a solidification crack due to the melt from the weld pool on the basis of the reconstructed temperature field for the real weld.

Fig. 10 shows the reconstructed mushy zone for the real weld produced on the Al-Si binary alloy containing 3 % of silicon. As can be seen, the length of the mushy zone is approximately 2.3 mm (Fig. 10a). The photographs of the microstructure represented in Figs. 10b–c were obtained at the centerline of the weld. One of them represents the region with the solidification crack, another one shows the region of the weld where crack growth was interrupted.

Due to the similar solidification conditions, the parameters of both observed microstructures are equal. The value of the primary dendrite arm spacing is in the range of 5 μm. It means that more than four hundred dendrites can be placed on the length corresponding to the distance between the isotherms of liquidus and solidus at the weld centerline.

It should be emphasized that during solidification the tips of dendrites growing from both sides of the weld meet each other under the temperature, which is close to the liquidus one, i.e. in the region of the liquidus isotherm. Such situation can be illustrated using the real microstructure as shown in Fig. 10b. Therefore, during solidification there must be a dense dendritic network consisting of several hundreds of dendrites in the mushy zone at the weld centerline.

This means that if healing would be even possible in the region close to the liquidus isotherm, it would surely be hampered by the dense dendritic net in the interior regions of the mushy zone. In this region, far from liquidus, tearing of the liquid film of segregates caused by the strain accumulation can freely take place, involving solidification cracking as illustrated by the real microstructure shown in Fig. 10c.

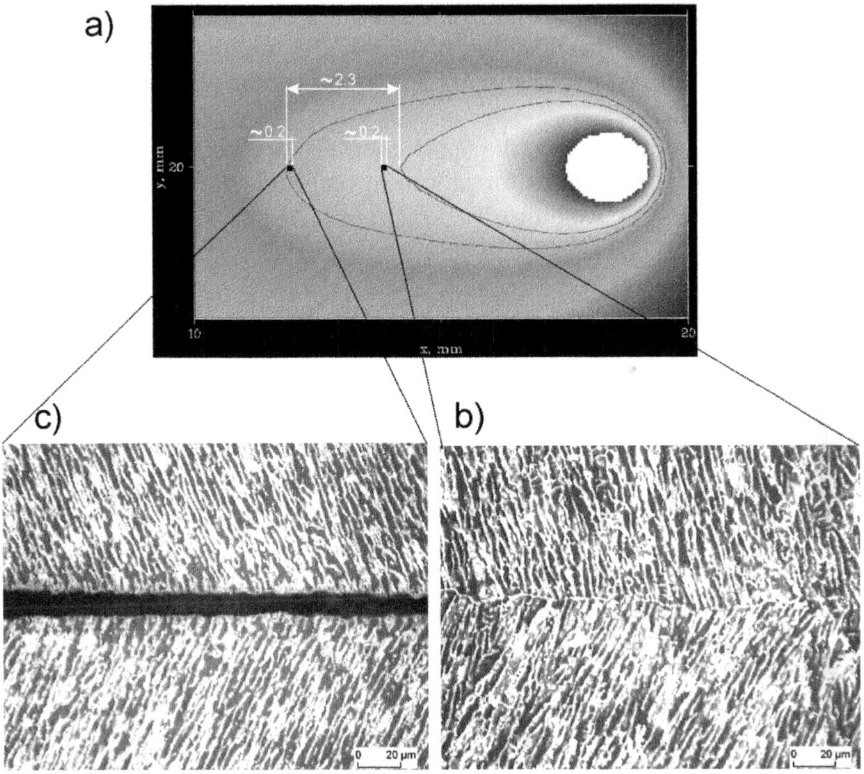

Fig. 10. To the role of the healing effect: **(a)** reconstructed shape of the mushy zone for the real weld produced in the binary alloy Al-3 % Si, **(b)** microstructure in this weld observed in the region where the crack growth was interrupted, and **(c)** microstructure of the same weld observed in a region with the solidification crack. Both microstructures illustrate the density of the dendritic network in the mushy zone (app. 200 dendrites at 1 mm length) which should hamper the healing effect

Example of Application: Multi-Beam Laser Welding

The idea to use additional heat sources as a means to prevent solidification cracking in welds was suggested in the 1970ies [32, 33]. On the basis of FEM-simulations it was demonstrated [32, 34] that introducing additional heating "in a right place and at the right time" [32] leads to "beneficial" compressive stress (or strain) in regions which are critical with respect to solidification cracking. Despite the apparent simplicity of the suggested idea, additional heating has still not found the expected industrial application. The determination of "the right place and the right time", i.e. the optimal position, the size and the power of the additional heating spots, seems to be the most important problem hindering the practical application of this technique.

As will be shown below, the application of the presented modeling approach can be very helpful for industrial realization of multi-beam techniques for crack-free welding.

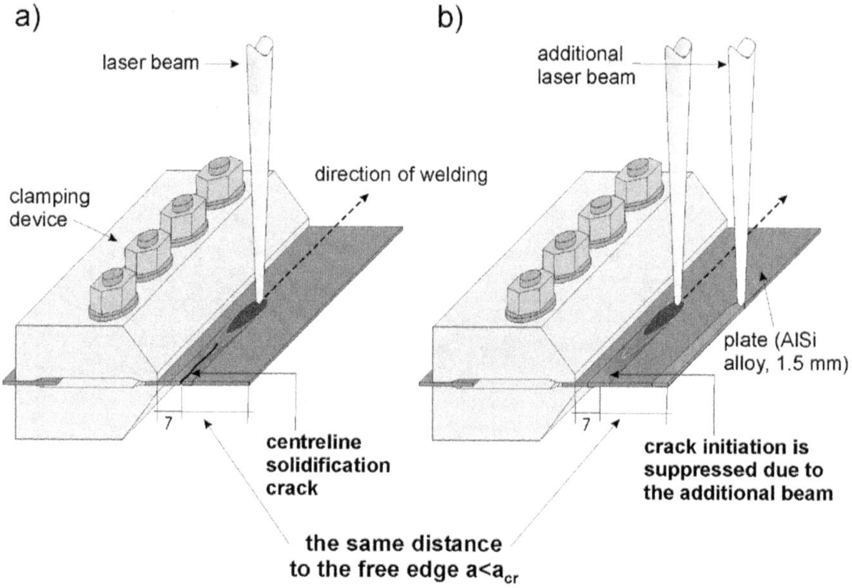

Fig. 11. Experimental validation of the multi-beam technique: **(a)** generation of solidification cracking due to the small distance to the free edge and **(b)** suppression of solidification cracking under the same conditions due to the application of the additional laser beam

The experimental laser welding close to the free edge described above was chosen for validation of the numerical calculations (Fig. 11). The idea was to define the distance from the free edge which assures the initiation and the stable growth of the solidification crack as shown in Fig. 11a.

Then, the optimal position and the power of the additional laser beam, that will ensure the suppression of crack initiation, were found using the numerical simulation (Fig. 11b). The experiments were carried out on Al-alloy AA6056 with a plate thickness of 2 mm.

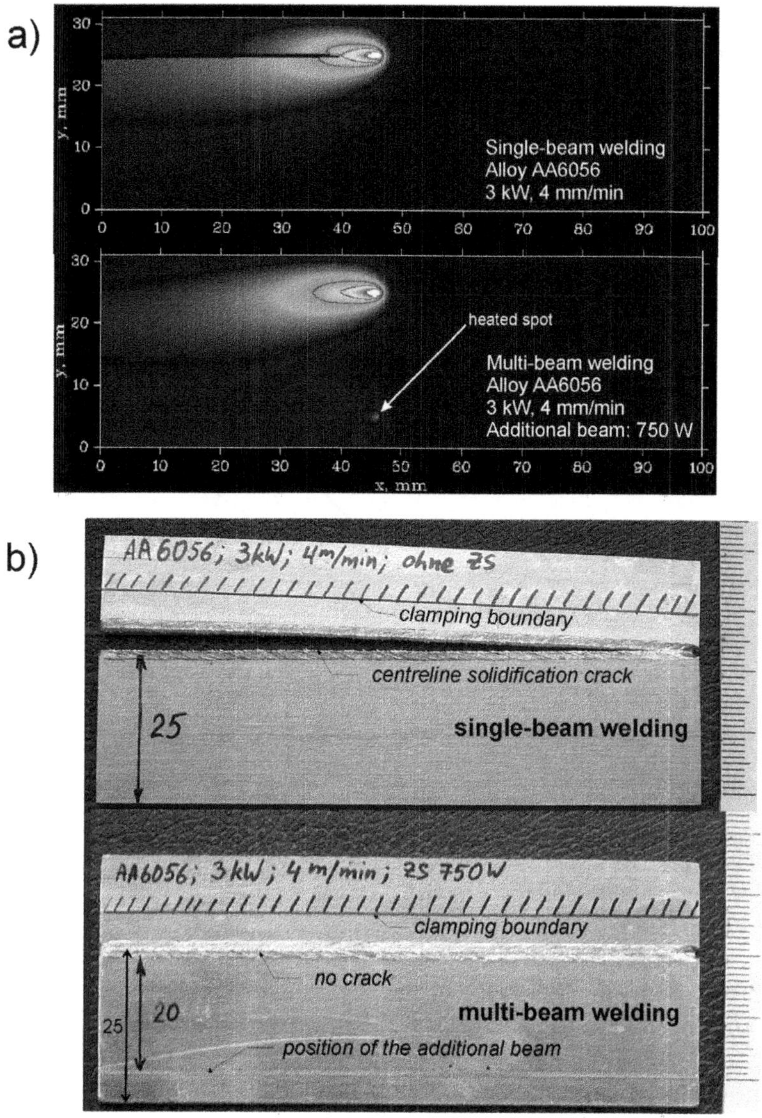

Fig. 12. Realization of the multi-beam technique: **(a)** preliminary numerical simulation of the temperature fields and **(b)** samples produced under the same conditions using single-beam (above) and double-beam (below) laser welding

Fig. 12 shows the results of the preliminary numerical simulations (temperature fields for single- and double-beam welding) together with the results of the experimental validation. As can be seen, the additional beam, which moves parallel to the main laser beam at a distance of 20 mm, can effectively be used for the suppression of crack initiation. The power of the additional beam is significantly lower than the power of the main beam. In this special case, the positive effect is achieved at the additional beam power which was higher than 40 % of the main beam power.

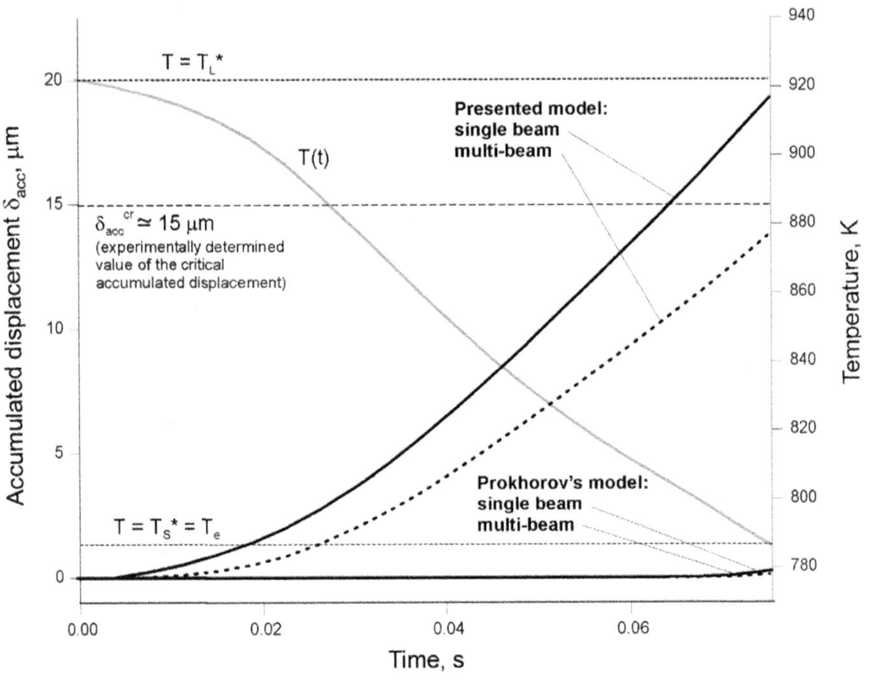

Fig. 13. Dynamics of strain accumulation: differences between the presented modeling approach and Prokhorov's model

Fig. 13 represents the dynamics of the accumulated displacement calculated for single- and multi-beam laser welding on the basis of the presented modeling approach as well as on the basis of the model of Prokhorov. The curves of the accumulated displacement were calculated for the initial point of the plate at the weld centerline, i.e. for the point where crack initiation can be expected. In all cases, the strain accumulation was calculated for the same temperature interval, from the real liquidus until the real solidus (the last one corresponds to the eutectic temperature).

The calculated maximal values were compared with the value of the critical displacement determined on the basis of the known critical distance from the free edge using the calibration of the experiment and the numerical simulation.

As can be seen from Fig. 13, according to the present approach, the use of the additional beam leads to a continuous reduction of the displacement accumulation. The maximal value of the accumulated displacement for multi-beam laser welding is less then the critical one.

Prokhorov' model neglects the effect of strain localization. For this reason, it predicts very small, physically not explainable values of the accumulated displacement needed for solidification cracking (single-beam welding). Although Prokhorov's model reveals a similar tendency to the reduction of a accumulated strain by multi-beam welding, the difference between both curves is not significant enough to be helpful for the prediction of the threshold from solidification cracking to crack-free welding.

Summary

Based on experimental observations, an approach to modeling of solidification cracking in welds is presented which integrates both the thermo-mechanical and the metallurgical aspects of the cracking phenomenon. The presented approach implies the solution of the thermo-mechanical problem taking into account the effects of the localization and the accumulation of displacements in the intergranular liquid film of the segregating alloying elements. The importance of proper preliminary evaluation of the real solidification range and the phase distribution within the mushy zone is emphasized. The high potential of the presented approach for the development of crack-free welding techniques in industry is demonstrated on the basis of the realization of multi-beam laser welding.

Acknowledgements

This work was carried out within the framework of the project "Optimisation of weldability" sponsored by the Bavarian Government, the Foundation of Oberfranken, Allianz-Zentrum für Technik GmbH, AUDI AG, EADS Deutschland GmbH, Linde AG und MTU Aero Engines GmbH. The authors would like to thank all partners for their active support as well as for their very useful discussions. Special thank is addressed to Dr. C. Heimerdinger (EADS Deutschland GmbH) for his creative co-working in the experimental realization of the multi-beam technique.

References

1. Singer ARE, Jennings PH (1946) Hot-Shortness of the Aluminium-Silicon Alloys of Commercial Purity. Journal of The Institute of Metals 73: 197–212
2. Singer ARE, Jennings PH (1947) Hot-Shortness of Some Aluminium-Iron-Silicon Alloys of High Purity. Journal of The Institute of Metals 73: 273–284
3. Jennings PH, Singer ARE, Pumphrey WI (1948) Hot-Shortness of Some High-Purity Alloys in the Systems Aluminium-Copper-Silicon and Aluminium-Magnesium-Silicon. Journal of The Institute of Metals 74: 227–248
4. Pumphrey WI, Lyons JV (1948) Cracking During the Casting and Welding of the More Common Binary Aluminium Alloys. Journal of The Institute of Metals 74: 439–455
5. Pellini WS (1952) Strain Theory of Hot Tearing Foundry, pp 125–133, 192, 194, 196, 199
6. Prokhorov NN (1952) Hot Cracking During Welding (in Russian). Mashgiz, Moscow
7. Bochvar AA, Rykalin NN, Prokhorov NN, Novikov II, Movchan BA (1960) The Question of "Hot" (Crystallisation) Cracks. Welding Production 10: 5–7
8. Prokhorov NN (1962) The Technological Strength of Metals While Crystallising During Welding. Welding Production 4: 1–8
9. Matsuda F, Nakagawa H, Nakata K, Kohmoto H, Honda Y (1983) Quantitative Evaluation of Solidification Brittleness of Weld Metal During Solidification by Means of In-situ Observation and Measurement (Report I). Transactions of JWRI 12 (1): 65–72
10. Matsuda F, Nakagawa H, Tomita S (1987) Quantitative Evaluation of Solidification Brittleness of Weld Metal During Solidification by Means of In-situ Observation and Measurement (Report IV). Transactions of JWRI 16(2): 90–95
11. Senda T, Matsuda F, Takano G, Watanabe K, Kobayashi T, Matzusaka T (1971) Fundamental Investigations on Solidification Crack Susceptibility for Weld Metals with Trans-Varestraint Test. Transactions of the Japan Welding Society 2(2): 45–66
12. Arata Y, Matsuda F, Nakata K, Sasaki I (1976) Solidification Crack Susceptibility of Aluminium Alloy Weld Metals I. Transactions of JWRI 5(2): 53–67
13. Arata Y, Matsuda F, Nakata K, Sinozaki K (1977) Solidification Crack Susceptibility of Aluminium Alloy Weld Metals II. Transactions of JWRI 6(1): 91–104
14. Jonsson M, Karlsson L, Lindgren LE (1984) Thermal Stresses, Plate Motion and Hot Cracking in Butt-Welding. In: Mechanical Behaviour of Materials IV. Pergamon Press, New York, pp 273-279
15. Dike JJ, Brooks JA, Li M (1988) Comparison of Failure Criteria in Weld Solidification Cracking Simulations. In: Mathematical Modelling of Weld Phenomena 4 (ed Cerjak). IOM Communications Ltd, London, pp 199–222

16. Feng Z (1994) A Computational Analysis of Thermal and Mechanical Conditions for Weld Metal Solidification Cracking. Welding in the World 33(5): 340–347
17. Zacharia T (1994) Dynamic Stresses in Weld Metal Hot Cracking. Welding Journal 73: 164-172
18. Dike JJ, Brooks JA, Krafcik JS (1996) Finite Element Modelling and Verification of Thermal-Mechanical Behaviour in the Weld Pool Region. In: Trends in Welding Research (ed Smartt HB et al). ASM Intern, pp 159–164
19. Feng Z, Zaharia T, David SA (1996) On the Thermomechanical Conditions for Weld Metal Solidification Cracking. In: Mathematical Modelling of Weld Phenomena 3 (ed Cerjak), pp 114–147
20. Feng Z, Zaharia T, David SA (1997) Thermal Stress Development in a Nickel Based Superalloy during Weldability Test. Welding Journal 76: 470–483
21. Makhnenko VI, Velikoivanenko EA, Rozynka GF, Musiyachenko AV, Pivtorak NI, Seiffarth P (1998) A Computer Program for Prediction the Zones with the Risk of Formation of Hot Cracks in Welding with Deep Penetration. The Paton Welding Journal 10(2): 57–65
22. Weise S (1998) Heißrißbildung beim Laserstrahlschweißen von Baustählen. BIAS-Verlag, Bremen
23. Herold H, Streitenberger M, Pchennikov A (2001) Modelling of the PRV-Test to Examine the Origin of Different Hot Cracking Types. In: Mathematical Modelling of Weld Phenomena 5 (ed Cerjak et al). IOM Communications Ltd, London, pp 783–792
24. Shibahara M, Serizawa H, Murakawa H (2000) Finite Element Method for Hot Cracking Analysis under Welding Using Temperature Dependent Interface Element. In: Modelling of Casting, Welding and Advanced Solidification Processes IX (ed Sahm PR et al). Shaker-Verlag, Aachen, pp 844–851
25. Shibahara M, Serizawa H, Murakawa H (2001) Finite Element Method for Hot Cracking Analysis Using Temperature Dependent Interface Element. In: Mathematical Modelling of Weld Phenomena 5 (ed Cerjak et al). IOM Communications Ltd, London, pp 253–267
26. Bergmann HW, Hilbinger RM (1998) Numerical Simulation of Centre Line Hot Cracks in Laser Beam Welding of Aluminium Close to the Sheet Edge. In: Mathematical Modelling of Weld Phenomena 4 (ed Cerjak). IOM Communications Ltd, London, pp 658-668
27. Hilbinger RM, Bergmann HW, Köhler W, Palm F (2001) Considering of Dynamic Mechanical Boundary Conditions in the Characterisation of a Hot Cracking Test by Means of Numerical Simulation. In: Mathematical Modelling of Weld Phenomena 5 (ed Cerjak et al). IOM Communications Ltd, London, pp 847–862
28. Hilbinger RM (2000) Heißrissbildung beim Schweißen von Aluminium in Blechrandlage. Universität Bayreuth, Bayreuth
29. Diepers HJ (2003) Simulation des Primärabstandes gerichtet erstarrter Dendriten mit der Phasenfeldmethode. Shaker Verlag, Aachen

30. Andersson JO, Helander T, Höglund L, Shi P, Sundman B (2002) Thermo-Calc & DICTRA. Computational Tools For Materials Science, Calphad, 26: 273–312
31. Saunders N, Miodownik AP (1998) CALPHAD (Calculation of Phase Diagrams): A Comprehensive Guide. Pergamon Materials Series vol 1. Elsevier Science, Oxford, New York, Tokyo
32. Akesson B, Karlsson L (1976) Prevention of Hot Cracking of Butt Welds in Steel Panels by Controlled Additional Heating of the Panels. Welding Research International 6 (5): 35–52
33. Shumilin VG, Karkhin VA, Rakhman MI, Gatovsky KM (1980) A Technique of Arc Welding. Patent No 1109280, 1980, USSR
34. Herold H, Streitenberger M, Pchennikov A, Makarov E (1999) Modelling of One Sided Welding to Describe Hot Cracking at the End of Longer Butt Weld Seams. Welding in the World 43(2): 56–64

Influence of the Weld Pool Geometry on Solidification Crack Formation

M. Wolf, H. Schobbert, Th. Böllinghaus

Federal Institute for Materials Research and Testing, Berlin, Germany

Abstract

Numerous externally loaded as well as self-restraint hot cracking test procedures have been developed over the years to evaluate soldidification crack phenomena. But the interpretation and the subsequent transfer of test results towards component weldments has still to be regarded as a decisive challenge. Some progress has been achieved by hypothesizing crack / no crack criteria referring to uniform and uniaxial loading of the solidification front during welding. But, as the present results show, solidification cracking significantly depends on the geometry of the solidification front, which has so far not been considered in the respective criteria. In this contribution, the position of solidification cracks is correlated with the position-dependent strain rates along the solidification front and with the position-dependent curvature of the weld pool. Especially at positions of high curvature, an increased rate of shrinkage can be predicted on the basis of a geometrical Rate of Feeding (ROF) Rate of Shrinkage (ROS) model, corresponding to the measured positions of the solidification cracks. The assumption of position-dependent strain and strain rates along the solidification front during Varestraint-Testing has been confirmed by means of numerical simulations.

Introduction

Solidification cracks originate and grow as interdendritic or intergranular material separations directly behind the moving weld pool and preferably occur in fcc-solidifying alloys. At the end of the welding process, their dimensions can reach from a few micrometers to the whole length of the weld.

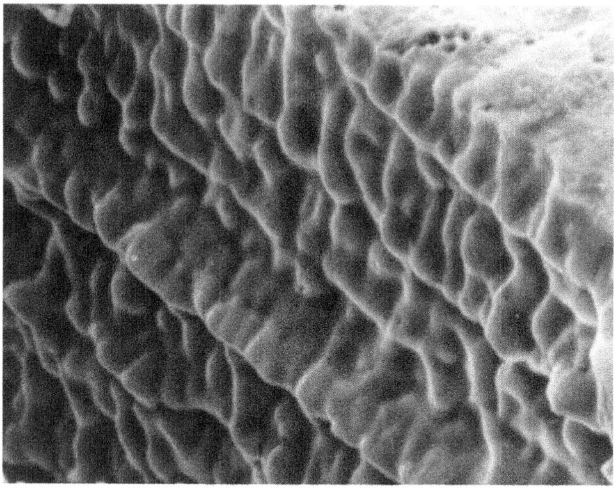

Fig. 1. Solidification crack of a Ni-base alloy (German mat.-no.: 2.4889 – 2000 x)

Fig 1 shows a photograph of a solidification crack taken by a scanning electron microscope. Typical of the surface morphology of solidification cracks is the visibility of dendrite tips.

The occurrence of solidification cracks is determined by the thermomechanical and metallurgical processes in the mushy zone surrounding the weld pool and corresponding to a temperature range, in which molten as well as solid material is coexistent. Solidification cracks are initiated, if the thermomechanical strains caused by the time-dependent temperature distribution during welding and the material-specific shrinkage during liquid/solid phase transition cannot be compensated by the ductility of the weld metal or by reflow of melt.

Although the most decisive critical factors such as the local strain rates, local shrinkage rates and the critical temperature range could be identified, the solidification cracking susceptibility of materials can still only be ranked in relation to other materials, referring to equal welding parameters and other boundary conditions. The reason for the difficulties in assessing the materials solidification crack susceptibility by absolute values is its concatenation to a complex system of feed-back processes. As for example technological parameters like the welding speed can simultaneously influence the thermomechanically induced material displacements in the vicinity of the weld pool as well as metallurgical processes like the morphology of the solidification front, solidification cracking can often not be related to a definite reason.

From the metallurgical point of view, solidification cracking can be ascribed to a great variety of different reasons. For example, some iron companions such as sulphur, phosphor or boron can lower the solidification cracking resistance of stainless steels drastically. Moreover, some stainless steels with a chemical composition nearby the eutectic rim can lose their relatively high solidification cracking resistance, if a critical solidification rate is exceeded and their primary solidification mode changes from ferritic to austenitic [1].

Although plenty of metallurgical factors favoring or reducing the tendency to solidification cracking are known, the hot cracking behavior can still very often not even approximately be prognosticated on the basis of the materials composition. For this reason and because of the non-linear processes of solidification crack initiation, hot cracking tests have to be performed, in which a material is exposed to welding conditions. For a subsequent transfer of hot cracking test results to component weldments, crack / no crack criteria are required, by which solidification cracking can be predicted on the basis of significant input parameters. However, universal solidification crack criteria could not have been established so far.

Two solidification cracking criteria have gained the most acceptance on international level: The criterion of the brittleness temperature range (BTR) related to the solidification cracking theory of Prokhorov [2] and the so called Rate of Feeding (ROF) Rate of Shrinkage (ROS) model with the criterion ROF < ROS [3].

The BTR criterion of Prokhorov, which has been applied and partly overworked, for example, by Matsuda et al. [4] and Herold et al. [5], bases on a continuum-mechanical approach. Prokhorov postulates a defined temperature range in which the ductility of the solidifying material is significantly low (comparable to the ductility dip of a material at much lower temperatures, causing the ductility dip cracks). Within this temperature range, which reaches from a temperature between liquidus and solidus to a temperature below solidus, the material is sensitive to solidification cracking. A material can thus be characterized according to the extension of this temperature range in combination with the critical strain or to a critical temperature-related strain increase $d\varepsilon/dT$.

Solidification cracking never occurs in a completely solidified state of the material, so that the designation "below solidus" refers to dispersed residual melts which might be present at the grain boundaries. According to the literature, different shapes of the ductility curve can be found referring either to an increase or a decrease of ductility when the solidification process reaches the solidus temperature. It can only be assumed in this connection that if bridging between dendrites or grains is not present during solidification, the minimum ductility is reached at the final stages of solidification. If cooling favors the bridging between dendrites or grains, then the ductility might finally increase at the solidus temperature. It becomes clear in this connection that for a characterization of the solidification cracking, not only the temperature range but also the shape and the absolute values of the critical strain curve have to be determined.

The ROF-ROS theory refers to the competitive and dynamic process of volume shrinkage caused by the liquid/solid phase transition and a leveling reflow of the melt from the weld pool (Fig 2). The balance between the material-specific volume shrinkage, which might moreover be superimposed by thermomechanically induced strain rates (ROS), and the backfilling by the melt (ROF) can mathematically be expressed in simplified terms by ROF = ROS. If the maximum melt reflow is not sufficient for filling the interdendritic voids or the interdendritic pressure becomes lower than the cavitation pressure, then a solidification crack is initiated. The corresponding solidification crack criterion can then be written as ROF < ROS.

A lot of theoretical work has been done for the calculation of ROF and ROS and a good agreement between the prognosis and the experimental data has been observed [6–8]. The ROF-ROS criterion has furthermore been extended by the interfacial effect of Marangoni forces influencing the ROF [9]. However it still has to be clarified, whether this criterion can be applied to all solidification crack phenomena.

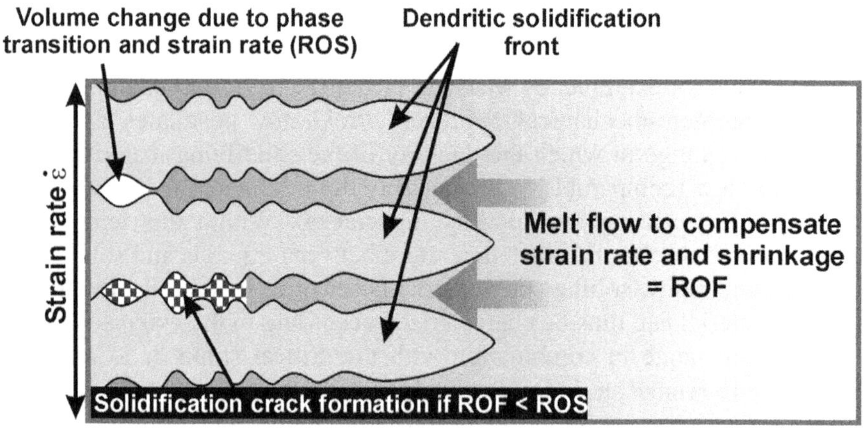

Fig. 2. Rate of Feeding (ROF) – Rate of Shrinkage (ROS) model

In a direct comparison, the BTR criterion seems to be better applicable to intergranular solidification cracking which is especially promoted by residual melt at the grain boundaries and a relatively low heat conductivity of the solid. Since the shrinkage caused by the solidification of residual melt is very low and also a potential reflow of the melt is implausible due to the length of the cracks, a continuum-mechanical approach like the BTR seems to be suitable for describing intergranular solidification cracking caused by residual melts.

The ROF-ROS criterion is more qualified for describing the solidification cracking, which bases on permanent interdendritic volume shrinkage during solidification directly at the solidification front. Both criteria have in common that strain and strain rates transverse to the solidification direction increase the tendency to solidification cracking. However, as an advantage of the proposed ROF-ROS mechanism, the local strain rate at the solidification front is accepted as the decisive factor whereas for the BTR the accumulated strain is considered.

Externally Loaded Hot Cracking Tests

The hot cracking resistance of materials is usually ranked according to the results of externally loaded hot cracking tests, in which a specimen is deformed during welding or a simulated time-temperature cycle (Hot Tensile Test) under defined conditions [10]. In order to minimize the material-specific internal (thermomechanical) loading, the geometry of the specimens is usually simple. One reason why the externally loaded hot cracking tests are widely spread in research is that defined strains and strain rates related to the mechanical loading can be applied onto the specimens. Therefore input variables comprising loading and welding parameters can be assigned to the measurable output data like number, length and position of solidification cracks which makes the results interpretable compared to the self-restraint hot cracking tests.

Numerous externally loaded hot cracking tests exist, for which the specimen loading (tensile loading, bending) as well as the interpretation of the test results can be quite different. Usual output data are for example the total crack length (TCL), the critical strain rate of the specimen and the length of the longest hot crack (MCL). Due to this multitude of variations, a comparison of the test results can only be done qualitatively and, as mentioned before, a direct transfer of test results to real component welds is still controversial.

Internationally, two of the most frequently applied externally loaded hot cracking tests are the Varestraint-Test (VAriable-REstraint-Test) [11] and the Transvarestraint-Test [12]. However, such tests do not account for a shrinkage restraint as it can only be applied in a self-restrained test. Both tests differ only in their bending direction.

For a Varestraint-Test (which is sometimes designated more precisely as Longitudinal Varestraint Test), a specimen is bent longi-tudinally over a mandrel within a short period of time during producing the test weld run. For a Transvarestraint-Test, the specimen is bent transversely to the weld direction. Both types of test have specific advantages: The Varestraint Test is suitable for testing of all types of hot cracks (solidification cracks, liquation cracks and ductility dip cracks). The results are interpreted by the total crack length (TCL), the maximum crack length (MCL), the number of all cracks or by the minimum strain at which hot cracks occur (ε_{min}).

The Transvarestraint-Test is preferably applied for the determination of the brittleness temperature range (BTR), since in most cases a centerline crack is engendered representing the maximum crack length. If the temperature regime in the centerline of the weld is determined by measurements or by numerical simulations, the BTR can directly be calculated by the length of the centerline crack. An advanced method for the assessment of the solidification crack susceptibility is the determination of the Solidification Cracking Temperature Range (SCTR), by which the maximum crack distance (MCD) is especially regarded [13].

Referring to the solidification cracking criteria, the loading rate of a hot cracking test has to be discussed in this context. In the Tranvarestraint-Test, for example, very high bending speeds in comparison to the welding speed are usually applied for determining the BTR. By this mechanical shock-loading, the length of the centerline crack represents directly the brittle material zone in the weld metal related to the momentary temperature distribution. The strain rate above a certain level does not influence the result anymore and can therefore be eliminated as a free parameter for Transvarestraint testing.

But besides a few exceptions which occur, for example, from a break open of tack welds, shock-loadings do usually not occur in real component welds, so that the question might be raised of which loading rate is realistic and comparable to the solidification crack formation in real component welds. As a compromise between shock-loading and a possibly uncritical loading rate an intermediate average strain rate of about 1–2 %/s is applied to the MVT-Test used in this investigation [14]. Using such a loading rate the BTR cannot be determined after the test by measurement of the longest solidification crack, whereas the intermediate loading rates are in particular suitable for testing the ROFROS criterion.

Fig. 3a–d shows a sequence of the solidification crack formation from a high speed video taken during an MVT-Test in Transvarestraint mode. It can be seen that the first fraction of the first crack (3.2 mm) forms up almost instantaneously within a time interval of 0.4 seconds (Fig. 3b–c) by an average growth speed of 8 mm/s.

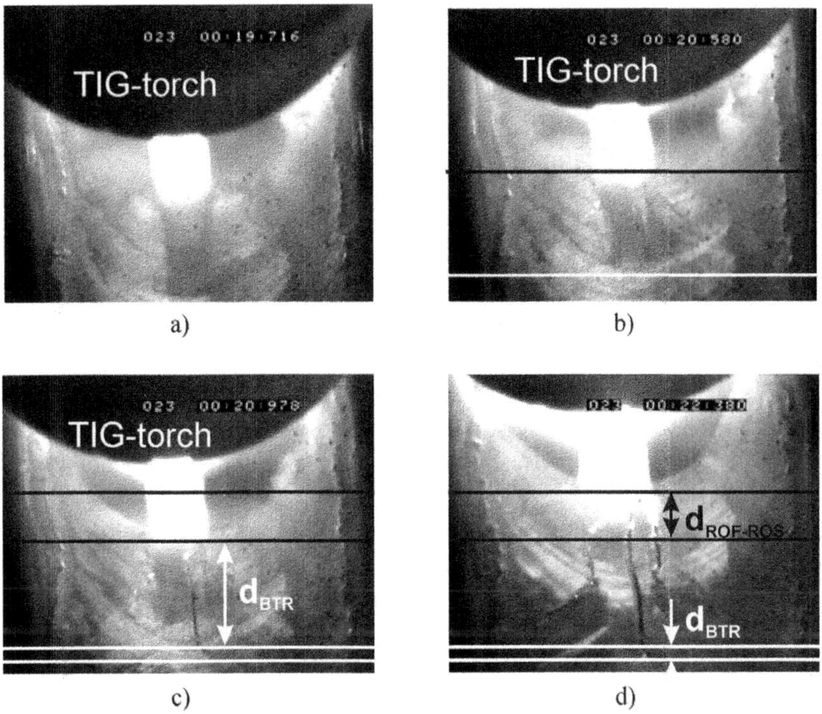

Fig. 3. Solidification crack formation in Alloy 602 CA during the MVT-Test; Transvarestraint mode (4 %); shielding gas: Ar + 5 % H2; v_{weld} = 1.4 mm/s, I = 171 A; U = 11.4 V

Obviously, a critical strain is exceeded at this instant, leading to an abrupt crack formation compared to the welding speed of 1.4 mm/s. After the sudden crack opening, the growing proceeds much slower. Within a period of 1.4 seconds between Fig. 3c and Fig. 3d, the crack growth rearward to the welding direction is very small, but the crack lengthens in forward direction with the speed of the solidification front until straining is over.

It can thus be concluded that the first fraction of the crack forms up depending on the brittleness of the material zone behind the weld pool, whereas a second fraction grows on account of insufficient melt reflow.

It has to be emphasized in this connection that this kind of crack formation is not necessarily representative of any alloy. But as a compromise for including all possibilities of crack formation and for a comprehensive test result, an intermediate bending speed seems to be most appropriate.

Objective

As a contribution to the ongoing standardization discussion correlated with the transfer of hot cracking test results towards component weldments and, in particular, to clarification of the mechanisms and crack criteria behind solidification crack phenomena, the influence of the weld pool geometry on the formation of solidification cracks has been investigated. For this purpose the Varestraint- as well as Transvarestraint-Test have been applied. The weld pool geometry is formed by material- and process-specific properties such as the heat conductivity of the solid, the melt flow within the weld pool, the energy distribution of the heat source and the solidification kinetics. It can be shown in this connection that the position-dependent probability of solidification crack formation can be correlated to geometrical parameters of the weld pool in combination with the global loading direction.

Besides the theoretical fundamentals of the numerous solidification cracking theories which mostly base only upon metallurgical effects measurements as well as computations of the position-, time- and direction-dependent strains and strain rates in the surrounding of the weld pool are very difficult due to the extreme temperature gradients and a complex high-temperature material behavior. This is the reason why the results of different test procedures might scatter broadly. But, it has to be emphasized in this connection that the strains and strain rates, respectively, occurring in the vicinity of the weld pool are generally the fundamental reason for solidification crack formation and represent the most explicit test parameter in every externally loaded hot cracking test to correlate laboratory results and practical cracking.

Therefore, the local material displacements leading to solidification cracks have to be quantified for a transfer of solidification cracking test results. On this account, in particular the second part of this contribution deals with numerical simulation of the position dependent material displacements in the surrounding of a weld pool during MVT testing.

Experimental Procedure

The following investigations have been carried out using the Modified Varestraint Transvarestraint-Test (MVT-Test) already mentioned above (Fig. 4). The major feature of the MVT-Test is to change the loading direction (transversely or longitudinally to weld direction) only by replacing the dies (Fig. 5).

Fig. 4. MVT-Hot Cracking Test

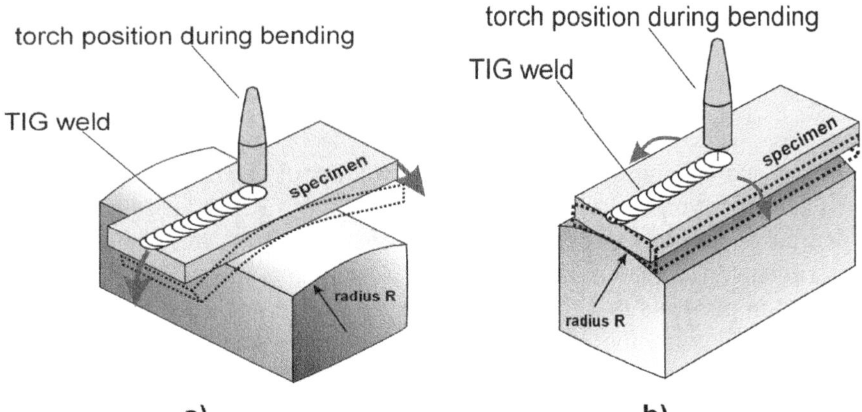

Fig. 5. MVT - Varestraint mode **(a)**; MVT - Transvarestraint mode **(b)**

Therefore, by means of the MVT-Test it is possible to perform alternatively Varestraint- or Transvarestraint Tests leaving relevant parameters like the geometry of the specimens or the timing of the test procedure unaffected.

The investigations have been performed based on the observation that under defined MVT-Test conditions, solidification cracks often form up symmetrically (Fig. 6). A symmetry can be observed preferentially for a small number of solidification cracks and therefore with minor crack formation interaction. From the symmetry it can thus be concluded that the weld pool geometry as well as the position-dependent strain or strain rates must be considered as a critical factor for solidification cracking, since their position correlated with the weld pool geometry is deterministic.

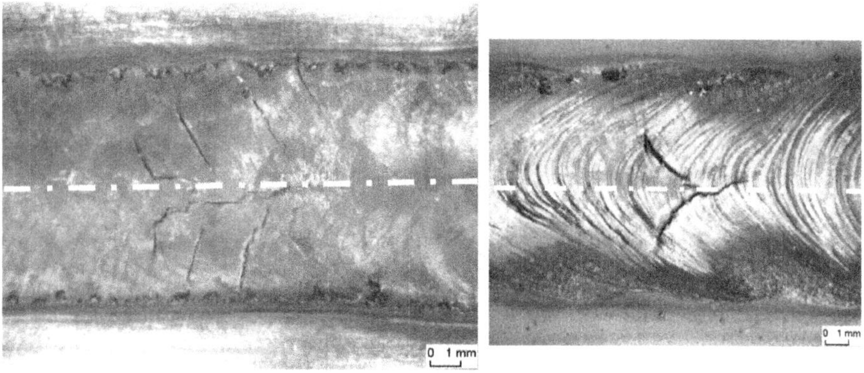

a) Alloy 602 CA (NiCr25FeAlY) **b)** AISI 316L

Fig. 6. Crack initiation site after MVT testing; TIG bead-on-plate weld; Varestraint mode; 4 % total strain

For quantifying the influence of the weld pool geometry on the position of solidification cracks, the welding parameters such as heat input per unit length and welding speed have been systematically varied in MVT experiments. The material used in these experiments was AISI 316 L. It was found in pre-tests that the shape of the end craters of bead-on-plate welds agree well with the solidification front at the surface during welding. Therefore the specific end crater shape of every weld has been used for the determination of the weld pool geometry and solidification front, respectively.

Fig. 7 shows exemplarily photographs of end craters and the resulting solidification cracks after MVT-testing. For these kinds of experiments, a quantification of the test results has been carried out by measuring the position of solidification cracks in correlation with the weld pool geometry. The results can be described as follows: For a slow welding speed, which means that the weld pool geometry is approximately circular, the solidification cracks preferably appeared nearby the fusion line (Fig. 7a).

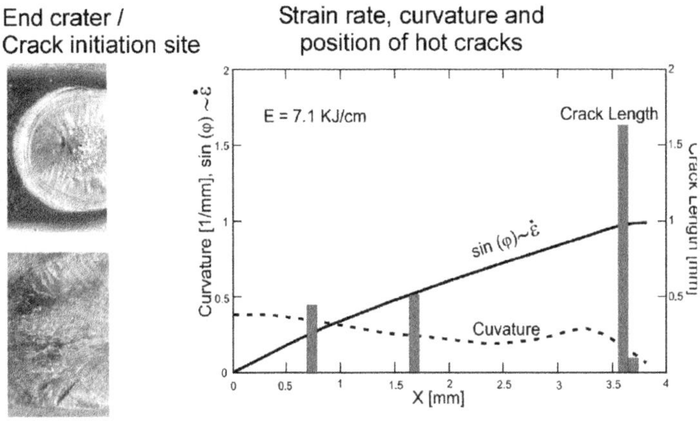

a) I = 110 A; U = 10.7 V; v_{weld} = 1.6 mm/s

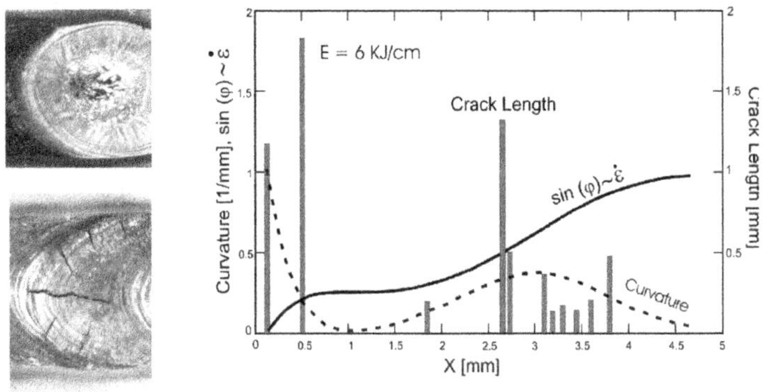

b) I = 130 A; U = 10.7 V; v_{weld} = 2.3 mm/s

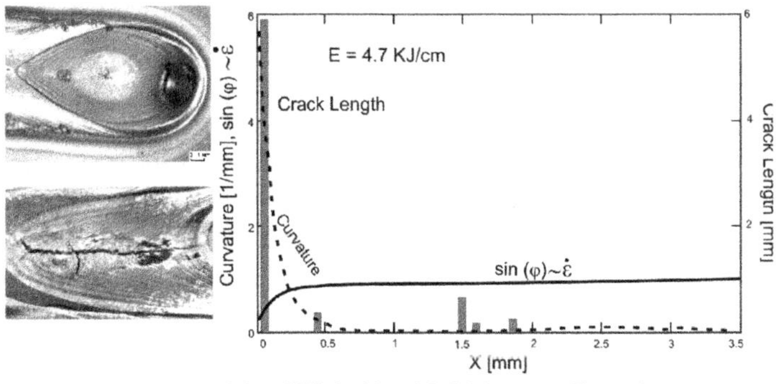

c) I = 255 A; U = 13.8 V; v_{weld} = 7 mm/s

Fig. 7. Alloy AISI 316 L; Correlation of weld pool geometry with solidification cracking; MVT Varestraint mode; total strain = 2.3 %; shielding gas: Ar + 5 % N2

When the welding speed was increased and the weld pool geometry became tapered, then the solidification cracks concentrated more and more to positions nearby the centerline, and even a small centerline crack occurred (Fig. 7b). When the weld pool geometry was sharply tapered and mathematically unsteady at the centerline, then a single centerline crack occurred (Fig. 7c). The geometries of the end craters were measured and fitted by mathematical functions basing on the ellipsoid equation and correction functions. By means of end crater functions, the curvature along the solidification front as well as the normalized strain rate function occurring during bending were calculated and plotted in a common diagram as shown on the right hand side in Fig. 7a to 7c.

It has to be remarked in this connection, that the position dependent strain rates were calculated transverse to the solidification direction, which means transverse to a potential crack direction, and base only on the geometry of the weld pool. Thermomechanical effects occurring during a weld process are not considered so far, but their influence will be discussed below. Moreover, the positions and lengths of the solidification cracks were charted in the respective diagrams. Following tentative conclusions can be drawn: Due to testing in Varestraint mode, which means that the specimen is bent longitudinally to the welding direction, the strain rates transverse to the potential crack are monotonic increasing from the centerline to the fusion line. The trend of the curvature functions highly depends on the geometry function of the solidification front. Correlating the positions of the solidification cracks with the position-dependent curvature and strain rate, it can be concluded that a solidification crack occurs preferably at positions, where a relatively high curvature and a relatively high strain rate coincide.

A significant correlation of the weld pool geometry in form of the curvature with the solidification cracking becomes in particular clear on the basis of the example illustrated in Fig. 7c. Although theoretically the strain rates at or in the direct vicinity of the centerline were very low, a single centerline crack occurred which originated slightly next to the centerline. The fact that straining longitudinally to the weld direction can cause centerline cracks, even though the mechanical loading is lowest at this position, means that the influence of the respective weld pool geometry on solidification cracking is determinant in some cases.

Fig. 8 shows an example of an MVT-Test result from the Transvarestraint mode, where the welding parameters correspond as far as possible to the example in the second row of Fig. 7. It can be seen that, due to the change of the bending direction, the strain rate function is oppositely directed and that the positions of the solidification cracks are shifted towards the centerline.

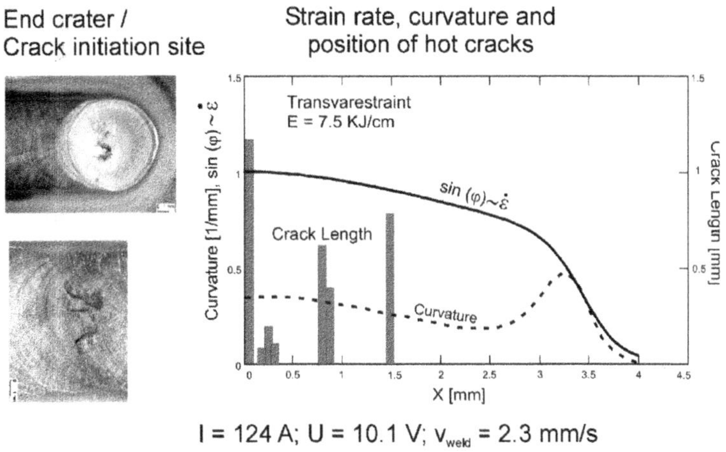

Fig. 8. Alloy AISI 316 L; Correlation of weld pool geometry with solidification cracking; MVT Transvarestraint mode; total strain = 2.3 %; shielding gas: Ar + 5 % N2

In this example, the strain rate has to be regarded as the determinant factor for solidification cracking, since highest strain rates occurred mathematically at the crack positions. It can thus be concluded, that both the position-dependent strain rate and the position dependent curvature influence solidification cracking. The effect of the curvature on solidification crack formation is discussed in the next paragraph.

Rate of Area Shrinkage Correlated with the Weld Pool Geometry

The influence of the weld pool geometry on the position of solidification cracks has been theoretically analyzed by a geometric model. In this two-dimensional model the rate of shrinkage by phase transition at the solidification front is related to the position-dependent growth direction and speed of the solidification front.

As a first approach it bases on strain rate free solidification. The function of the growth path (Fig. 9), which represents the idealized trace of an arbitrary point at the solidification front shifting from the fusion line to the centerline, can be calculated from the end crater function representing the solidification front at an arbitrary point in time, as described below. It is assumed that cells with a length directly corresponding to the local extension of the solidification interval, protrude into the melt.

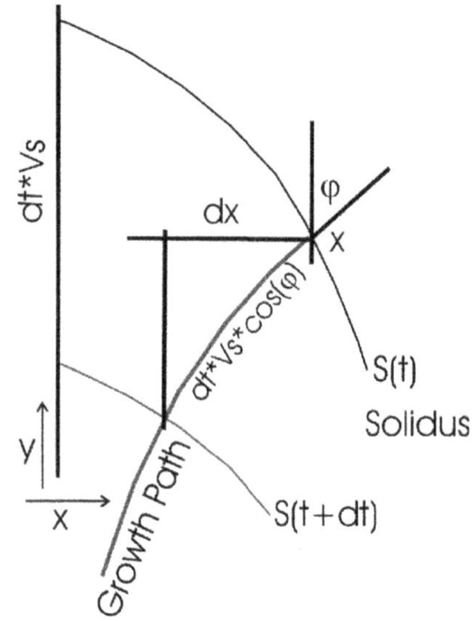

Fig. 9. Calculation of the time-dependent solidification front evolution; welding direction -y

The root point, for example, of such a cell is differentially shifted by:

$$ds = v_S \cdot \cos(\varphi(x)) \cdot dt \qquad (1)$$

perpendicular to the solidification front, whereby $\varphi(x)$ is the position dependent angle between the welding direction and the normal of the solidification front and v_s is the welding speed.

From Eq. 1 follows:

$$v_S \cdot dt = 2 \cdot dx \, / \, \sin(2 \cdot \varphi(x)) \qquad (2)$$

By integration of Eq. 2 the time-dependent shifting function of the relevant root point as well as of the tip of a cell can be calculated.

For the determination of the area change per time correlated with the proceeding of the solidification front, the change of the area of liquid between two adjacent cells caused by the phase transition involving a change of the specific density $\rho_l \Rightarrow \rho_s$ has to be calculated (Fig. 10).

As a first approximation this area shrinkage is set as a measure ROS_γ for the ROS. Two arbitrary adjacent cells start growing with an offset of λ in x-direction.

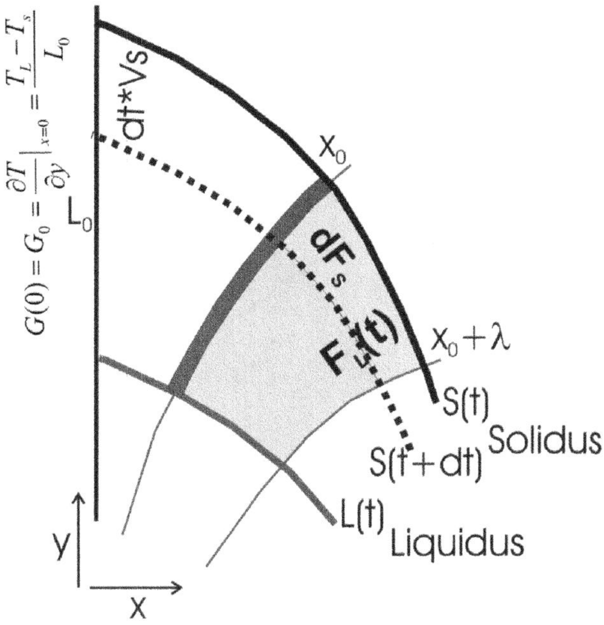

Fig. 10. Change of area by the proceeding of a bent solidification front

The functions of the liquidus and solidus isotherms y = y(x) are set as equal but they are shifted by the solidification interval moving with the speed of welding in y-direction.

After the solidification of dF_S within the period dt, there exists a residual liquid portion $F_L(t)$ - γdF_S, whereby $\gamma = \rho_s/\rho_l$. With the shrinkage of dF_S in correlation with the corresponding liquid portion, it follows for the change of area:

$$\mathrm{ROS}\gamma\, dt = (\gamma - 1)\cdot dF_S(t) \tag{3}$$

and

$$\mathrm{ROS}\gamma = (\gamma - 1) \cdot dF_S/dt \tag{4}$$

For a verification of the geometric ROF-ROS model, the end crater function of Fig. 7b has been used which exhibits a significant curvature at a distance of about 3 mm from the centerline.

The position-dependent ROS function, for which a value for the cell distance λ was estimated, is shown in Fig. 11.

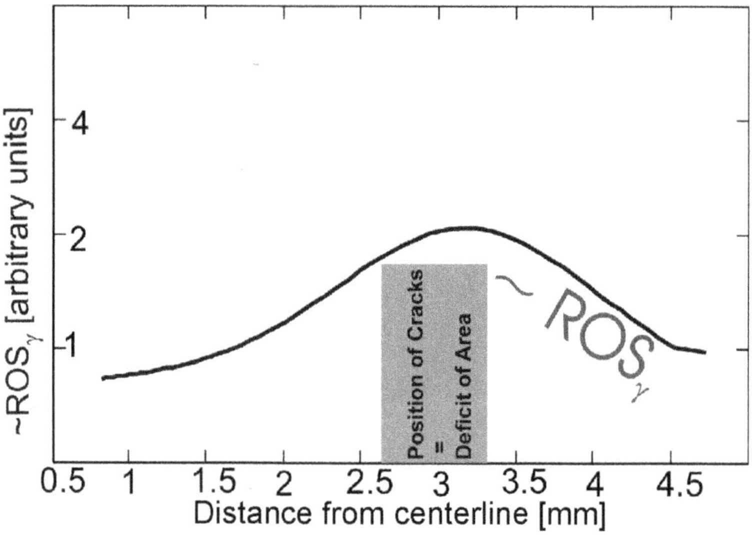

Fig. 11. Position-dependent area shrinkage per time ~ ROS referring to Fig. 5b

The ROS function shows a peak at the position where the solidification front exhibits a high curvature and, moreover, this position corresponds to a section of increased solidification cracking. Referring to the ROF-ROS model it can thus be concluded that in particular at positions of high curvature the probability of ROF < ROS is increased. As a first approximation of the ROF, the position-dependent function of the opening length of the cells representing the geometric obstacle for the melt flow was calculated. This function increases monotonically from the centerline to the fusion line, but a quantitatively critical intersection to the ROS function can so far not be deduced.

From these calculations it can be concluded that at highly bent solidification fronts an increased rate of shrinkage occurs requiring an increased ROF. Since the ROF can be assumed to be only slightly linearly increasing along the solidification front, the probability of solidification cracking must be increased at positions of high curvature.

Numerical Simulation of the MVT-Tests

Solidification cracks generated in an externally loaded hot cracking test usually result from a superimposition of local thermomechanical and global mechanical strains and strain rates, respectively. For a correlation of the position-, time- and direction-dependent material displacements leading to solidification crack formation, numerical simulations are most appropriate. With the help of computer technology, non-linear material properties, heat source characteristics and experimental boundary conditions, for example, can be considered. The software chosen for the numerical analysis of the MVT-Test was Ansys 6.0.

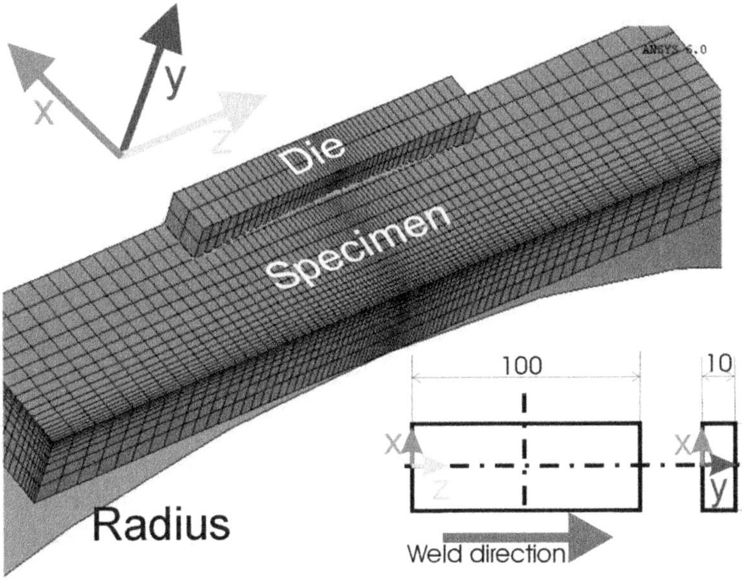

Fig. 12. MVT model of the relevant components: specimen, radius and die

Fig. 12 shows the discretized model for the numerical simulation of the MVT-Test in Varestraint mode. In order to calibrate and verify the material properties, the numerical model, the experimental boundary conditions and the modeled heat transfer between arc and specimen the temperature regimes at defined positions were measured by means of thermocouples for different conditions and compared to the results of the numerical simulations.

Fig. 13 represents such a comparison, in which the temperature regimes at four positions at the surface of a MVT specimen (Alloy 602 CA) were simultaneously measured during a TIG weld run without any mechanical loading. By an iterative variation of the energy efficiency used in the numerical simulations the computed curves could be adapted to the measured curves by which an average energy efficiency of $\eta = 0.66$ was determined for the TIG-weld process of the MVT test.

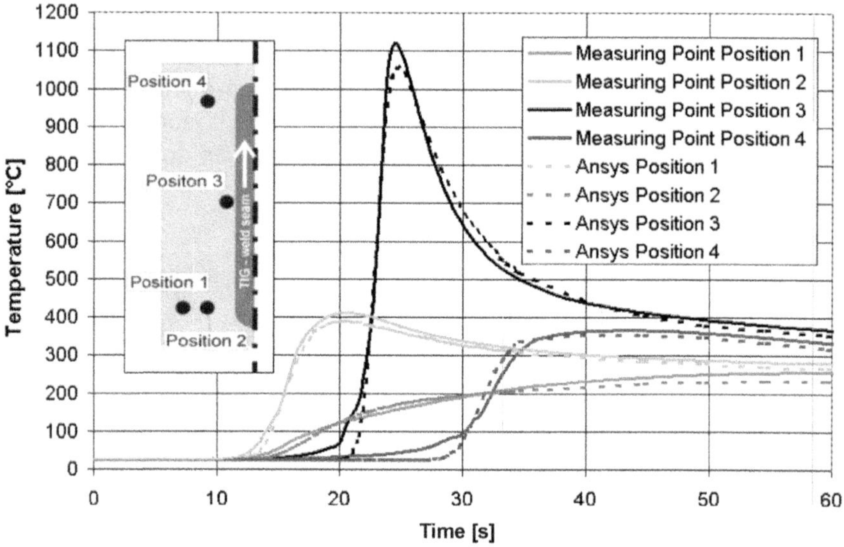

Fig. 13. Temperature regimes at the specimen surface compared to the numerical simulation by means of Ansys; Alloy 602 CA; I = 205 A; U = 11.5 V; v_{weld} = 3 mm/s; $\eta = 0.66$

It should be noted in this connection that the energy efficiency refers directly to the welding process and that the energy loss of the welding apparatus is not considered.

Fig. 14 depicts another comparison between experimental and simulated temperatures. The thermocouples were placed in a row 5 mm below the surface in the symmetry line parallel to the welding direction. The temperature regimes were recorded during a real MVT-Test in order to take account of the increased heat dissipation taking place via the dies during bending. For quantifying the influence of the heat flow via the dies, numerical simulations were performed with and without consideration of heat dissipation via the contact area.

The results show that for Alloy 602 CA and a specimens thickness of 10 mm significant deviation of the simulated temperature regimes occurs only at a temperature below 700 °C. Since solidification cracking occurs at much higher temperatures, the heat dissipation via the dies can be neglected for the numerical simulation and has no direct influence on the MVT-Test results from standard size specimens.

The simulation of the time-dependent temperature distribution represents the basis for subsequent thermomechanical simulations in which the strains and strain rates in the vicinity of the weld pool can be computed.

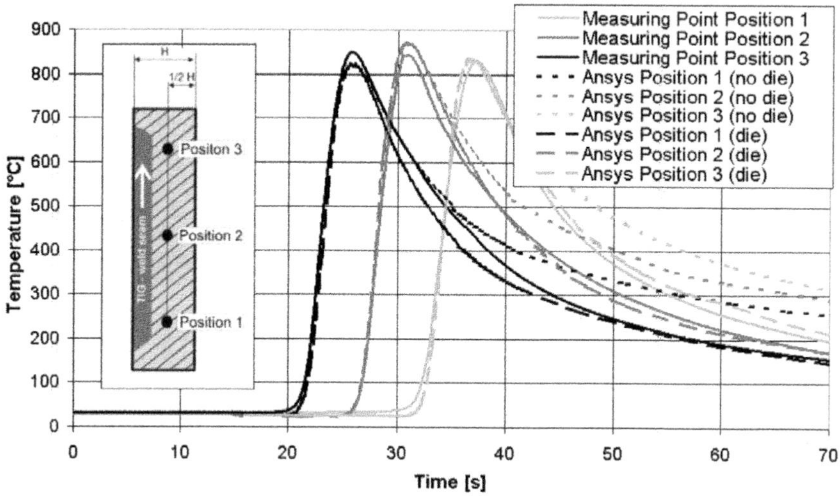

Fig. 14. Temperature regimes 5 mm below the surface; with / without heat dissipation via the dies; Alloy 602 CA; I = 194 A; U = 10.8 V; v_{weld} = 3 mm/s; $\eta = 0.66$

For the numerical simulation of the MVT-Test, the test procedure is divided into three steps: In the first step, the TIG bead-on-pate welding is simulated from the edge of the specimen to the middle without any external loading. In the second step, for which the load stepping must be well adjusted in the simulation, the specimen is bent while the TIG welding continues.

Bending is finished within 1.5 to 3.5 seconds, depending on the respective radius. In a third step, the heat source proceeds until the final position is reached without any further external loading. In order to reduce the computation time, the third step, which is irrelevant to solidification cracking, is omitted in the numerical simulations.

Fig. 15a–b shows the strain distribution in x- and z-direction at the specimen surface before bending. The weld pool is thermomechanically represented by a moving void generated by dynamical "kill" and "rebirth" of elements. The illustrations show that the weld pool is flanked by an area of compressive strains caused by the heating of the respective material zones. The weld metal exhibits compressive strains, too, which correlated with the appearance of the TIG-weld after an MVT-Test, which was slightly overcut.

Fig. 15c–d shows the strain distribution in the vicinity of the weld pool directly after the bending process in x- and z-direction.

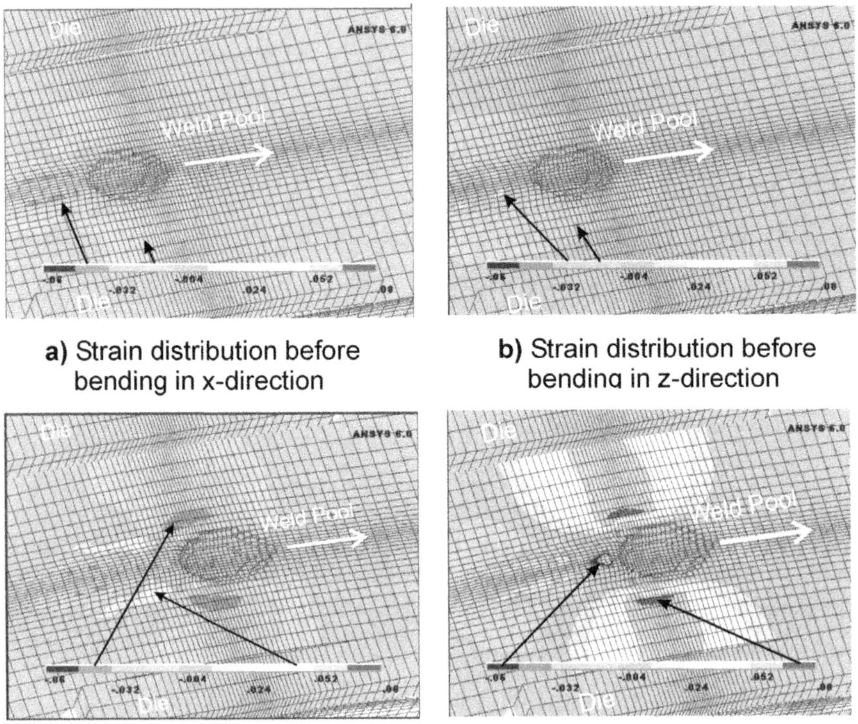

a) Strain distribution before bending in x-direction

b) Strain distribution before bending in z-direction

c) Strain distribution after bending in x-direction

d) Strain distribution after bending in z-direction

Fig. 15. Simulated transient strain distribution on the specimens surface; material properties according to Alloy 602 CA; I = 225 A; U = 12.5 V; v_{weld} = 1.8 mm/s; η = 0.66; radius 125 mm (total strain 4 %); Varestraint mode

The strain distribution becomes highly inhomogeneous by the mechanical loading compared to that before. For solidification cracking close to the fusion line, the strain distribution in z-direction is relevant representing the strain direction transverse to a potential solidification crack.

The strain distribution in z-direction reveals that the maximal strain of 7.2 % occurs not directly at the solidification front but in a distance of 2.8 mm. Obviously, the global tensile strain by external loading is superimposed by compressive strains present in the direct vicinity of the weld pool leading to a slight decrease of the total strain. According to the numerical simulation, the area close to the centerline directly behind the weld pool experiences compressive strains in z-direction but compressions in z-direction are not relevant for solidification crack formation at this position and might be correlated with the transverse contraction of the specimen by bending in longitudinal direction.

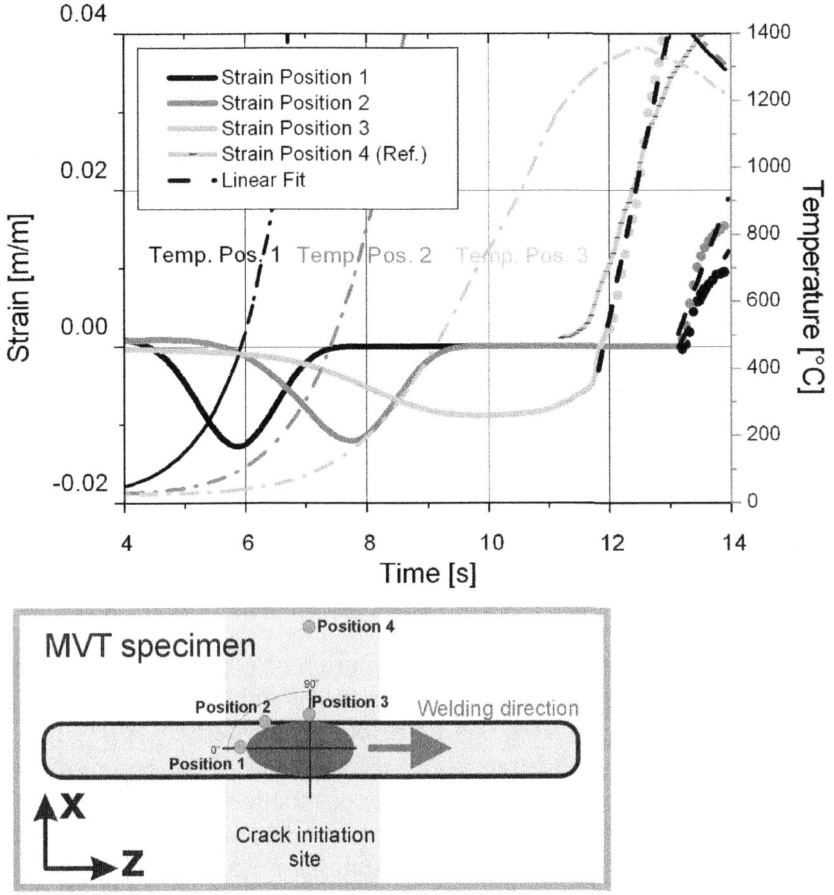

Fig. 16. Position-dependent strain increase during bending at 3 positions correlating to the weld pool geometry and a reference position at the specimens edge

Fig. 16 shows time-dependent strain curves and the corresponding temperature regimes at different positions on the specimen surface located at the crack initiation site which is the middle zone of the MVT specimen. The positions were chosen according to the weld pool geometry based on the respective thermal simulation, so that the weld pool runs over position 1 and position 2, whereby position 3 is only touched.

The strain functions refer to a strain direction transverse to a potential crack, which means, that the function of position 1 represents the strain in x-direction, positions 2 is a vectorial sum of the respective strains in z and x-direction and position 3 represents solely the strain in z-direction. When the weld pool approaches towards these positions, the material is firstly under compressive strains. Then the weld pool runs over position 1–2, the material becomes liquid, and no strain or stresses emerge. About two seconds before the material becomes solid again at position 1 and position 2, the mechanical bending process begins which can be seen by the strain increase at position 3 and position 4. The latter one is located nearby the upper edge of the specimen representing the function of the mere mechanical strain increase since nearly no thermal influences occur at this position.

From Fig. 16 following conclusion can be drawn: At position 1 the strain rate (0.015 /s) as well as the total strain after the bending process (0.009 m/m) is lower compared to position 2 at which a strain rate of 0.021 /s and a total strain of 0.015 m/m occurs. At position 3 the total strain accumulates most (0.068 m/m) and also the strain rate is highest (0.038 /s) since the material is not melted and the relevant strain direction is also the mechanical loading direction. The total strain at position 3 and the strain rate at position 2 and 3 are higher than the total mechanical strain (0.041 m/m) and the strain rate (0.018 /s) at position 4 since the material is exposed to high temperatures in the vicinity of the weld pool leading to a locally decrease of the mechanical strength.With respect to the Varestraint mode of the MVT-Test and under the condition of a steady weld pool geometry as shown in Fig. 15 it follows that strain and strain rate transverse to a potential crack are lowest but not zero at the centerline. At the fusion line, strain and strain rates transverse to a potential crack are absolutely highest, which has been verified by further calculation of strains and strain rates transverse to a potential crack between the fusion line and the centerline along the solidification front. The assumption, that the normalized position-dependent strain rate function can approximately be calculated by the geometry of the weld pool and by the loading direction, as shown above, can thus be confirmed by the results of the numerical simulations, although the strain rate is not zero at the centerline.

Regarding the transfer of experimental results to component welds, absolute values of critical strain and strain rate in the vicinity of a weld pool are required, which can most precisely be obtained by numerical simulations. Such values are given depending on the position along the weld pool geometry. Since the influence of the melt flow on the weld pool shape has not been considered in the numerical simulation, so far and the high-temperature material properties can only be assumed, a non-restricted transfer of the numerical results is still difficult. However, the weld pool shape of the presented numerical simulation is best comparable to the weld pool shape shown in Fig.7a. Since solidification cracks were preferably found at position 2 and position 3, the critical strain rate can be approximated between 0.021 /s and 0.038 /s.

Conclusions and Perspectives

The geometric analysis of MVT-results confirmed that there has to be expected significant dependency of the solidification crack formation on the geometry of the weld pool. The solidification crack position has been correlated with the local curvature of the solidification front and with the local strain rates.

By means of the end crater functions, which represent the solidification front during welding at the surface, theoretical solidification paths can be calculated which are the basis for a geometric ROFROS-model. It can be demonstrated that high curvatures of the solidification front cause a high rate of shrinkage. This means that a high probability of crack formation can be predicted at these positions, verified by MVT-results. By means of FEM, the strains in the vicinity of the weld pool during MVT-Testing in Varestraint mode have been analyzed. The results confirm the assumption of a monotonically increased strain and strain rate distribution around the weld pool during bending, but highest strains related to the MVT specimen do not occur directly at the solidification front.

The correlation of solidification cracking and weld pool geometry can especially be used for hybrid welding technologies, where by the specific geometric arrangement of at least two heat sources the weld pool shape as well as the thermomechanical strain and strain rates around the weld pool can be varied within broad limits. Such optimization tasks can be performed most conveniently by numerical simulations for which an appropriate solidification cracking criterion is a prerequisite. Therefore, further investigations are being pursued with a view to determining a solidification crack criterion especially matched to the requirements of numerical simulations.

References

1. Schobbert H, Böllinghaus Th, Wolf M (2003) Hot cracking resistance of laser and hybrid welded austenitic stainless steels. Trends in Welding Research 6: 76–81

2. Prokhorov N (1962) The technological strength of metals while crystallizing during welding. Welding Production 9(4): 1–8
3. Fuerer U (1977) In: Proc Int Sym Eng Alloys, Delft, The Netherlands, p 131
4. Matsuda F, Hashimoto T, Senda (1969) Fundamental investigation on solidification structure in weld metal. Trans Nat Res Inst Metals 11: 43–58
5. Herold H, Streitenberger M, Pechnnikov A (2001) Prevention of centreline solidification cracking during one-side welding. IIW-Doc IX-2000-01
6. Rappaz M, Drezet J M, Gremaud M (1999) A New Hot-Tearing Criterion. Met Mater Trans 30A (2): 449–455
7. Rindler W et. al (2002) A Modified Hot Tearing Criterion for Steels. In: Mathematical Modelling of Weld Phenomena 6, pp 819–835
8. Kannengiesser Th, McInerney T, Florian W, Böllinghaus Th, Cross C E (2002) The Influence of Local Weld Deformation on Hot Cracking Susceptibility. In: Mathematical Modelling of Weld Phenomena 6: 803–818
9. Olson DL, Edwards R (1998) Interfacially driven mass transport in joining and coating technologies. In: Marangoni and Interfacial Phenomena in Materials Processing (Book 692), The Royal Society, Cambridge, pp 113–126
10. prCEN ISO/TR 17641-3 Destructive tests on welds in metallic materials – Hot cracking tests for weldments – Part 3 – Externally loaded tests11. Savage W, Lundin C (1965) The Varestraint Test. Weld Res Suppl 10: 433–442
12. Mc Keown D (1970) Versatile Weld Metal Cracking Test. Met Constr 8: 351–352
13. Lippold J.C, Lin, W (1996) Weldability of commercial Al-Cu-Li alloys. In: Proc of ICAA5, Transtec Publications, pp 1685–1690
14. Wilken K, Kleistner H (1982) The MVT-Test – A novel procedure for testing of the hot cracking susceptibility during welding (in German). Material und Technik 1: 3–10

IV Testing and Standardization

Recent Developments in Weldability Testing

J.C. Lippold

Welding Engineering Program,
The Ohio State University, USA

Abstract

The term "weldability" can be used to describe a wide variety of material characteristics when a material is subjected to welding. These may include the physical and mechanical properties of the material, the ease with which welding can be accomplished from a practitioner's standpoint, or the ability of the material to avoid metallurgical degradation, usually assessed by its susceptibility to cracking during welding or subsequent heat treatment. A number of weldability tests have been developed over the years to evaluate and quantify material weldability. Many of these test techniques have focused on the phenomenon known as "hot cracking". This paper will review the basic concepts associated with hot cracking and other forms of elevated temperature cracking associated with welds and describe some recent advances in the use of testing approaches to quantify susceptibility to these forms of cracking. This description will include the use of the Varestraint test, the cast pin tear test, and the Gleeble[1] thermo-mechanical simulator for quantifying cracking susceptibility and providing comparative measures of weldability among alloys.

Introduction

The cracking of welded construction during fabrication has been a problem since the first welding processes were first widely adopted in the early 20th century. Cracking occurs for two reasons, 1) the presence of tensile stress, and 2) a susceptible microstructure in the weld metal or heat-affected zone.

---

[1] Gleeble is a registered trademark of DSI. Inc, Troy NY, USA.

Since the elimination or control of stresses during welding is usually quite difficult, a better approach is often control of the weldment microstructure.

The various forms of cracking are generally grouped by the temperature range over which they occur. "Hot cracking" is associated with the presence of liquid films along grain boundaries or elsewhere in the structure and includes weld solidification cracking, HAZ liquation cracking, and weld metal liquation cracking. "Warm cracking" occurs in the solid state at temperatures between the solidus and approximately half the melting temperature of the material and may occur either during fabrication or subsequent postweld heat treatment. Various forms of warm cracking include ductility-dip cracking, reheat cracking, strain-age cracking, and lamellar cracking. Finally, "cold cracking" occurs at or near room temperature and is usually associated with the presence of hydrogen and hydrogen-assisted cracking mechanisms.

Over the past 50 years, numerous studies of weld cracking have been published and various theories proposed to describe the cracking mechanisms. In addition, a number of test techniques (well over 200) have been developed to study and quantify cracking susceptibility.

In this paper, four types of weld cracking are reviewed and test procedures described to quantify susceptibility, namely, 1) weld solidification cracking, 2) HAZ liquation cracking, 3) ductility-dip cracking, and 4) strain-age cracking. All three of these cracking phenomena have been observed in austenitic stainless steels and Ni-base alloys, and often plague the fabrication of these materials.

Weld Solidification Cracking

Weld solidification cracking occurs during the final stages of solidification when tensile shrinkage stress accumulates and liquid films still persist along solidification grain boundaries in the structure. If the imposed shrinkage strain exceeds the inherent ductility of the solidifying weld metal, cracking will occur.

The temperature range over which this occurs has been defined by Prokhorov [1], Matsuda [2], and others as the Brittle Temperature Range (BTR), where the ductility of the weld metal shows a minimum value where weld solidification cracking is possible.

As a general rule, the wider the BTR, the more susceptible the material is to solidification cracking. Thus, the ability to measure the BTR during welding should provide some approximation of a material's susceptibility to weld solidification cracking.

The Transverse Varestraint Test

The Varestraint test, in a variety of forms, has been used since the 1960's to quantify susceptibility to "hot cracking", i.e. weld solidification and liquation cracking. A variety of methods have been used for quantification, including total crack length, maximum crack length, total number of cracks, threshold strain for cracking, and others. Recently, Lin et al. [3] developed a new methodology for evaluating cracking susceptibility using the transverse Varestraint test that provides a measure of the temperature range over which cracking occurs. This has been termed the solidification cracking temperature range (SCTR) and is a subset of the BTR. Samples are tested over a range of augmented strain and the maximum crack distance (MCD) in the fusion zone is measured. A schematic of the transverse Varestraint test is shown in Fig. 1.

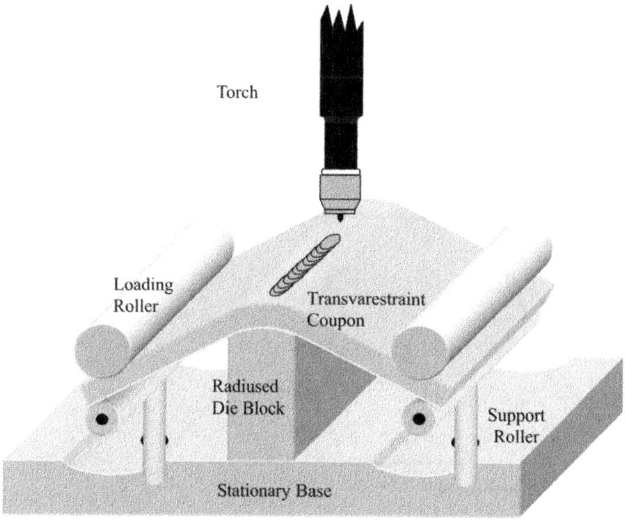

Fig. 1. Schematic of the transverse Varestraint test for evaluating weld solidification cracking susceptibility

Above a critical strain level, designated the saturated strain, the MCD does not increase with increasing strain. This indicates that the solidification crack has propagated the full length of the crack susceptible region. Representative cracking as it appears at the trailing edge of the weld pool is shown in Fig. 2 for a Type 310 sample tested at 5 % strain. By testing over a range of augmented strain, an MCD versus strain plot such as that shown in Fig. 3 can be generated.

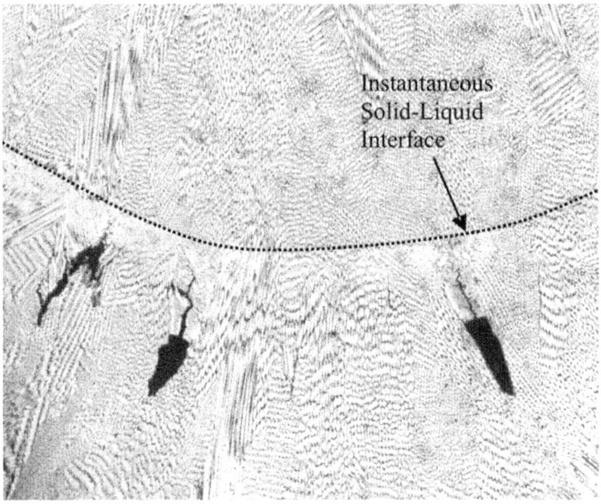

Fig. 2. Weld solidification cracking in a Type 310 stainless steel transverse Varestraint specimen

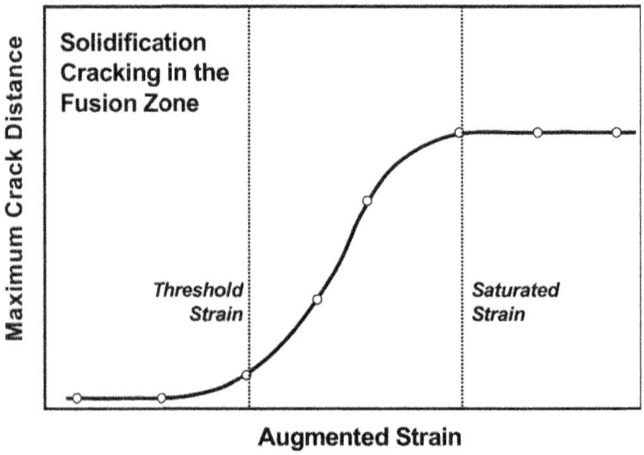

Fig. 3. Maximum crack distance in the fusion zone versus applied strain during transverse Varestraint testing

In this manner, the threshold strain for cracking to occur and the saturated strain above which the MCD does not increase can be identified. Typical strain ranges over which samples are tested are 0–7 %. Most fully austenitic weld metals stainless steels and Ni-base alloys exhibit saturated strain levels between 5 and 7 %. Threshold strain levels are generally in the range from 0.5 to 2.0 %.

Although, the threshold strain may, in fact, be an important criterion for judging susceptibility to weld solidification cracking, the MCD at or above saturated strain is much easier to determine and provides a measure of the SCTR.

In order to determine SCTR, the cooling rate through the solidification temperature range is determined by plunging a thermocouple into the weld pool. The time over which cracking occurs is approximated by the MCD above saturated strain divided by the solidification velocity. Using this approach, SCTR can be calculated using the following relationship, where V represents the welding velocity.

$$SCTR = [Cooling\ Rate] \times [MCD/V]$$

The concept for determining SCTR using this approach is shown in Fig. 4. By using a temperature rather than a crack length as a measure of cracking susceptibility, the influence of welding variables (heat input, travel speed, etc.) can be eliminated. SCTR then represents a metallurgically significant, material-specific measure of weld solidification cracking susceptibility.

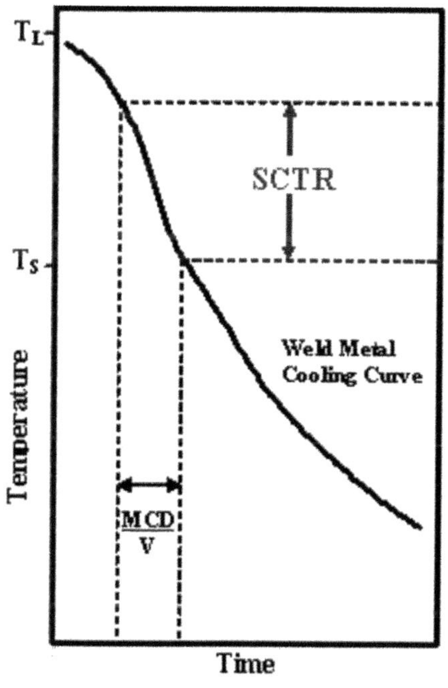

Fig. 4. Method for determining the solidification cracking temperature range (SCTR) using the cooling rate through the solidification temperature range and MCD at saturated strain

A bar chart showing the SCTR values for a number of austenitic and duplex stainless steels is shown in Fig. 5. Alloys that solidify as primary ferrite (duplex stainless steels 2205 and 2507, and Types 304 and 316L) have low SCTR values, typically less than 50 °C. Alloys that solidify as austenite exhibit SCTR values above 100 °C. Alloy A-286, which is notoriously susceptible to solidification and liquation cracking, has a very high SCTR value.

The SCTR data allows a straightforward comparison of cracking susceptibility. These values may also allow alloy selection based on restraint conditions. For example, in high restraint situations, SCTR values below 50 °C may be required to prevent cracking, while for low restraint weldments 150 °C may be sufficient.

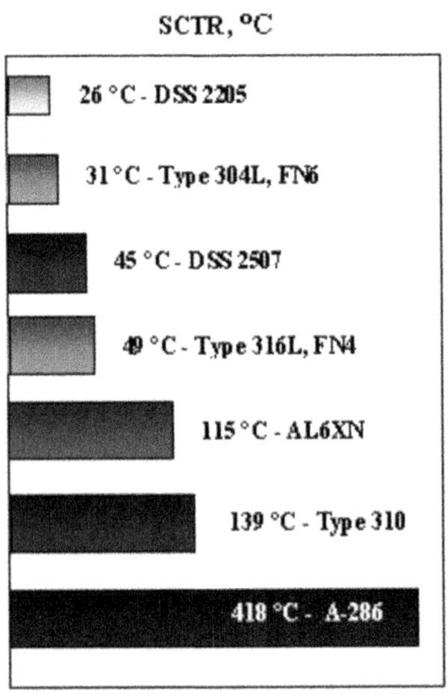

Fig. 5. SCTR values for a number of austenitic and duplex stainless steels

Recently, Finton and Lippold [4] used a statistical approach to evaluate the variables associated with transverse Varestraint testing. This study used both austenitic stainless steels (Type 304 and 310) and Ni-base alloys (Alloys 625 and 690) to determine the statistical importance of different variables and to establish variable ranges in which testing should be conducted to give reproducible results. Based on this study, they recommended the variable ranges in Table 1 for use with stainless steel and Ni-base alloys.

These ranges are not necessarily appropriate for other alloy systems. More research is required to relate the SCTR to the local restraint conditions required for cracking. As a minimum, the overall transverse Varestraint approach described here appears to provide a good relative measure of solidification cracking susceptibility and has been found to work well in predicting the behavior in other systems, including structural steels and aluminum alloys.

Table 1. Variables and variable ranges for transverse Varestraint testing of stainless steels and Ni-base alloys

Arc Length Range:	0.05–0.15 in.
Maximum Voltage Changes:	± 1–1.5 volts
Minimum Specimen Length:	3.5 in.
Minimum Specimen Width (parallel to welding direction):	3.0 in.
Current Range:	160–190 amps
Travel Speed Range:	4–6 in./min
Arc Length Range:	0.05–0.15 in.
Maximum Voltage Changes:	± 1–1.5 volts

The Cast Pin Tear Test

Although the Varestraint test has great utility for assessing the solidification cracking susceptibility of most structural alloys, the test may be too severe for evaluating some of the highly-alloyed Ni-and Co-base alloys used for repair of turbine engine components, since these alloys may crack at very low strain levels. For these materials, a modified version of the cast pin tear (CPT) test, originally introduced by Hull [5], can be used. Using this test, small charges of the material of interest are melted in a copper crucible using a gas tungsten arc welding (GTAW) torch under argon shielding. This charge is then dropped through the bottom of the crucible into a copper mold. A range of mold diameters and lengths are used to control the restraint in the solidifying pin. This procedure and apparatus design is described in detail elsewhere [6].

Using the CPT test, a plot of percent cracking versus mold size is developed, as shown in Fig. 6. 100 % cracking represents the situation where cracking occurs completely around the diameter of the pin or there is complete separation of the pin. Note that for Alloys 3 and 5 only small increases in mold length result in large changes in cracking susceptibility, while Alloy 1 and Alloy 625 are resistant to cracking until relatively long mold lengths are used.

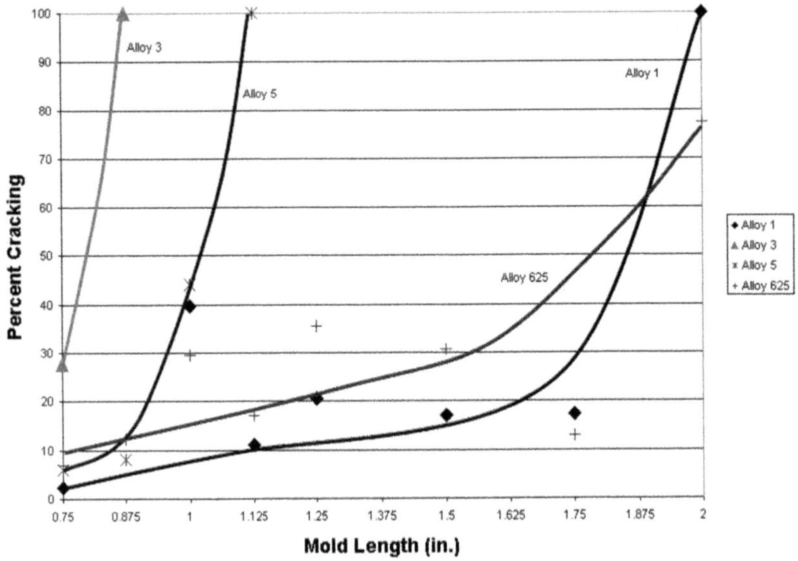

Fig. 6. Cast pin tear test data for Ni- and Co-base alloys. Alloy 1 is Co-20Cr-10Ni-W-4Al, Alloy 3 is Ni-10Co-8Cr-10W-5Al, and Alloy 5 is Ni-12Co-7Cr-5W-6Al

Other distinct advantages of this test are that virtually no sample preparation is required and very little material is used. An entire curve, such as those shown in Fig. 6 can be generated with about 200 grams of material. Additionally, testing is not time intensive. For a given material, testing and analysis can be completed in just a few hours. The cooling rates achieved, based on evaluation of solidification substructure size, are equivalent to those in arc welds.

The test is under further development and refinement at Ohio State University. An improved molten metal delivery system is being tested that will greatly facilitate mold filling and allow a wider range of mold geometries to be used. This should allow alloys with only moderate solidification cracking susceptibility to be tested.

Heat-affected Zone Liquation Cracking

Heat-affected zone liquation cracking is associated with the partially-melted zone (PMZ) that forms in the HAZ just adjacent to the fusion boundary. This form of cracking occurs along grain boundaries that liquate at elevated temperature due to a variety of mechanisms.

In many alloys, this liquation occurs simply due to the segregation of impurity and/or other melting point depressant elements to the grain boundary during the HAZ thermal cycle. If the concentration of these elements and the time above the boundary melting temperature are sufficient, local boundary melting will occur.

Another mechanism for PMZ formation is known as constitutional liquation [7]. Constitutional liquation occurs when a "constituent" particle, such as a carbide, reacts with the surrounding matrix resulting in local melting at the particle/matrix interface. The particle itself does not melt, rather the reaction zone at the interface undergoes eutectic melting. If a grain boundary then intersects this liquid region, the boundary can be wet by the liquid and becomes susceptible to liquation cracking. These mechanisms have been described in some detail elsewhere [7]. Susceptibility to HAZ liquation cracking can be quantified using both the Varestraint test and the hot ductility test, as described by Lin et al. [8].

The Spot Varestraint Test

The Varestraint test technique for HAZ liquation cracking differs from that described previously for weld solidification cracking, since it uses a stationary spot weld to generate a stable HAZ thermal gradient and microstructure.

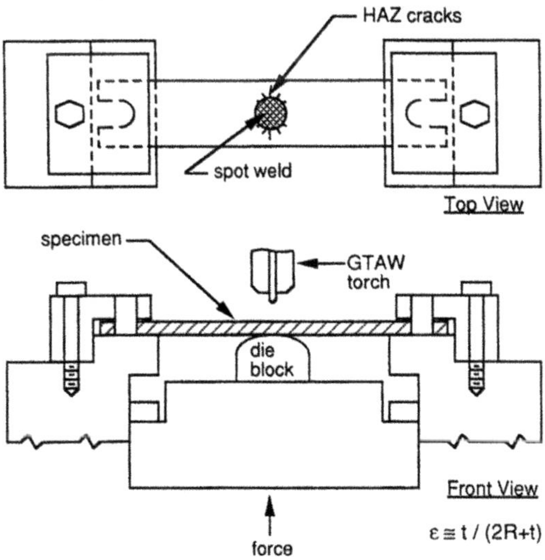

Fig. 7. Schematic illustration of the spot Varestraint test [5]

The technique developed by Lin [9] used both an "on-heating" and "on-cooling" approach to quantify HAZ liquation cracking. A schematic of the test is shown in Fig. 7.

The on-heating test is conducted by initiating the GTA spot weld, ramping to the desired current level, and then maintaining this current level until the weld pool size stabilizes and the desired temperature gradient in the HAZ is achieved. For the Type 310 and A-286 stainless steels evaluated by Lin et al. a spot weld surface diameter of approximately 12 mm was achieved after a weld time of 35 seconds. The arc is then extinguished and the load immediately applied to force the sample to conform to the die block. By using no delay time between extinction of the arc and application of load, HAZ liquation cracks initiate at the fusion boundary and propagate back into the HAZ along liquated grain boundaries (Fig. 8).

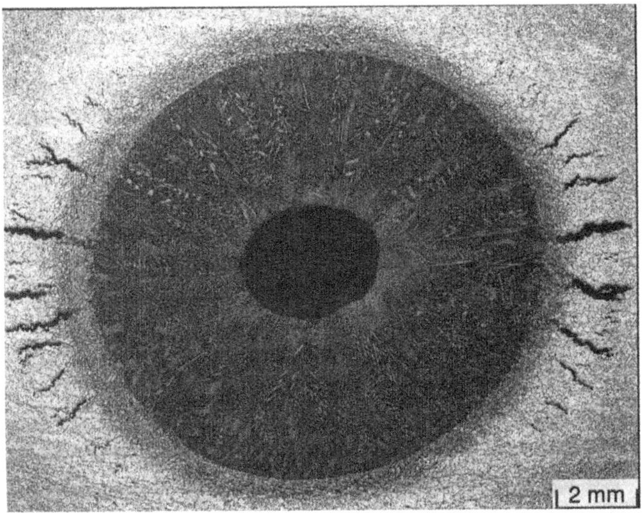

Fig. 8. Spot Varestraint test sample (plan view) of A-286 stainless steel tested at 5 %strain

By plotting maximum crack length (MCL) versus strain, a "saturated strain" can be determined which defines the strain above which the maximum crack length does not change. Examples of on-heating MCL versus strain plots for Type 310 and A-286 stainless steels are shown in Fig. 9A. Note that a threshold strain for cracking cannot be identified for A-286 and that both materials achieve saturated strain at 3 %.

For the on-cooling test, the same procedure as described above is used, but after the arc is extinguished there is a delay before the sample is bent.

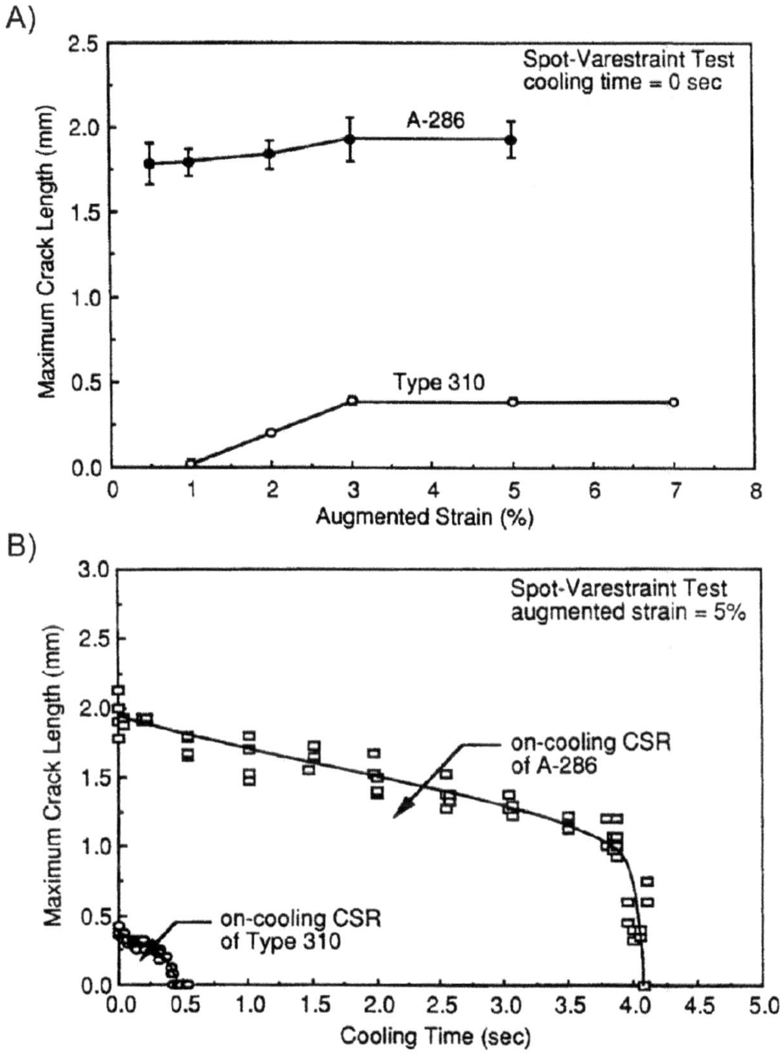

Fig. 9. On-heating **(A)** and on-cooling **(B)** HAZ liquation cracking as measured by the spot Varestraint test [5]

By controlling the delay, or cooling, time, the weld is allowed to solidify and the temperature in the PMZ drops until eventually the liquid films along grain boundaries are completely solidified. By plotting MCL versus cooling time, the time required for liquid films in the PMZ to solidify can be determined. This is shown in Fig. 9B.

Note that over 4 seconds is required before cracking disappears in A-286, indicating that the grain boundary liquid films persist to quite low temperatures. By measuring the temperature gradient in the HAZ using implanted thermocouples, it is possible to determine the thermal crack susceptible region (CSR) surrounding the weld within which HAZ liquation cracking is possible. As described by Lin [9] this is done by converting the MCL at saturated strain for both on-heating and on-cooling spot Varestraint tests to a temperature by multiplying the MCL by the temperature gradient in the HAZ. Using this approach, it is possible to describe the region around a moving weld pool that is susceptible to HAZ liquation cracking.

Plots of the thermal CSR for A-286 and Type 310 are shown in Fig. 10. Note that the width of the CSR at the periphery of the weld is 222 °C for A-286 and only 61 °C for Type 310. It is also interesting that, based on the on-cooling data, the PMZ in A-286 does not fully solidify until the temperature reaches 1035 °C while for Type 310 it is 1296 °C. This technique provides a quantifiable method to determine the precise temperature ranges within which cracking occurs and to allows differences in HAZ liquation cracking susceptibility among materials to be readily measured.

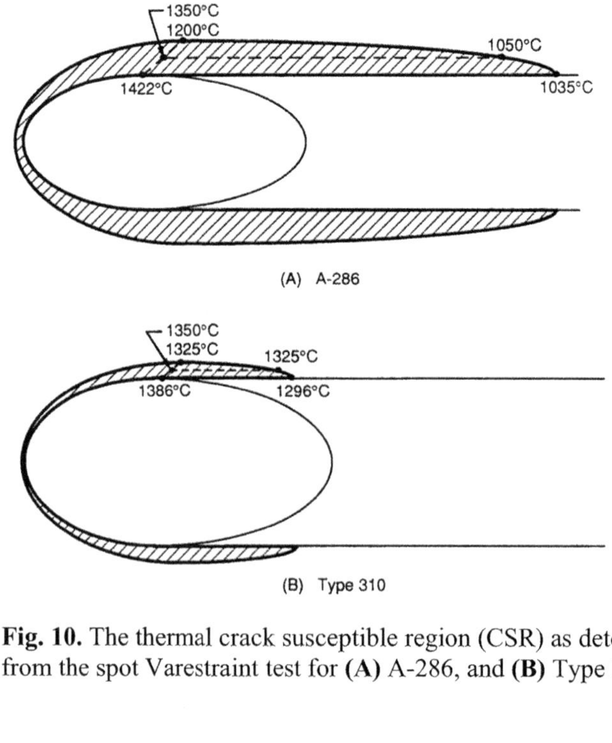

Fig. 10. The thermal crack susceptible region (CSR) as determined from the spot Varestraint test for **(A)** A-286, and **(B)** Type 310 [5]

Ductility-Dip Cracking

Ductility-dip cracking (DDC) refers to elevated temperature, solid-state cracking that results from a sharp drop in ductility at temperatures above approximately half the melting temperature of the material [10, 11]. It can occur in wrought alloys, castings, and in the HAZ and fusion zone of highly restrained weldments. Characteristically, it is associated with single phase austenitic alloys with large grain size, and is intergranular in nature.

Considerable work has been conducted over the last few years to understand the nature of DDC in welded austenitic stainless steels and Ni-base alloys [12, 13, 14, 15]. This work has included investigation of austenitic stainless steels (Types 304 and 310, and AL6XN), Ni-base alloy 690, and Ni-base filler metals 82 and 52.

A typical ductility-dip crack in a Ni-base weld metal is shown in Fig. 11.

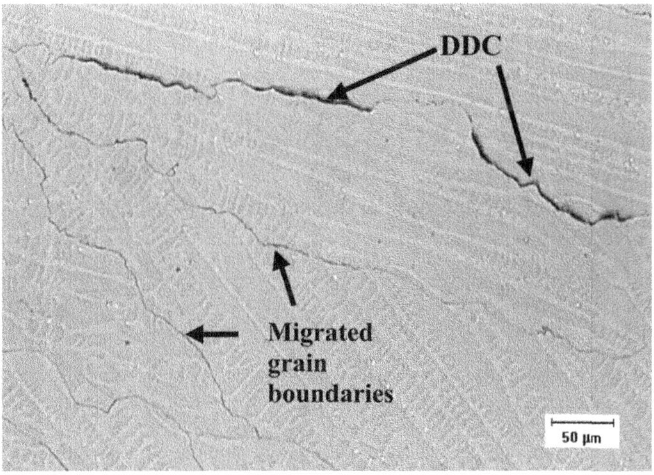

Fig. 11. Ductility-dip cracking along migrated grain boundaries in fully austenitic weld metal

In weld metals, DDC always occurs along migrated grain boundaries (MGBs). These are crystallographic, high-angle boundaries that have migrated away from their parent solidification grain boundaries during cooling below the solidification temperature range and/or during reheating in multipass welds. A detailed description of these boundaries can be found elsewhere [16].

Weld metal DDC in stainless steels and Ni-base alloys has been found to be a strong function of grain size, grain boundary character, and precipitation behavior. Weld metals exhibiting large grains with straight MGBs and few grain boundary precipitates tend to be the most susceptible. An increase in grain boundary tortuosity resulting from local pinning by precipitates or second phases that form at elevated temperature will decrease susceptibility to DDC.

This occurs by a grain boundary locking effect that resists grain boundary sliding. While not directly linked to DDC, impurity segregation to the MGBs tends to further increase susceptibility to this form of cracking. In Ni-base filler metals, the addition of hydrogen to the shielding gas increases susceptibility to DDC [14].

In order to quantify susceptibility to DDC, the strain-to-fracture (STF) test was recently developed by Nissley at Ohio State University [12]. The STF test employs a "dogbone" tensile sample with a GTA spot weld applied in the center of the gage section. The spot weld is made under controlled solidification conditions using current downslope control.

This results in an essentially radial array of migrated grain boundaries within the spot weld. Samples are then tested in a Gleeble™ thermomechanical simulator at different temperatures and strains. Temperature and strain ranges are typically 650–1200 °C and 0–20 %, respectively. After testing at a specific temperature-strain combination, the sample is examined under a binocular microscope at 50X to determine if cracking has occurred. The number of cracks present on the surface is counted.

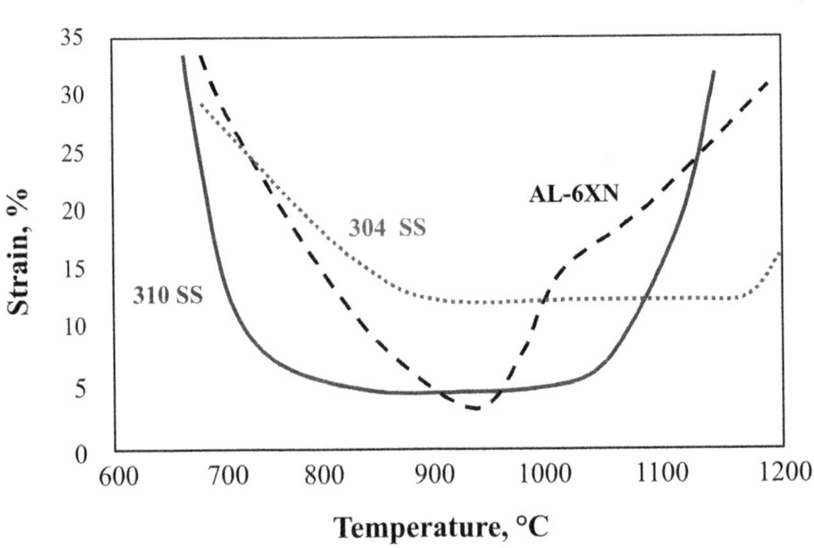

Fig. 12. Strain-to-fracture test results for three austenitic stainless steels [9]

Using this data, a temperature vs. strain envelope is developed that defines the regime within which DDC may occur. Both a threshold strain for cracking (ε_{min}) and ductility-dip temperature range (DTR) can be extracted from these curves. Temperature-strain curves are shown in Fig. 12 for Type 310, Type 304, and the super-austenitic alloy AL6XN. Based on these curves, Type 310 would be expected to have the highest susceptibility to DDC since the DTR at 15 % strain is 400 °C and ε_{min} is approximately 5 %.

This test has been shown to be remarkably sensitive to the onset of grain boundary cracking in the DDC range and should prove to be a valuable tool for studying elevated temperature embrittlement in the weld metal and HAZ. Work is ongoing at Ohio State University to further optimize the test. This test is also being using the study the composition and metallurgical variables that affect susceptibility to ductility-dip cracking.

Strain-age Cracking

Strain-age cracking (SAC) is a form of postweld heat treatment cracking that is associated with Ni-base superalloys. The term "strain-age" is derived from the fact that cracking occurs in the temperature range were extrinsic and intrinsic strain accumulation overlaps the onset of precipitation hardening, or aging. In most Ni-base superalloys, this form of cracking is closely related to the precipitation of gamma-prime, $Ni_3(Ti,Al)$. Alloys with higher Ti + Al contents tend to be more susceptible to SAC. [17]. Grain size and impurity content also influence susceptibility. Materials with much finer HAZ grain size or those with low levels of sulfur, phosphorus, and boron are more resistant to SAC. A relationship between SAC susceptibility and grain boundary liquation has also been reported [18].
SAC is usually associated with the HAZ of either wrought alloys or castings and occurs along grain boundaries in close proximity to the fusion boundary. A strain-age crack in Waspaloy and the corresponding fracture surface are shown in Fig. 13. From a mechanistic standpoint, cracking occurs along the grain boundary due to *intra*granular strengthening by precipitation and the corresponding formation of a precipitate-free zone at or near the grain boundary. Upon the application of sufficient strain, cracking occurs through this weakened region. There is an ongoing debate about the validity of this mechanism and further research is required to resolve this issue.

As with other forms of cracking, numerous tests have been developed to determine susceptibility to SAC. To date however, there is no standardized test for quantifying the susceptibility of an alloy to SAC.

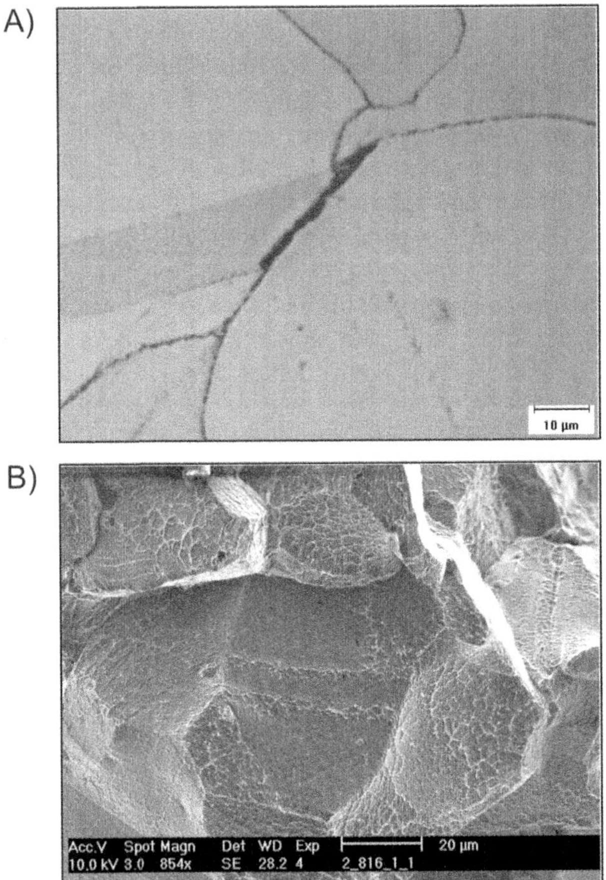

Fig. 13. Strain-age crack in Waspaloy, **(A)** optical micrograph, **(B)** SEM fractograph

Many of the test techniques that have been developed use a Gleeble™ thermo-mechanical simulator. The problem with these tests is that they do not accurately simulate the thermo-mechanical history of a weld. Most tests do not adequately simulate the development of residual stresses in a weldment as it cools after the weld metal is deposited. Those tests that impose stresses on cooling from the peak temperature to simulate weld residual stress do not allow relaxation of the stresses in subsequent PWHT simulations.

The approach used by Norton [19] at Ohio State University attempted to more closely simulate the actual conditions experienced by the HAZ during welding. The Gleeble was used to impose a simulated thermal cycle on the specimen. After reaching the peak temperature, the sample was restrained in the Gleeble jaws and allowed to cool to room temperature. This resulted in the buildup of considerable stress in the sample due to thermal contraction. The sample was then heated to an appropriate PWHT temperature and held at temperature for a predetermined time (0 to 4 hours).

Upon reheating to the PWHT temperature, considerable stress relaxation occurs and then stress begins to build in the sample as aging occurs, as shown in Fig. 14. The starting stress for both alloys has been subtracted so that that the stress buildup relative to each other can be shown. Note that the increase in stress is more rapid in Waspaloy due to the more rapid aging response associate with gamma-prime precipitates relative to gamma double prime in Alloy 718. After the prescribed hold time the sample is pulled to failure at the PWHT temperature and the ductility measured. The hot ductility following PWHT was used to develop a multivariate polynomial for calculating the ductility as a function of PWHT temperature and time.

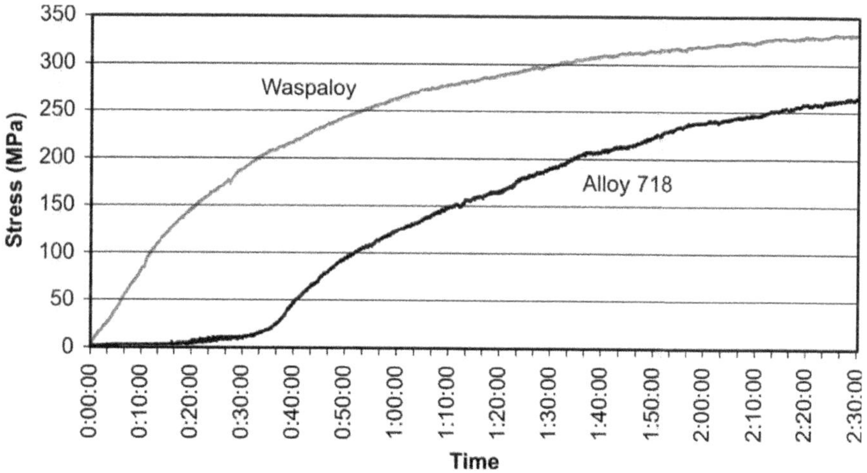

Fig. 14. Stress increase during hold at constant displacement during post-weld heat treatment cracking test. The initial stress for both alloys has been normalized to zero to allow for a better comparison of their hardening response

The collected data appeared to have a parabolic curve, so the model was chosen to be a second order polynomial. A spreadsheet for both alloys was created with factors of time, temperature, and the interactions between the two in a second order equation. A line was fit to the ductility (reduction in area) measurements. The resulting output gave the intercept and coefficient for each of the five variables as well as the coefficient of determination for the fit.

An example of ductility versus temperature curves for Waspaloy and Alloy 718 determined using this method is shown in Fig. 15. The coefficients of determination (R^2) for the Waspaloy and Alloy 718 surface plot polynomials are 0.92 and 0.91, respectively. The regression models show good fit to the measured data over the range of tested times and temperatures both by their high coefficients of determination and the ability to predict the ductility of samples. The curves in Fig. 7 are consistent with actual experience in welding these alloys. Alloy 718 is generally quite resistant to SAC, while Waspaloy is considered moderately susceptible.

Additional details of this test and its potential importance for determining susceptibility to PWHT cracking can be found elsewhere [19]. It should be noted that this test is not limited to the evaluation of Ni-base superalloys, but can also be applied to other materials that are susceptible to PWHT cracking, such as Cr-Mo-V steels and stainless steels (Type 347).

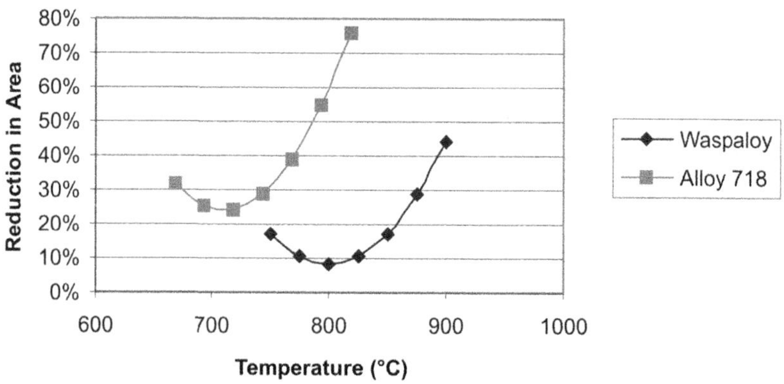

Fig. 15. Comparison of elevated temperature ductility of Waspaloy and Alloy 718 regression models for 3 hours of PWHT [19]

Summary

The weldability of engineering materials continues to be a topic of considerable interest and relevance in the manufacturing community. Despite the many studies that have been conducted and the hundreds of weldability tests that have been proposed, the quest to quantify weldability with the goal of avoiding (or at least predicting) susceptibility to weld cracking remains elusive. This elusiveness arises from the fact that most of the cracking mechanisms are still not well understood and there are relatively few standardized tests to measure weldability.

This paper has described tests for weld solidification cracking, HAZ liquation cracking, ductility-dip cracking, and postweld heat treatment cracking that have carefully considered the metallurgical variables that influence cracking in the development of the test techniques. These tests are useful not only for quantifying susceptibility, but for studying the fundamental mechanisms that promote weld cracking in engineering materials. While none of these tests are yet "officially" standardized, they each provide sufficient detail regarding sample preparation, testing procedure, and analysis that laboratory-to-laboratory variability in performing the tests should be minimized.

The goal to develop standardized weldability tests has been an elusive one and will probably not be realized until international cooperation and collaboration is achieved. It is hoped that this conference will be a first step in realizing that cooperation.

Acknowledgements

This paper represents the work of many students and colleagues over the last 15 years. I would particularly like to thank Dr. Wangen Lin for his insight and diligence in the development of the transverse and spot Varestraint test techniques, Mr. Daniel Ryan for developing the modified cast pin tear test, Mr. Nathan Nissley and Mr. Matthew Collins for the development and testing associated with the strain-to-fracture test, and Mr. Seth Norton for the evolution of the strain-age cracking test. It is my privilege to have had these individuals work in my research laboratory and I wish to acknowledge their "blood, sweat, and tears" that made these advancements possible. I would also like to recognize my research sponsors, including Edison Welding Institute, the American Welding Society, and BWXT, Inc. for providing the resources to support students and conduct the research presented here.

References

1. Prokhorov NN (1962) The technological strength of metals while crystallizing during welding. Welding Production 9 (4): 1–8
2. Matsuda F, Nakagawa H, Tomita S (1986) Quantitative evaluation of solidification brittleness of weld metal during solidification by in-situ observation and measurement (Rpt. 3). Transactions JWRI 15 (2)

3. Lippold JC, Lin W (1996) Weldability of commercial Al-Cu-Li alloys. In: Proc. of ICAA5, Aluminum Alloys – Their Physical and Mechanical Properties, eds.. Driver JH et al., Transtec Publications, pp 1685–1690
4. Finton T, Lippold JC (2004) Standardization of the Transvarestraint test, EWI Summary Report No 04–XX
5. Hull FC (1959) Cast-pin tear test for susceptibility to hot cracking. Welding Journal 38 (4): 176s–181s
6. Ryan D (2003) Development of a Modified Cast Pin Tear Test for Ni-base Alloys. Master's Thesis, The Ohio State University
7. Pepe JJ, and Savage WF (1967) Effects of Constitutional Liquation in 18 Ni Maraging Steel Weldments. Welding Journal, Vol. 46: 411s–422s
8. Lippold JC, Varol I, Baeslack WA (1992) An investigation of heat-affected zone liquation cracking in austenitic and duplex stainless steels. Welding Journal 71(1): 1s–14s
9. Lin W, Lippold JC, Baeslack WA (1993) An investigation of heat-affected zone liquation cracking, part 1 – a methodology for quantification. Welding Journal 71(4): 135s–153s
10. Hemsworth B, Boniszewski T, Eaton NF (1969) Classification and Definition of High Temperature Welding Cracks in Alloys. Metal Construction & British Welding Journal, February: 5–16
11. Rhines FN, Wray PJ (1961) Investigation of the Intermediate Temperature Ductility Minimum in Metals. Transactions of the ASM, Volume 54: 117–128
12. Nissley NE, Lippold JC (2003) Ductility-dip cracking susceptibility of austenitic alloys. 6th International Trends in Welding Research, ASM International, pp 64–69
13. Nissley NE, Guaytima G, Lippold JC (2002) Development of the strain-to-fracture test for evaluating ductility-dip cracking in austenitic alloys. IIW Doc #IX-2037-02
14. Collins MG, Lippold JC, Kikel JM (2003) Quantifying ductility-dip cracking susceptibility in nickel-base weld metals using the strain-to-fracture test. 6th International Trends in Welding Research, ASM International, pp 586–590
15. Kikel JM, Parker DM (1999) Ductility dip cracking susceptibility of Filler Metal 52 and Alloy 690. 5th International Trends in Welding Research, ASM International, pp 757–762
16. Lippold JC, Clark WAT, Tumuluru M (1992) An investigation of weld metal interfaces. In: The Metal Science of Joining. The Metals, Minerals and Materials Society, Warrendale, PA, pp 141–146
17. Prager M, Shira S (1968) Welding of precipitation-hardening nickel-base alloys. WRC Bulletin No 128, February
18. Duval DS, Owczarski WA (1969) Studies of postweld heat treatment cracking in nickel-base alloys. Welding Journal 48 (1): 10s–22s
19. Norton SJ, Lippold JC (2003) Development of a Gleeble-based test for postweld heat treatment cracking susceptibility. 6th International Trends in Welding Research, ASM International, pp 609–614

Hot Cracking Tests – The Route to International Standardization

J.C.M. Farrar

Introduction

Hot cracking in weld metals and parent material heat affected zones (HAZ) has been a problem for as long as steels and alloys have been fusion welded. In most codes and standards, detectable cracks of any type are not permitted and so efforts have been directed to the development of weld metals and parent alloys which are not sensitive to hot cracking and to the design of welding procedures which reduce or eliminate the risks of hot cracking.

A natural consequence of this has been the need for tests, which can be used to assess the sensitivity of weld metals and parent alloys, compare the sensitivity of different compositions, and predict possible occurrence in real fabrications.

Many tens of different tests have been developed over the years, some of which are specific to individual groups of materials, some to specific welding processes and some to individual pieces of test equipment at one or more laboratories. The complexities of the actual cracking mechanisms combined with the diversity of individual tests means that a form of standardization is difficult, and true international standardization is particularly challenging.

Nevertheless, considerable progress has been made in the last 20 years and this paper attempts to identify those issues, which need to be resolved, if an international standard is to be developed in the coming years. It concentrates on two specific standards. The first of these is the Russian Standard namely GOST 26389 – 84 [1], which was developed twenty years ago and has been widely used by countries of the former Soviet Union and by a number of countries in Eastern Europe which have strong technical links. The second is the draft International and European Union standard prEN/ISO 17641 [2], which has been developed within CEN TC121 Working Group 13, with considerable support and assistance from IIW sub commissions II-C and IX-H. A major review of the scientific basis of possible harmonization of Russian Federation and European Union Weldability Tests has been carried out at the Bauman University Moscow [3].

This paper also recognizes the major project carried out at BAM, Berlin and reported within Commissions II and IX of IIW. The final report, presented in 1999, not only described the results of a large worldwide 'Round Robin' test programme carried out by members of IIW but also highlighted the very real problems of test consistency and reproducibility between different laboratories [4].

Fundamentals of Hot Cracking and Definitions

Hot cracks arise in weld metals and in the HAZ of parent materials when the strains developed during cooling of welded joints, or imposed externally, exceed the ductility of a particular part of the joint. They range in size from very small liquation cracks (< 1 mm in length) in HAZ's or multipass welds up to large solidification cracks, which may extend along the complete length of a welded joint.

Hot cracks are not confined to any particular alloy system and can occur in steels, stainless steels, nickel-base alloys, copper and aluminium based alloys. The detailed reasons for the incidence of hot cracks are many and complex and fall outside the scope of this paper. In general terms, they occur when localized ductility is insufficient to support any imposed strains.

The lack of ductility can result from the presence of liquid metal, micro structural features, orientation (relative to the strain) and in some cases upon the presence of brittle impurities and low melting point (or liquated) films. In this respect some alloy systems are particularly sensitive to the presence of impurity elements such as sulphur, phosphorus and lead etc. It is widely recognized that fully austenitic single-phase microstructures, particularly weld metals, are susceptible to hot cracking of one form or another. Impurity levels strongly influence the incidence of cracking in such microstructures.

Simple definitions, taken from prEN ISO 17641-1 are given below:

- *Hot Cracks* are material separations occurring at high temperatures along the grain boundaries (dendrite boundaries), when the level of strain and the strain rate exceed a certain level. Small cracks, which are visible only at magnifications of more than about 50x, are often described as *micro fissures.*

The general term is further sub-divided into three specific types:

- *Solidification Cracks* are formed during solidification from the liquid phase of weld metals. They can sometimes be visualized as dendrites, which are growing more slowly than the strains being imposed, so that they fail to fuse on the weld center line and leave a cavity or centerline crack. They usually extend up to the surface of the weld metal, but sometimes remain subsurface.

- *Liquation Cracks* are formed in areas of liquation in the HAZ of a parent material or in multipass welds where weld metal is reheated during the deposition of subsequent weld beads. They are often small (micro fissures) and often subsurface in multi pass welds.
- *Ductility Dip Cracks* are formed in regions of reduced hot ductility in either parent HAZ or in multipass welds. They are associated with specific temperature ranges and may result from diffusion related compositional changes.

Types of Test

From the above brief commentary and definitions it can be seen that the successful design of a single all-embracing test is difficult.

A universal test would have to take in account all of the following possible variables:

1. Many different alloy types, steels, aluminium alloys etc.;
2. The three different cracking types, as described above;
3. Weld metal cracking in single and multipass welds;
4. Parent material HAZ cracking in single and multipass welds;
5. Visible and subsurface cracks;
6. Effects of both variable strain and variable strain rate.

It is therefore perfectly reasonable that a whole variety of tests should have evolved over the last fifty years or so. Some tests have been developed to assess a specific type of cracking in a specific material type. Some have been concerned with the relative susceptibility of similar alloys whereas others have been more concerned with the prediction of whether or not cracking would occur in a real situation.

There are tests, which are primarily used for the optimization of alloy compositions, and others are used to develop and optimize weld procedures. Some tests are highly quantitative and are designed to give a numerical indication of susceptibility, others tend be go/no-go or crack/no-crack types of test. However it is possible to broadly classify the various tests into two major groups, with a simple subdivision of one of the groups.

Self Restrained Tests

These are tests where the specimen loading is produced by stresses developed during welding of a restrained weldment. There are no externally applied loads and in the Russian standard these are described as 'technological' tests.

A number of specimen designs have been developed with varying degrees of restraint and a range of weld geometries, but there appears to be general agreement on the use of restrained fillet weld tests. The Russian standard also includes a circular patch type test with various design options to vary the level of restraint.

A major difference between the two standards is the EN/ISO use of weld procedure butt welds where longitudinal tensile or bend test specimens are extracted and subjected to additional external straining. It must be emphasized that this additional ambient temperature straining does not induce any new hot cracking but only serves to help detect cracks already existing in the weld. In this respect the tests are probably most useful in locating small cracks and micro fissures in weld metal.

Externally Loaded Tests

These are tests that require the use of equipment, which is capable of imposing a strain on the test specimen whilst welding or some form of thermal cycle is in progress. In the Russian standard these are described as 'machine' tests. Attempts have been made to standardize both test equipment and test methods and two Russian standards, namely GOST 15533-80 and GOST 7855-84 are available.

However many different laboratories world wide, have either built their own testing equipment or have developed specimen designs and procedures to suit existing test equipment. This was the main reason for the lack of between-laboratory reproducibility in the hot cracking Round Robin.

It is also the reason that part 3 of the EN/ISO standard is to be published first as technical report and not as a full standard. This document is intended to provide information and guidance on the use of externally loaded tests in the hope that future developments can lead to reproducibility of test methods and the procedures considered essential for an international standard.

Description of Tests

Restrained Weldments

Whatever the detailed design, (restrained fillet, patch test etc) these tests are intended to assess the solidification cracking susceptibility of weld metal in a single pass restrained fillet weld. Various types of test are available, with increasing restraint being provided by increasing thickness and/or stiffness of the plates used in the test assembly. The tests were originally designed for use with arc welding processes and filler metals, although they might prove useful with some of the newer processes (such as hybrid laser/MAG welding). A typical test assembly is shown in Fig. 1.

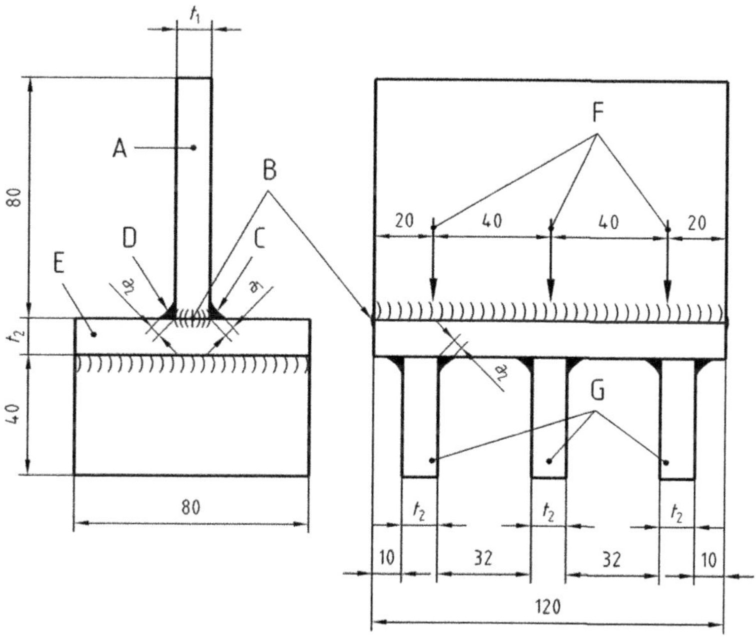

Fig. 1. One example of a self restraint T-joint. (From [2], Part 2.)

Assessment is based upon the length and position of cracks (if any) in the test weld. Assessment is essentially qualitative since no direct measurement of strain is available and it is not possible to directly link the incidence of cracking in the test with the risk of cracking in a real full size fabrication. However the test can be useful in ranking weld metals and in investigating variables such as heat input or welding speed.

Weld Metal Tensile Test

This test is designed to assess the susceptibility of weld metals to liquation and ductility dip cracking. Applying a load to rupture a cylindrical all-weld metal specimen taken from a butt weld opens cracks initiated during the welding operation. The surface regions adjacent to the fracture surface are examined and any cracking can be detected and measured, as shown in Fig. 2. This test can be combined with normal butt weld procedure tests, to qualify a particular consumable and/or procedure. Alternatively it can be used to investigate the effect of welding process variables, such as heat input, and even the influence of weld restraint if thicker test plates are used and welding is carried out in a restraining jig.

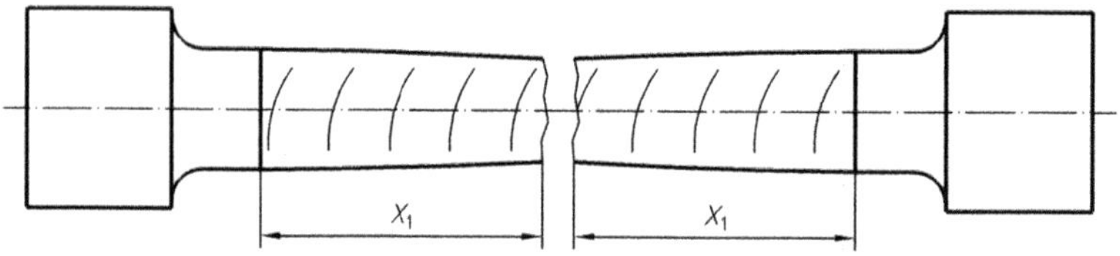

Fig. 2. Fractured all-weld metal tensile specimen, showing region examined for hot cracking. (From[2], Part 2)

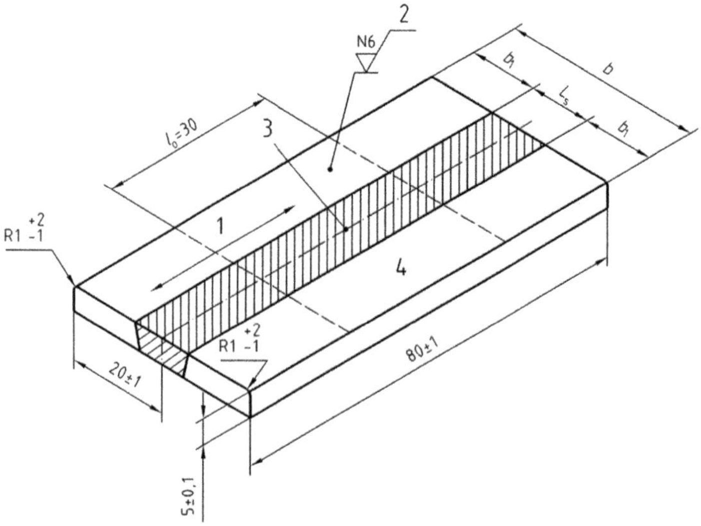

Fig. 3. Arrangement and dimensions of the Longitudinal bend Test (LBT) prior to bending, showing the area to be examined - 4 (From [2], Part 2)

Longitudinal Bend Test

This test is designed to assess the sensitivity to solidification, liquation and ductility dip cracking in all-weld metal butt welds. Bending of a test specimen taken longitudinally from a butt weld opens up any cracks initiated during welding as shown in Fig. 3.

Possible applications and variables are similar to the weld metal tensile test. In principle the positioning of the specimen can be such as to include parent metal HAZ, so that liquation cracking in this region can also be investigated.

Externally Loaded Tests – Background

These tests are designed to produce quantitative rather than qualitative results and to closely simulate, under controlled conditions, the important variables, which relate either directly or indirectly to hot cracking in real welds.

It is generally recognized [5] that hot cracking is a result of the exhaustion of plasticity/ductility at high temperatures and is controlled by the simultaneous influence of different key factors for example the size of the brittle temperature range (BTR), the ductility of the material within the BTR and the magnitude and rate of straining arising from weld metal shrinkage and overall structural restraint.

Many of the externally loaded tests attempt to combine the effects of the main parameters by:

1. Starting straining at the liquid state and finishing at full solidification, thus covering the complete BTR.
2. The strain rate in the BTR is similar to that encountered in real welds, thus allowing diffusion, stress relaxation during solidification and self healing of cracks by liquid metal, to take place during the course of the test.
3. Reproducing both local shrinkage strains and external strains up to the point of crack formation.

The four tests described in the draft EN/ISO technical report prCEN TR 17641-3 together with some of the tests from the Russian standard will be briefly reviewed in relation to their ability to meet the above criteria.

Hot Tensile Test

Hot cracking susceptibility is determined by carrying out a tensile test whilst at the same time heating the specimen to simulate a welding thermal cycle. The specimen can be extracted from parent material or from all-weld metal. In principle, imposing an appropriate prior thermal cycle on the parent material can be used to test HAZ material. These tests are usually carried out in specialized equipment such a Gleeble® machine which enables precise control of thermal cycles and imposed strains. Different procedures and specimen dimensions are used to assess susceptibility to solidification cracking from those used to assess susceptibility to liquation cracking:

1. To simulate solidification cracking the specimen is heated to the melting temperature and the equipment jaws holding the specimen are then held fixed so that the shrinkage strains are able to induce cracking.
2. To simulate liquation cracking, specimens are heated to a range of temperatures just below the solidus. A number of specimens are strained to failure and susceptibility is based on the production of a hot ductility curve.

Assessment is based on establishing the BTR as a fundamental characteristic of the material. In this respect, the test is useful for material selection, ranking and approval but as has been explained above the test does not necessarily cover all the key variables, which affect hot cracking, particularly procedural variables in weld metals and welded joints.

Weldments with Externally Imposed Strain

The basis of these tests is to impose a bending or tensile strain on a weld, whilst the weld is being deposited. Bending offers the potential advantage of inducing hot cracking in the surface regions of the test specimen which enables a more reliable assessment to be carried out. The weld may be an autogenous TIG melt run on parent material or it may be an arc welding process with filler metal. It is usual to employ a specially designed test machine and different design philosophies have been adopted in different countries. In Russia the emphasis has been on progressively increasing external strain, using arrangements similar to those shown in Fig. 4 and included in the GOST standard.

In the USA and parts of Europe machines based on the Varestraint or Transvarestraint Test have been developed, but not fully standardized (Fig. 5).

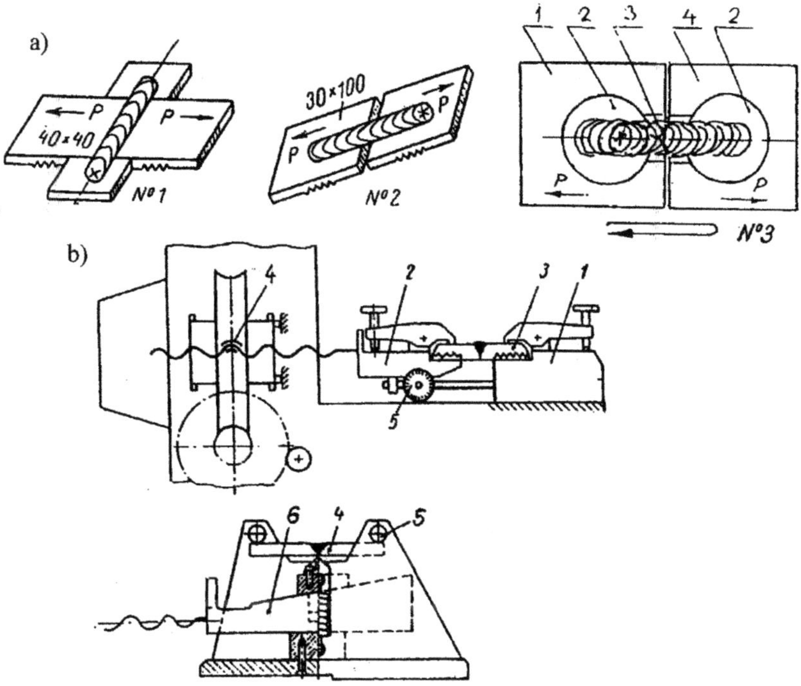

Fig. 4. Diagrams taken from the GOST standard [1] showing general arrangements for tensile and bend straining of externally loaded specimens

In these tests, the load to induce the strain is imposed at a predetermined point in the production of the weld. Yakushin [3] has expressed severe reservations about the use of so called 'shock' loading in some of these tests and claims that it does not reproduce the conditions experienced by real welds when strains develop progressively throughout the complete BTR during solidification and cooling.

Assessment of such tests is usually based on some measurement of cracks once the test is complete. The current consensus appears to favor using the total visible crack length, although some cracks may be subsurface particularly in the root region. It still needs to be confirmed that total crack length is a true and rapid measure of the hot cracking sensitivity.

These tests are widely used for both parent material and weld metal selection and approval. They are also useful for examining the effect of welding procedural variables. They are certainly used successfully within a given laboratory where sound protocols have been established and where a database correlating test results with service experience exists.

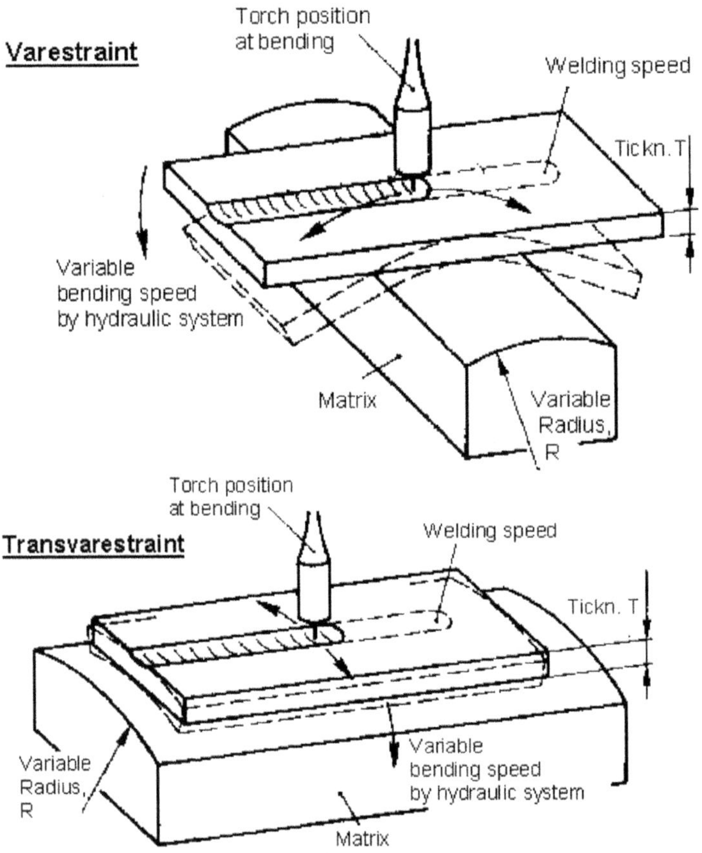

Fig. 5. General arrangement of the Varestraint and Transvarestraint Tests [4]

However, inter-laboratory trials have shown relatively poor reproducibility and this has been attributed largely to differences in equipment and test procedures [4].

The Flat Tensile Test

This test, sometimes known as the Controlled Deformation Cracking Test or PVR Test, was developed to give a relatively simple test, using more widely available and less specialized equipment. It is capable of quantifying the hot cracking susceptibility of base materials, weld metals and a range of conventional arc welding processes.

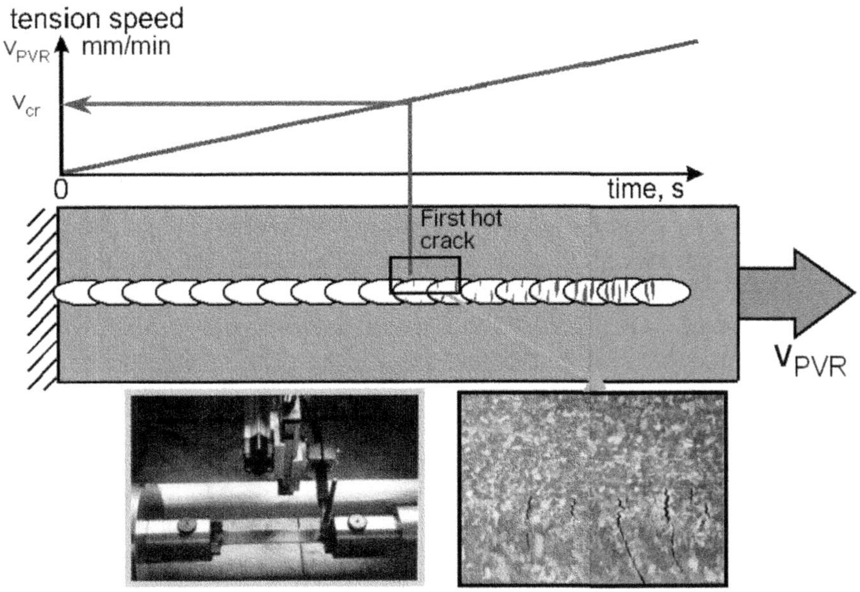

Fig. 6. General arrangement of the Flat Tensile Test or PVR-Test showing position of first hot cracking [6]

This test is carried out using a single flat tensile specimen, which is strained in a horizontal tensile testing machine capable of being programmed with a linearly increasing strain rate. The test differs fundamentally from the Varestraint or Transvarestraint tests in that a linearly increasing tension speed, rather than instantaneous loading imposes the strain. The test is normally carried out with the strain being imposed in the same direction as the weld run. However the test can be carried out with the weld being made transverse to the applied strain. The test can be performed both with and without filler materials, and in the latter case, base metals can be assessed using a TIG melt run and standardized welding parameters.

Filler materials can be assessed using welding conditions recommended by the consumable supplier, or actual welding conditions can be applied to reproduce a practical situation. Welding procedures can be optimized by variation of welding conditions e.g. electrical parameters, travel speed, shielding gas mixtures, flux-wire combinations etc. in order to minimize the risk of hot cracking.

All three types of hot cracking (solidification, liquation, and ductility dip) can be reproduced in a single test, and in principal, only a single test is needed to characterize a material, provided that some reference data are available for comparison.

It is claimed that the test is reproducible, with low scatter and is capable of good discrimination between the three main types of hot cracking [6].

The test procedure in its standard form is shown in Fig. 6. Specimen dimensions can be modified to suit the capacity of available testing equipment. Welding is carried using a constant welding speed and the tension speed is programmed to increase from zero at the start of the test to 70 mm/min at the end of the test.

On completion of the test, the test specimen is examined at about 25x magnification to identify the first hot cracks. The tension speed corresponding to the appearance of the first crack (more than one type may be present) is defined as the critical tension speed v_{cr}, and is a measure of hot cracking susceptibility. Streitenberger [6] has carried out a comprehensive comparison of two externally loaded tests namely the Modified Varestraint Test (MVT) and the PVR test.

Discussion

The main advantage of self-restrained tests is that they are relatively simple to carry out and require no specialized testing equipment. The major disadvantage is that the results are qualitative, or possibly semi quantitative in the case of the longitudinal bend test.

To assess the effect of welding variables a large number of tests are required and these can be both time-consuming and costly. However, they may have a place in a workshop environment where some reassurance of avoidance of hot cracking is required. The tests are not really suitable for the investigation of critical variables or material assessments in the laboratory.

The main general disadvantage of the externally loaded tests is the requirement for specialized equipment and procedures. Previous work has shown that reproducibility between laboratories is poor and that it is probably unreasonable to expect good agreement between a number of different types of test.

If progress is to be made towards an a truly international standard for externally loaded hot cracking tests, then one test should be chosen for further development, establishment of procedures and widespread validation and verification.

Such a test should meet the following requirements:

1. Be geometrically simple with controlled straining. Relatively complex bending and 'shock' loading should be avoided;
2. Utilize readily available, relatively inexpensive test equipment;
3. Be capable of reproducing all forms of hot cracking;
4. Be suitable for testing parent materials and weld metals;
5. Be suitable for investigating welding procedural variables;
6. Be quantitative;
7. Be reproducible both within and between laboratories.

Of the tests described in both the prEN/ISO and GOST standards , the one, which most clearly meets many of the above criteria, is the Flat Tensile Test or PVR Test although it should be noted that the actual strain in the molten is not measured. In addition, this test, because of its relative geometrical simplicity is the one most amenable to mathematical modeling. However the thermo-mechanical interactions are highly complex and much further work is required. Modeling could prove useful in further development of the test and in future validation [7]. It is therefore suggested that consideration should be given to carrying out the work necessary for this test to be developed as an international standard test.

It is believed that the following will be required:

i. A very clear and precise procedure for performing the test should be developed;
ii. If possible, the dimensions of the test piece should be standardized and fixed;
iii. The minimum specification for the testing machine, consistent with ii) should be defined;
iv. Evidence of within laboratory reproducibility should be published;
v. Evidence of between laboratory reproducibility should be published and if this is not available, then consideration should be given to carrying out an appropriate Round Robin exercise once items i) to iii) have been completed.

Conclusions

A comprehensive and critical review of the European International prEN/ISO and Russian GOST standards on hot cracking tests has been carried out. It is concluded that:

1. There is good agreement between the two standards on self-restrained tests.

2. It should be relatively easy to produce a common test (or tests), but it is recognized that such tests have limited application and are at best semi-quantitative in their output.
3. There is a need for a relatively simple, quantitative, externally loaded and programmable 'machine' test.
4. The test best suited, at the present time, to meet the needs of an International standard is the Flat Tensile or PVR Test.
5. More work is needed on testing procedures and demonstrations of reproducibility before the test can be standardized.

References

1. Nikolayev G, Yakushin B, Saharov Y, Misurov A, Deev A. Welded Joints – Test methods on resistance to shrinkage crack formation under fusion welding. State Standard of Soviet Union, GOST 26389-84
2. Destructive tests on welds in metallic materials – Hot cracking tests for weldments – Arc welding processes. Draft European International Standard. PrEN ISO 17641
   Part 1: General
   Part 2: Self restraint tests
   Part 3: Externally loaded tests (technical report)
3. Yakushin BF (2003) The Scientific Base for Russian Federation and European Weldability Tests Standards Harmonisation. IIW doc IX-H-573-03
4. Wilken K (1999) Investigation to Compare Hot Cracking Tests – Externally Loaded Specimens. IIW doc IX-1945-99
5. Yakushin BF (2003) Analysis of GOST Standard 26389-84.
   IIW doc IX-H-582-03 (translated by Peter Bernasovsky)
6. Streitenberger M (2002) Comparison of hot cracking test methods: PVR-test and MVT- test. IIW doc IX-H-527-02
7. Herold H, Streitenberger M, Pchennikov A (2000) Modelling of the PVR-test to Examine the Origin of Different Hot Cracking Types.
   IIW doc IX-H-474-00

Value of Different Hot Cracking Tests for the Manufacturer of Filler Metals

H. Heuser

Böhler Thyssen Schweisstechnik Deutschland GmbH, Germany

Introduction

For material evaluation and quality assurance an efficient, flexible and profitable test concept for estimation of welding qualification with regard to hot cracking susceptibility is needed. It has to provide the possibility for evaluating base materials, filler metals and welded joints. There is a need of both test methods for permanent evaluation during production of filler metals and for special tests in critical cases as for instance hot cracking problems the user is faced with. Here, comparative analyses are necessary to some extent. Adequate hot cracking tests have to be available which can be used to determine the effects of batch and welding conditions. Manufacturers of filler metals therefore need different test methods.

For quality control, normally self-restraint test methods (tensile test, double fillet weld test, cylinder test and ring segment test) are used [1]. For material optimization and for the elimination of hot cracking problems, externally loaded tests such as the Modified Varestraint Test (MVT) or the Hot Deformation Rate Test (HDR test) [1] are necessary.

In the following report, fields of application and the benefit of different test methods are described.

Self-Restraint Tests

Tensile Tests

A test method that is used every day by manufacturers of filler metals is the tensile test – a tensile specimen of the all-weld-metal. Fig. 1 shows a sample for the tensile test of austenitic weld metal of a stick electrode. The weld metal shows a strong deformation (high elongation) and strong Reduction of Area (RoA). There is no visible evidence of hot cracks. Fig. 2 shows in contrast to Fig. 1a distinctly different appearance. Weld metal alloy 230 shows a poor deformability (elongation 25 %, Reduction of Area (RoA) 24 %) and many small cracks near the fracture zone. These are hot cracks. Hot cracks occur during the welding process, when a critical temperature range is passed, and are therefore part of the all-weld metal sample.

Cracks near the surface of a tensile test sample – cut out along the welded seam – will open towards the surface during the tensile test. Cracks are often L-shaped and therefore called "hook cracks".

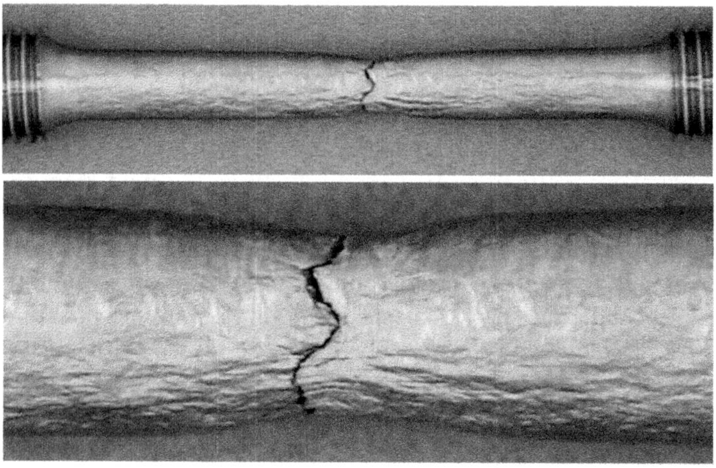

Fig. 1. Hot cracking test with the tensile specimen (all-weld metal Thermanit 19/15 W; E Z 20 16 3 Mn N L R 1 2; no hot cracks)

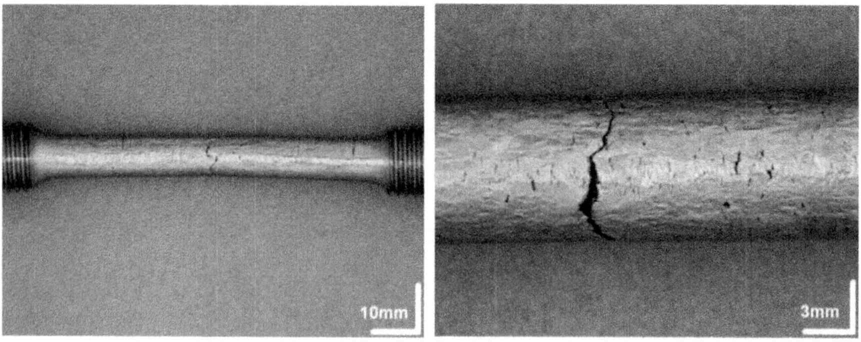

Fig. 2. SMAW all-weld metal alloy 230 (22 Cr, 2 Mo, 14 W, Co, Al, La, remain Ni) open cracks are hot cracks; El. = 25 %, RoA = 24 %

Prior to fracture is always such a crack. The RoA is considerably smaller than with crack-free samples. But not every surface crack of a tensile test sample points to a hot crack.

Fig. 3 shows a crack caused by a pore near the surface. The sample shows an excellent ability of elongation at a RoA of 78 %. Some experience is required to derive assured findings regarding the hot cracking susceptibility of weld metal from a tensile test. A weldment that shows a RoA of a minimum of 40 % is, despite small isolated hot cracks, unobjectionable.

Fig. 4 shows the dependency of RoA on the number and size of hot cracks at a CrNiMo weld [2].

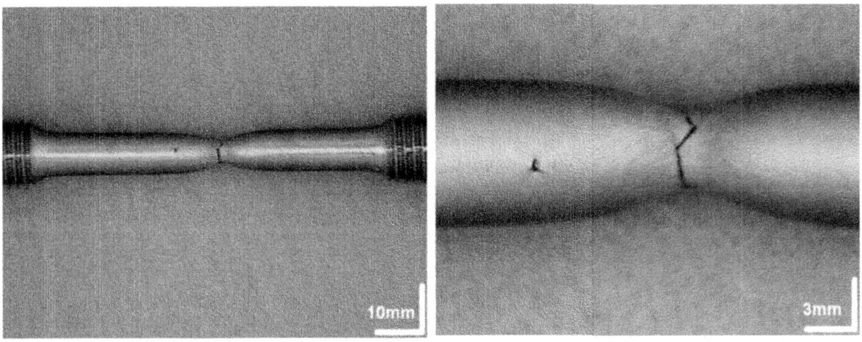

Fig. 3. GTAW all-weld metal ER 80 S-Ni 2 (3,5 % Ni) open cracks caused by a pore, no hot crack; RoA = 78 %

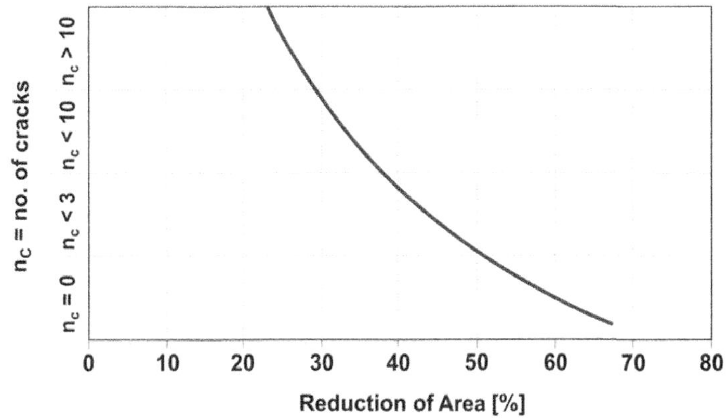

Fig. 4. Relationship between number of hot cracks (>1 mm) and RoA at the tensile specimen for austenitic weld metal

But RoA is not the overall indicator for all alloys. Filler metals with application temperatures > 800 °C for petrochemistry with 0.4 % C, 25 % Cr, 1 % Si, 1.5 % Nb and 35 % Ni produce elongation and RoA values < 10 % at room temperature (RT). With this little deformability of the weld hot cracks covered under the surface will not be opened in every case during tensile testing. Fig. 5 shows a sample for the tensile test of such a weld bead.

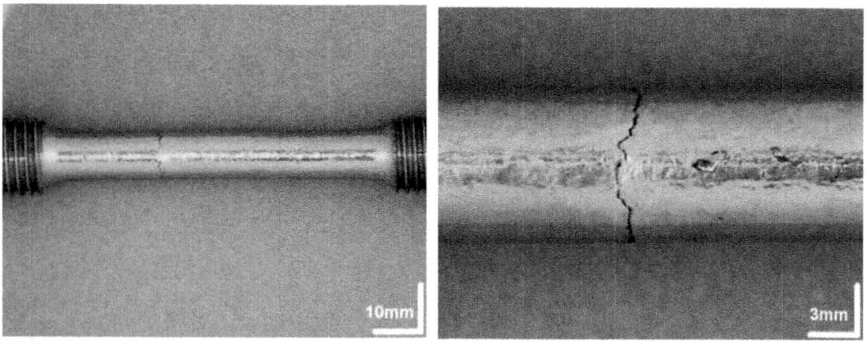

Fig. 5. GTAW all weld metal; W Z 25 35

T-Joint Test

The T-joint test is often required in specifications of Classification Societies and also included in the draft version of prEN 17641, part 2 [1]. This test method is very often used for tests of filler metals for welding dissimilar base materials. Here, the base plate should be made of low-alloyed materials.

Fig. 6 shows such a sample. The informational value of that test depends first and foremost on the deformation resistance of the base metal plate. It has to be thick enough or stiffened to avoid shrinking. This is described in prEN ISO 17641, part 2 [1].

Unless otherwise agreed, filler materials of the following diameters are used:

Stick electrode:	4 mm
Filler rods (TIG):	3 mm
Wire electrodes:	1.2 mm

Samples should be welded down hand. The first fillet weld should be welded at a consistent thickness (A-measure) of about 5 mm.

Fig. 6. Standardized hot cracking test required in various specifications, T-joint

Not later than 20 s after welding the first weld, the test weld (sealing run) in reverse direction (each without interruption) is to be produced. The test weld should be about 20 % thinner. After cooling down, a dye penetration test is carried out. If cracks are detected in the first weld, the hot crack test is not valid. The test can be carried out quickly and easily and is particularly useful for an optimal choice of a filler metal for single-pass fillet welds on dissimilar metals.

Cylinder Sample Test

Beside the T-joint test, there is another sample test standardized in the German Standard DIN 50129 [3] – the cylinder sample test – which is also suitable for performing hot cracking testing quickly and easily. Two cylinder samples with diameters of 50 mm and of 60 mm lengths, respectively, are necessary. Both are firmly clamped and a double-pass weld with the filler metal to be tested is produced. Afterwards, the actual test weld will be carried out (Fig. 7). Due to the existing prevention from shrinkage filler metals susceptible to hot cracking will show throat cracks which will be visible after cooling down in a surface fracture test. The result of that test depends very much on the welding parameters and the welding procedure. A reliable result can only be expected after many preliminary tests in which the welding parameters and the hot cracking susceptibility of the filler metal are synchronized. A disadvantage of this test is the fact that similar base materials are normally not available as bar stock.

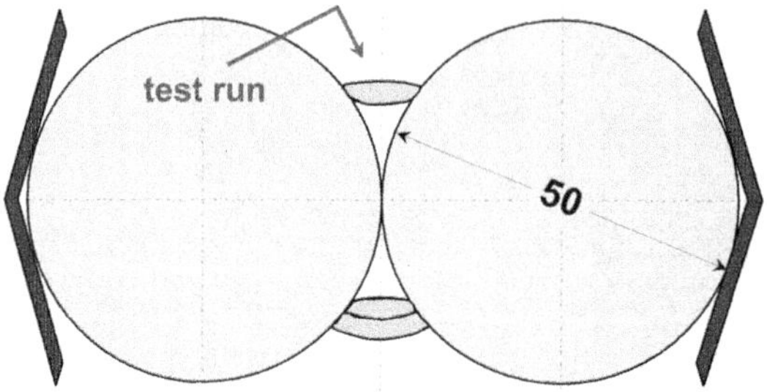

Fig. 7. Hot cracking test on cylindrical specimen (length: 60 mm)

Ring Segment Test

This test is – in compliance with the German Standards – mandatory for high-alloyed filler metals and filler metals on a Ni-base which are used in nuclear fields of application.

The test sample (Fig. 8) consists of four quadratic segments of the same size into which, after double-sided tack welding, a single-sided groove is cut. Base material with a thickness of 25 mm matching the filler metal to be tested should be used. This often cannot be realized. Therefore, compromises have to be found with the certifying organization and the customer.

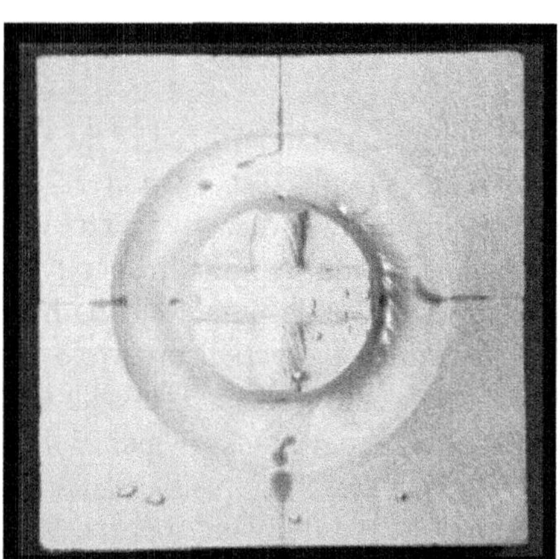

Fig. 8. Standardized hot cracking test required in various specifications, ring segment

This test applies to filler metals for GTAW, GMAW and SMAW. Type of current, polarity and inert gas have to meet the welding conditions under which the test filler metal should not be susceptible to hot cracking.

A weld has to be carried out without oscillation or interruption starting at about 30° before a butt joint and ending at about ¾ of circumference. After cooling down to room temperature and cleaning of the weld surface and the groove (brushing), the remaining length (¼) has to be welded in the same direction. After cooling down of the sample, the cleaned ring weld has to be analyzed for cracks using the dye-penetrant test.

The informational value of that test is very small. Improper preparation of the sample and welding performance can cause hot cracks that cannot be attributed to the filler metal used. In the future European Standard prEN 14532 dealing with qualifying examinations of filler metals, this test is not considered. It should be carried out only if mandatory (production in compliance with Standards 1408 of KTA [Nuclear Safety Standards Commission]).

Cracking Test According to Stamicarbon

In many cases the processing fabricator does not rely on the manufacturer's standard tests but requests also a filler metal check according to his own specifications. Stamicarbon (licensor of urea synthesis facilities) developed a particular specification to verify the crack resistance of filler metals with large surface cladding. Here, large surface double-layer band claddings (SAW or RES) are welded with austenitic strips to an unalloyed support material. Cladding joints are welded to specific gaps and laterally with a fully austenitic stick electrode (E Z 25 22 2 L B 2 2, 25 % Cr, 22 % Ni, 2 % Mo).

Fig. 9 shows the set-up and the samples to be taken. Their surfaces will be leveled by mechanically grinding. Afterwards, the surface will be pickled and checked for defects under a stereomicroscope with 20-fold magnification. Defects have to be identified as pores, metal inclusions or hot cracks. This procedure will be repeated three times after 1 mm of the surface has been ground off each time. After each grinding, only a certain number of defects is accepted. This test is very demanding in regard to the hot cracking susceptibility of filler metal. Particularly the weld metal of the stick electrode reacts very sensitively to this test.

Fig. 10 shows an overview of an analyzed ground surface, a metal inclusion (tolerable) and a hot crack. Both failure types are marked with different symbols. This test shows the highest sensitivity of the hot cracking tests presented so far.

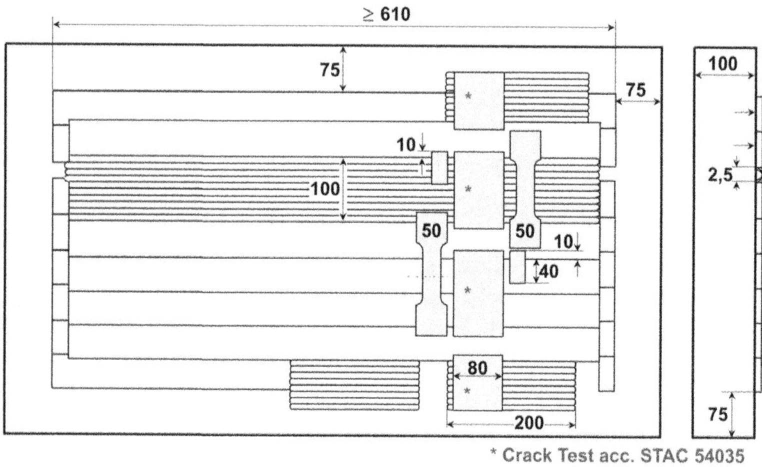

Fig. 9. Stamicarbon Test, test plate layout (urea synthetic plant)

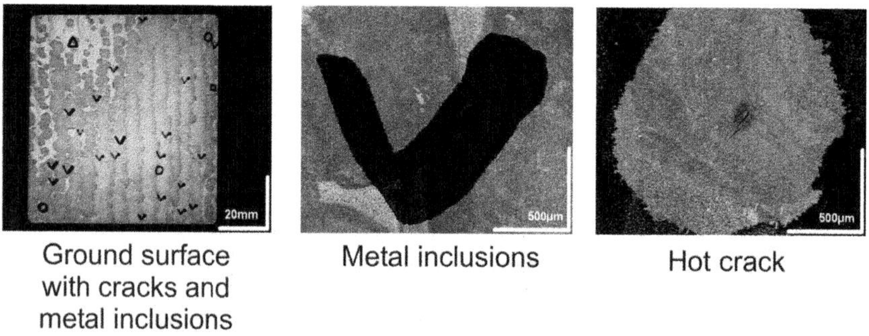

Fig. 10. Stamicarbon Test (54035), filler metal: E Z 25 22 2 L B 2 2

Comparative studies on all-weld metals between tensile test, side bend and longitudinal bend test with the Stamicarbon test (STAC 54035) showed for the three different electrodes no cracks in the tensile test and side bend test, isolated cracks in the longitudinal bending test and an unacceptable number of cracks in the Stamicarbon test (Table 1 and Table 2).

Table 1. Chemical composition (all-weld metal); El-∅: 4.0 mm

No.	C	Si	Mn	P	S	Cr	Mo	Ni	N
A	0.034	0.27	4.68	0.015	0.007	24.69	2.28	21.60	0.15
B	0.033	0.30	5.31	0.011	0.007	25.04	2.26	22.08	0.13
C	0.042	0.66	5.64	0.012	0.006	24.57	1.97	20.55	0.14

Table 2. Results of different hot cracking tests; SMAW: E Z 25 22 2 L B 22 (all weld metal)

No.	Tensile Specimen	Side Bend Test (180 °C)	Longitudinal Bend Test	STAC 54035
A	No cracks; RoA 60 %	No defects	4 cracks	Not accepted
B	No cracks; RoA 58 %	No defects	1 crack	Not accepted
C	No cracks; RoA 61 %	No defects	4 cracks	Not accepted

Externally Loaded Tests

Hot cracking tests during which the sample is stressed with controlled external load are used for basic considerations in view of optimizing the hot cracking behavior and analyzing hot cracking problems the processing fabricator is faced with.

The most important tests in the perspective of a filler metal manufacturer will be outlined below and their relevance will be described considering performed analyses [1].

Programmable Deformation Cracking Test (PVR Test)

For a programmable deformation cracking test, a tensile test with a defined rising elongation rate is carried out during welding. After exceeding the critical value for hot crack formation, all three types of hot cracks (solidification cracks, liquation cracks and ductility dip cracks (DDC)) can occur.

Fig. 11 shows the principle of approach. The achieved deformation rate and deformation of the sample at the point in time at which the first forced hot crack appears is a measure for evaluating the hot cracking susceptibility.

Fig. 12 shows the effect of the δ-ferrite content on the critical elongation rate. In the examined alloy (19 % Cr, 9 % Ni), up to about 3 FN primary ferritic solidification is present, which causes a high deformation rate without crack formation. Below 3 FN, the critical elongation rate decreases strongly [4] with diminishing δ-ferrite.

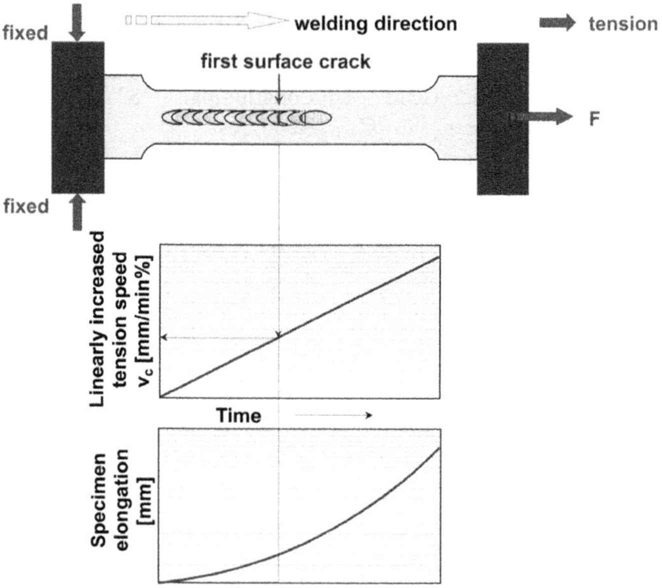

Fig. 11. PVR Test: Hot cracking test procedure of the controlled deformation cracking test (externally loaded test)

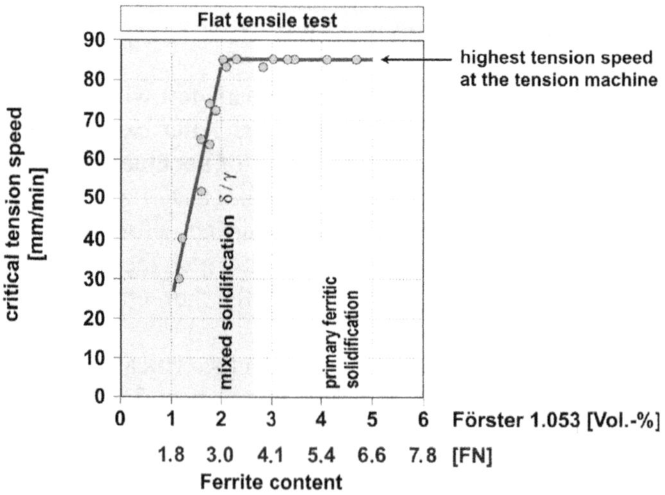

Fig. 12. Controlled deformation cracking test of weld metal CrNi 19 9 L; influence of the ferrite content and the solidification mode on the critical tension speed [4]

Short arc (~ 3 mm) 135 Amps; 26 Volts			Long arc (~ 8 mm) 135 Amps; 35 Volts		
C	Si	Mn	C	Si	Mn
0.011	1.20	1.10	0.011	1.10	1.00
Cr	Ni	N	Cr	Ni	N
17.05	10.80	0.065	16.60	10.80	0.11

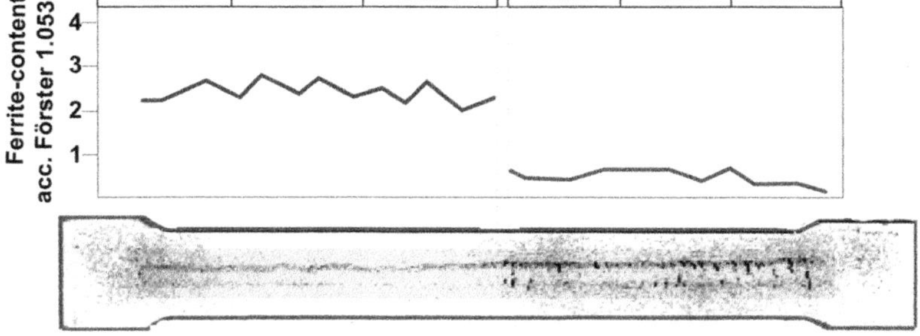

Fig. 13. Influence of the arc length on the hot cracking sensitivity of austenitic weld metal (PVR test) [4]

Fig. 13 shows what effect a welder can exert on the hot cracking behavior. On the left part of the tensile test sample the welder assured a short electric arc (low content of N). This resulted in a ferrite content of about 3 %. On the right part of the tensile test sample the welder adjusted a long electric arc that leads to a N pick-up (0.11 %). As a result, the ferrite content decreased to less than 1 %. That necessarily leads to a higher hot cracking susceptibility.

Hot Deformation Rate Test (HDR Test)

During welding in the HDR test, a pre-defined transverse deformation rate will be applied which varies from test to test until the first or last hot crack is forced (Fig. 14) [5]. This test is used to determine the susceptibility to solidification cracking. It is particularly useful for welding parameter optimization and appropriate selection of filler metals regarding hot crack prevention, because the effects of the base material and of the weld geometry as well as further variables can also be analyzed.

Fig. 15 shows the dependency of the hot cracking susceptibility of a fully austenitic SA weld (19 % Cr, 15 % Ni, 3 % Mo, 0.15 % N) on the welding parameters, the welding amperage, the welding voltage and the welding speed. The motor revolutions per minute are plotted versus the welding parameters which can be used, on the basis of the geometric data of the test facility, to determine the critical transverse deformation rate.

These analyses are carried out to prevent hot cracks during SA welding of a liquid petrol gas container. The problem could be minimized by reducing welding amperage and welding speed. By additional analyses using the Modified Varestraint Test, the hot cracking susceptibility could be totally prevented (see next section).

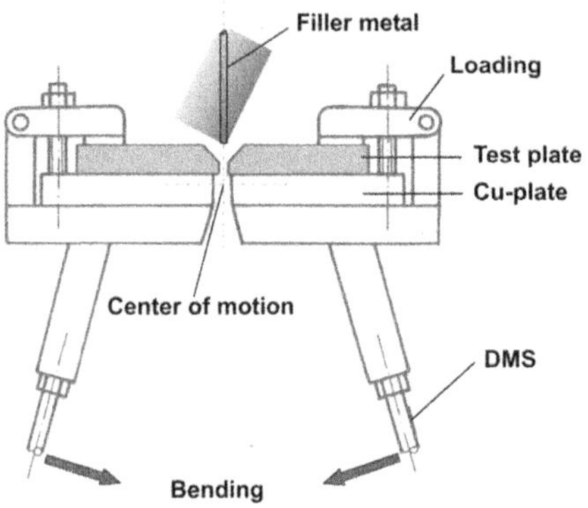

Fig. 14. Hot Deformation Rate Test

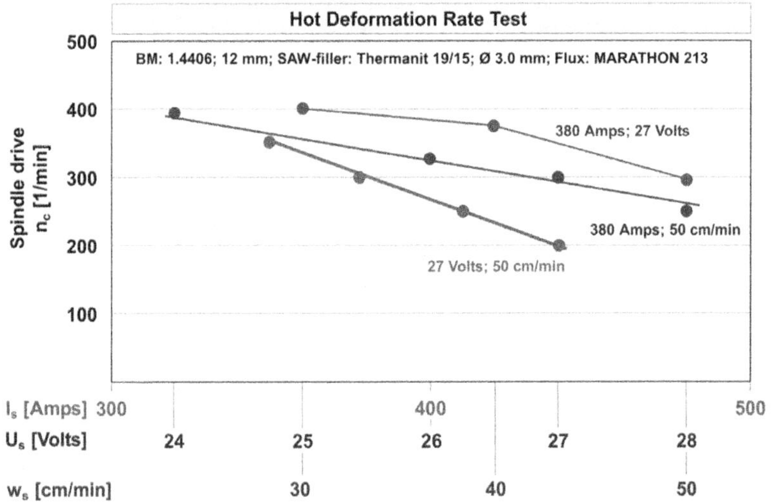

Fig. 15. Influence of welding amperage, voltage and welding speed on the hot cracking susceptibility (solidification cracking); as higher the spindle drive during the HDR-test without forcing cracks, as higher is the hot cracking resistance of the proved material

Modified Varestraint Transvarestraint Test (MVT Test)

In the MVT Test, a defined bending deformation is applied to the test sample during welding (Fig. 16) [1]. After exceeding the critical value for hot cracking, all three hot crack types (solidification cracks, liquation cracks and DDC) can occur. Base materials, filler metals, welding conditions, effects of welding auxiliary materials (flux, gas) and welding procedures can be clearly distinguished and evaluated in their effects on hot cracking susceptibility. The test weld bead will be analyzed under the stereomicroscope (25-fold), i.e. the individual crack lengths will be recorded, added up and inserted into a chart showing the applied elongation by variation of mandrel radius (Fig. 16).

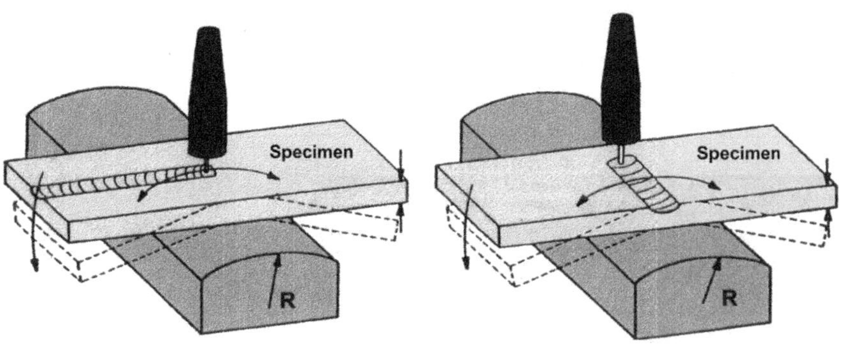

Fig. 16. MVT Test

This test is used to optimize filler metals, meet customer specifications and detect reasons for hot cracking problems. Some examples will be given further on.

Liquid Petrol Gas Container Made of Material 1.4406

Hot cracks occurred during SA welding of a liquid gas tank (Fig. 17) made of the material 1.4406. The wall thickness was 15–22 mm. Welding was carried out with a fully austenitic filler metal (1.4455) and a highly basic flux (degree of basicity according to Boniszewski B = 2.7).

Fig. 18 shows a weld cross-section of a joint. A typical hot crack of that joint is shown in Fig. 19.

Partially such hot cracks originated from Mn-silicate lines in the base material (Fig. 20).

Fig. 17. Base metal: 1.4406;
SAW-filler: S 20 16 3 Mn N L acc. EN 12072 (ER316L mod.)

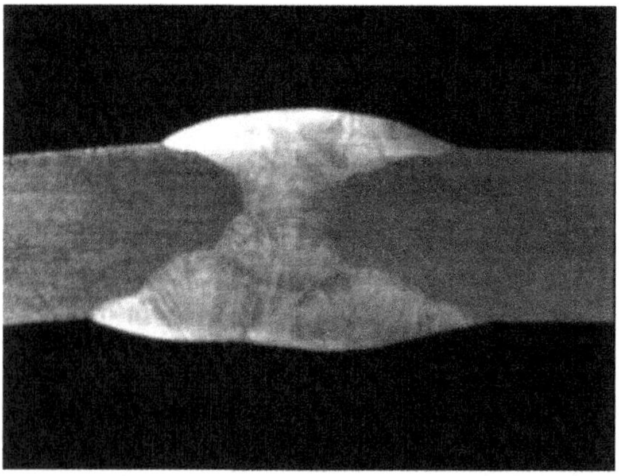

Fig. 18. SA-weld of LG-Tank; Base metal: 1.4406 (X2CrNiMoN17-11-2); filler metal: S 20 16 3 Mn N L acc. EN 12072

MVT Tests showed that a flux with a lower index of basicity (B = 1.3) resulted in a much better hot cracking behavior. The problem could be solved by using this flux and by adjusting welding parameters (see section about the HDR test). Fig. 21 is divided into three segments:

Test results located in Section I (weldable) indicate that hot cracks will hardly occur in practice. Section II (restricted weldable) means that special measures may be necessary to prevent hot cracks (welding parameters). Section III (not weldable) is an area that does not permit crack-free welding. This classification relies on a huge field of experiences of the Bundesanstalt für Materialforschung und –prüfung, Berlin (BAM) that carried out this test in regard to many thousand research results with this test and practical knowledge.

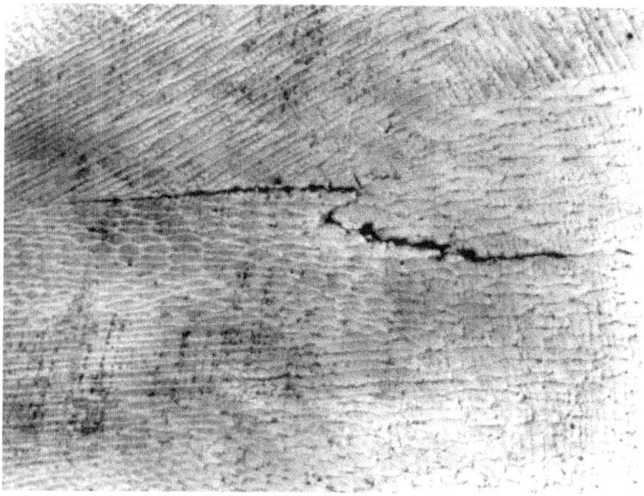

Fig. 19. Hot crack in SA-weld of S 20 16 3 Mn N L (ER316L mod.)

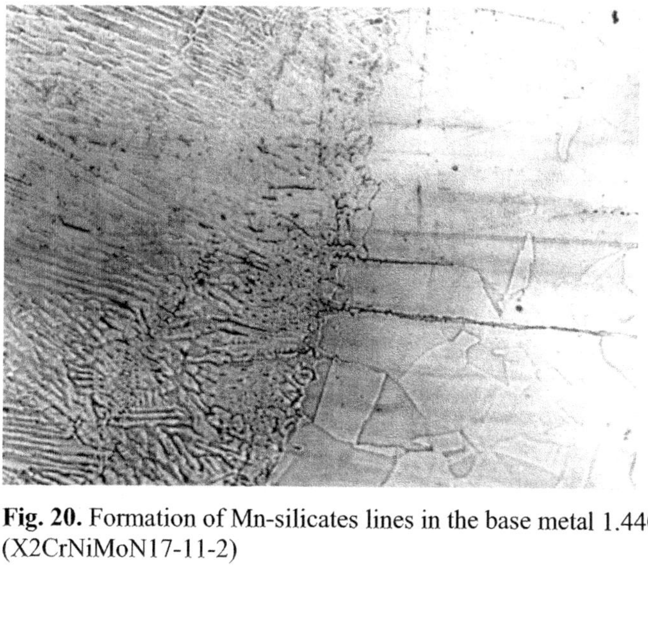

Fig. 20. Formation of Mn-silicates lines in the base metal 1.4406 (X2CrNiMoN17-11-2)

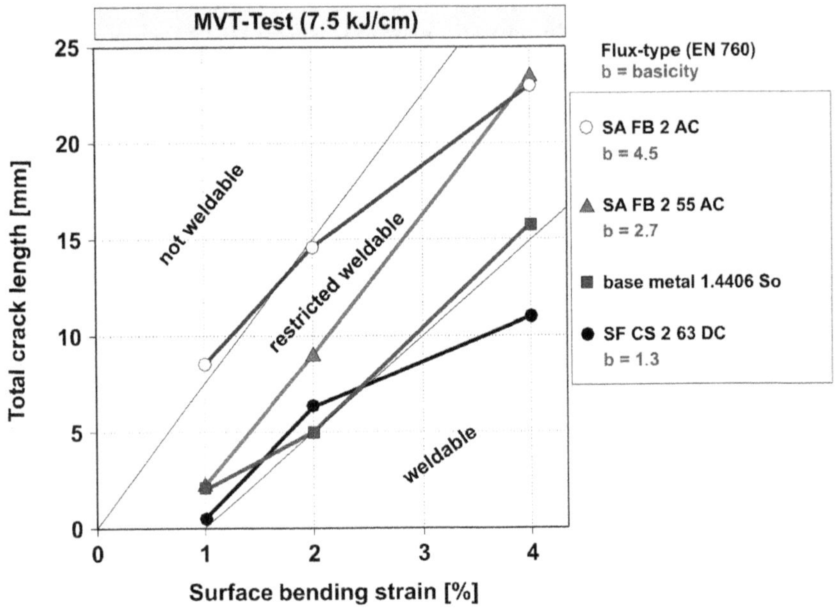

Fig. 21. SA-weldment; BM: 1.4406 (X2CrNiMoN17-11-2) ; filler metal : S 20 16 3 Mn N L (ER 316L mod.)

Hot Cracks during SA Welding of Heavy-Walled Components of Duplex Steel 1.4462

Duplex steel and matching filler metals are considered as unsusceptible to hot cracks, because they solidify as ferrite. But in practice, hot cracks are sometimes problematic when using this material during SA welding of heavy-walled components (wall thickness > 20 mm).

In the present case, circumferential welds were welded on pipes of 42 mm wall thickness and 300 mm of diameter. Welding procedure and welding parameters can be found in Fig. 22.

Flux with an index of basicity B = 1.8 was used for welding.

Fig. 23 shows a hot crack along the original ferritic grain boundary. Even by varying the welding parameters, no crack-free welds could be achieved. Cracks were revealed in side bend tests. The MTV test helped to determine optimal wire / flux combinations. In this case, a highly basic flux B = 2.7 was successfully used.

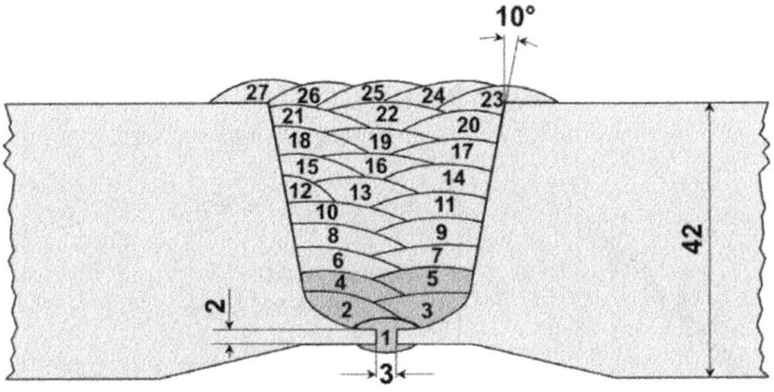

Welding parameter:

Pass	Welding process	EL- Ø [mm]	I [A]	U [V]	T_i [°C]
1	GTAW	2.0	100		
2 - 3	SMAW	3.2	100		≤ 150
4 -5	SMAW	4.0	140		≤ 150
6 - 27	SAW	3.2	380/400	27/29	≤ 150

Fig. 22. Welding sequence and parameters of circumferential welding on DSS 1.4462 (UNS S 31803); matching filler metal

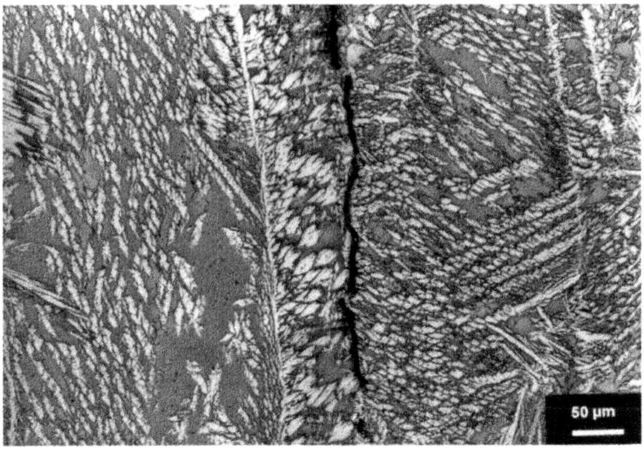

Fig. 23. Hot crack in the DSS weld (SAW; 1.4462, UNS S 31803), crack proceeding along the former ferrite grain boundaries

Fig. 24 shows the differing hot cracking susceptibility of different SAW fluxes for this application. A reduction of the amperage to 300 A or even 250 A also generated advantageous effects (Fig. 24). In addition to this, MVTs proved that the hot cracking susceptibility diminishes with an increasing ferrite content in all-weld metals (Fig. 25) [6].

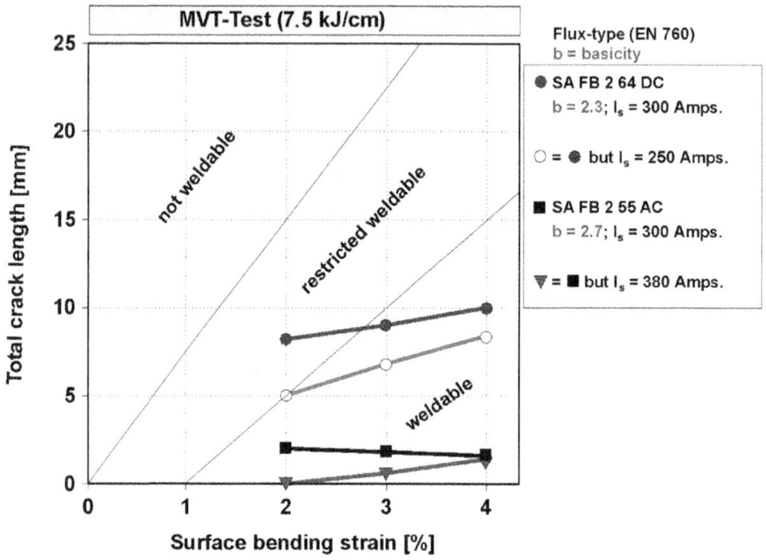

Fig. 24. SA-weldment; BM: DSS 1.4462 (UNS S 31803); filler metal: S 2209 (ER2209)

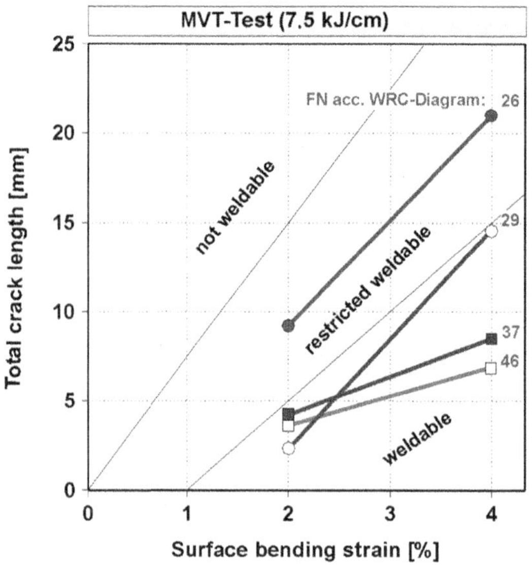

Fig. 25. SA-weldment; BM: DSS 1.4462 (UNS S 31803) matching filler metals; from different heats with different ferrite contents

SA Welding of Alloy 617

As a rule, nickel base filler metals are more susceptible to hot cracking than stainless steel ones. Therefore, a smaller wire diameter is used for SA welding. During SA welding with a wire matching to alloy 617, increasing formation of hot cracks occurred. A wire with a diameter of 1.6 mm and a highly basic flux (Marathon 444) was used for welding.

Tests with different current densities but constant heat input led to a prevention of hot cracks. Instead of the wire with a diameter of 1.6 mm, a wire with a greater diameter was used, whereas the amperage of the 1.6 mm wire was kept, that means current density was reduced because of the greater wire diameter. This leads to colder droplets, and hence to lower heat input. Fig. 26 shows this dependency in the MVT.

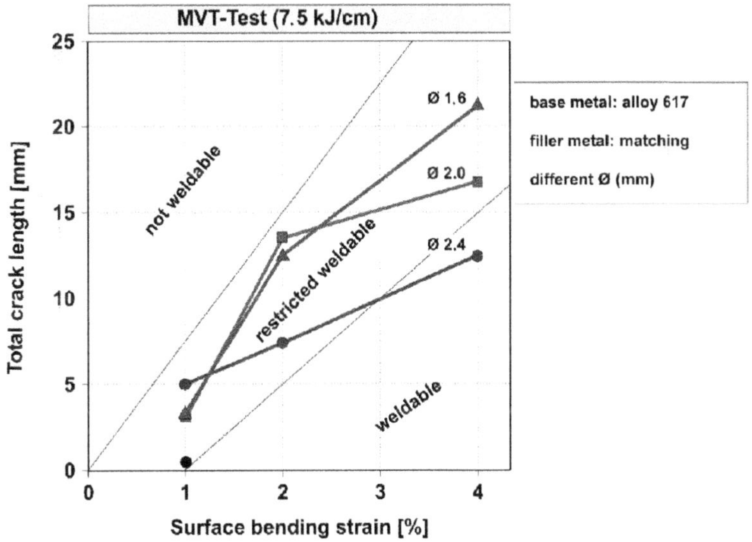

Fig. 26. Influence of wire-∅ in the SAW-process on the hot cracking susceptibility of alloy 617

Hot Cracks during Welding of Alloy 690

The alloy 690 is characterized in the references as very sensitive to hot cracking. This material is said to be particularly susceptibile to ductility dip cracking (DDC). This crack appearance is not characteristic in comparison to other hot crack appearances, because it occurs in alloys with a low degree of contamination. Hot cracks in alloy 690 welds are caused by grain boundary precipitation as illustrated in Fig. 27.

Fig. 28 illustrates that different batches of wire show a different hot cracking behavior.

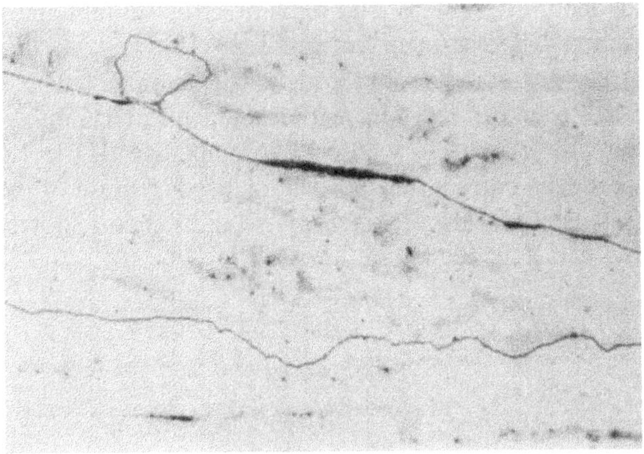

Fig. 27. GTA-weld of alloy 690; interdendritic cracks (hot cracks); grain size segregation

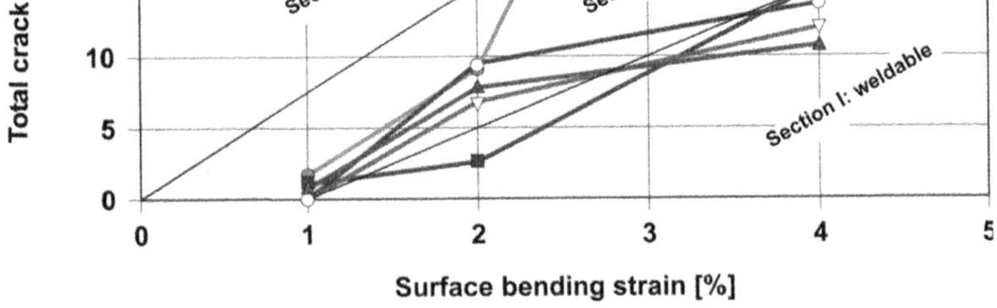

Fig. 28. Hot cracking behavior of different wire heats of alloy 690

Analyses underlying those MVT's are shown in Table 3. A comparison of the different analyses, however, gives no hint of different hot cracking susceptibilities.

Table 3. Different alloy 690-heats; wire analysis (in wt.-%); heats with different hot cracking behaviors in the MVT Test

heat	C	Si	Mn	P	S	Cr	Mo	Ni	Fe	Al	Ti
A	0.029	0.25	0.33	0.006	0.004	29.4	0.01	rem.	9.11	0.60	0.30
B	0.017	0.23	0.29	0.003	0.003	29.4	0.02	rem.	9.37	0.74	0.30
C	0.032	0.25	0.33	0.008	0.003	29.3	0.09	rem.	9.85	0.77	0.30
D	0.030	0.22	0.27	0.003	0.002	29.4	0.01	rem.	8.70	0.51	0.39
E	0.030	0.20	0.25	0.003	0.002	29.4	0.01	rem.	9.14	0.68	0.31

It seems that different effects play a role, e.g. the Ti carbon nitrides that can be recognized in Fig. 29 and that occur in a linear shape in the hot rolled wire (starting product for filler metals). Those wires led to an increased hot cracking susceptibility.

Fig. 29. Alloy 690; Ti-carbonitrides in the wire

The potential hot cracking susceptibility of these alloys can be seen in Fig. 30. Beside the weld in the HAZ, several transversal cracks can be distinguished which are classified as DDC. An evaluation of MVTs is shown in Fig. 31. The high susceptibility to DCC is clearly shown. But also hot cracks within the GTAW-remelted bead have a great length (section II: restrictedly weldable).

The hot cracking susceptibility of matching filler metals of this alloy can be reduced to a higher extent by adjustment of welding parameters or also of the welding process than by a restriction of the chemical composition of the hot rolled wire. The production process of the hot rolled wire seems to have an effect, too.

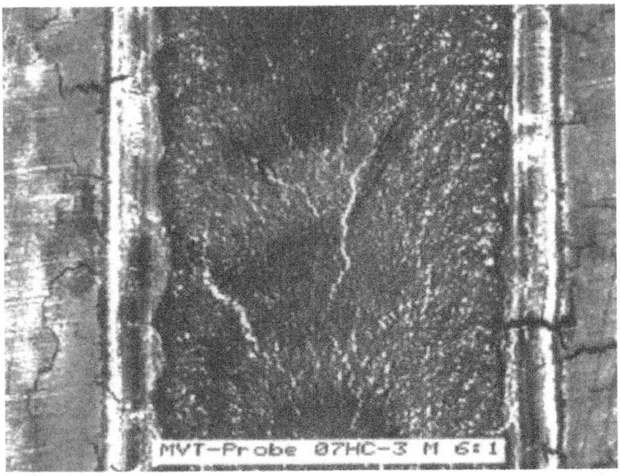

Fig. 30. Solidification cracks in the weld of alloy 690; Preheat Solidification Cracks and Ductility Dip Cracks in the HAZ; heat input: 7.5 kJ/cm; elongation 2 %

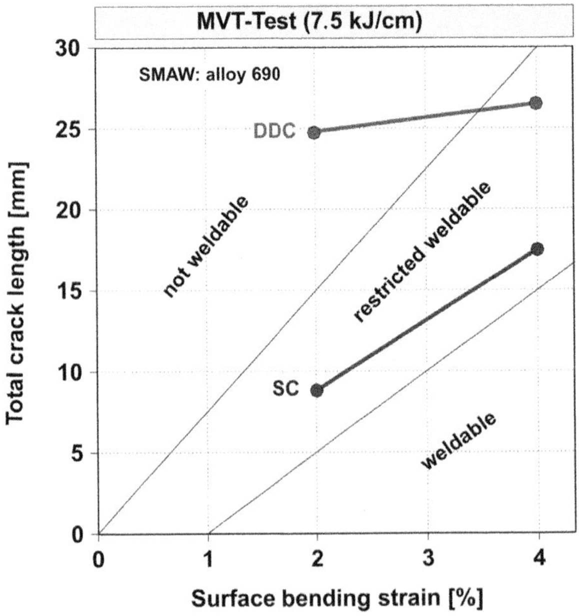

Fig. 31. Solidification Cracks (SC) and Ductility Dip Cracks (DDC) in SMA-weld of alloy 690

Conclusions

Manufacturers of filler metals use several hot cracking tests, particularly for filler metals being susceptible to hot cracking. Self-restraint tests are not always informative. Therefore, in critical cases and mainly for the disclosure of reasons for hot cracking at a component, externally loaded hot cracking tests are necessary.

In addition to metallurgical examinations, externally loaded tests such as the PVR or MVT tests should also be used to determine the hot cracking behavior and to be able to take appropriate measures. Here, the base material should be taken into consideration. If a base material is susceptible to hot cracking, one cannot expect that a matching filler metal shows a considerably lower hot cracking behavior. For that reason, welding parameters or welding procedures have to be adjusted to the hot cracking susceptibility of the material. To do this, externally loaded test are the appropriate tools.

References

1. prEN ISO 17641: Destruction tests on welds in metallic materials – Hot cracking tests for weldments – Arc welding processes
   Part 1: General
   Part 2: Self-restraint tests
   Part 3: Externally loaded tests
2. Pohle C (1985) Möglichkeiten für die Beurteilung der Heißrissanfälligkeit von austenitischem Schweißgut. Schweißen und Schneiden 37, Heft 2: 498–502
3. DIN 50129: Prüfung metallischer Werkstoffe; Prüfung der Rissanfälligkeit von Schweißzusatzwerkstoffen. Beuth Verlag
4. Tösch J, Schabereiter H, Perteneder E, Rabensteiner G (1997): Bedeutung und praktische Beeinflussbarkeit des Ferritgehaltes bei der Schweißung austenitischer Stähle. Schweiß- und Prüftechnik Wien, Heft 2: 18–26
5. Wilken K, Kleistner H (1985) Remarks on the classification and evaluation of hot cracking tests. IIW-Doc IX-1379-85
6. Heuser H, Groß V, Ladwein T (1997) Hot cracking problems with submerged arc welding of heavy wall components from 22 % chromium duplex stainless steels. Duplex Stainless Steels 97 – 5th World Conference 1997, Maastricht, The Netherlands; KCI Publishing BV; D97–034

Influence of the Deformation Rate of Different Tests on Hot Cracking Formation

H. Herold, A. Pchennikov, M. Streitenberger

Institut für Füge- und Strahltechnik,
Otto-von-Guericke-Universität Magdeburg, Germany

Abstract

Referring to the ISO standardization of hot cracking test procedures with externally loaded specimens, three different and fundamental test procedures are assessed with the help of experiments and finite element analyses to find out the influence of different deformation rates on the test results of three well known stainless steels.

A comparison is made between the hot tensile test carried out with help of the Gleeble system using the Moscow procedure of the Standard of Russian Federation, the American instructions for the Hot Ductility Curve as well as the PVR-test (controlled deformation crack test) created in Austria. These main test procedures have generally been used for more then 30 years.

While within the American Hot Ductility Curve Test a round tensile specimens is torn after heating by thermal resistance to a test temperature some degrees below liquation at a tension speed of about 50 mm/s, the Russian method heats all round tensile specimens to the same maximum temperature just above the melting point and loads each specimen at another tension speed between 0.01 and 0.1 mm/s for approximation of the critical tension speed.

For tensile test comparison, the PVR test is carried out on round specimens with a diameter of 10 mm. A TIG melt run is simultaneous welded longitudinally at a superposed tension speed linearly increasing from zero to 1 mm/s in welding direction. The length position of the first crack appearing at the specimen defines the critical tension speed.

Three different stainless steels have been subjected to these test procedures. The test results of the different procedures precisely correlate within the materials ranking, but differ in the amount of evaluated test criteria.

Introduction

Hot cracking tests are conducted with the following different procedures using self restraint specimens or externally loaded specimens. Most of them relate to the theories of Prokhorov [1]and Matsuda [2]. They are described in Drafts of ISO standards [3–5].

While the theoretical hot cracking research is focused on local thermo-mechanical kinetics and metallurgical dynamics during solidification, the practically applied hot cracking test procedures are based on global concepts with measurable constitutional, mechanical, thermal and time-affected test criteria. Their development began with descriptions of hot cracking events by yes/no criteria (test procedure with self loaded specimens [4]) to quantify the hot cracking sensitivity by thermo-physical simulations, and finally to tests direct in welds during the welding process [5].

For example, the hot tensile test (Gleeble test) quantifies by the hot ductility curve the Brittleness Temperature Range (BTR), where the ultimate tensile strength and the reduction in area are plotted dependent on both the heating and the cooling process as a function of the test temperature. Both types, the Varestraint Test and the Transvarestraint Test, assess the hot cracking sensitivity by a diagram of total crack length versus applied surface bending strain longitudinally or transversally to the welding direction of a TIG-bead-on-plate weld. The PVR test (controlled deformation crack test) defines the hot crack initiation by a critical tension speed for the first hot crack of each hot cracking type.

Former comparisons of the results of different hot cracking tests with externally loaded specimen showed by [6, 7] that there is a good consistency in the cracking sensitivity of the selected four materials tested, although the test procedures and test criteria are very different.

The hot cracking theories by Prokhorov and Matsuda are based on a thermomechanical concept using the deformation rate, evaluating the equilibrium between initiation of and resistance against solidification cracking. The deformation rate is the mathematical derivation of the internal deformation per temperature within the Brittleness Temperature Range (BTR). Its dimension [%/°C] is the percentage of the changed deformation per temperature. Hot cracking appears when this deformation rate $d\varepsilon/dT$ during solidification within the BTR is higher than the critical deformation rate.

The preference of test procedures with external load is their possibility of quantifying the hot cracking sensitivity of high-alloyed materials and of deposited filler metal as well as the reactions of applied welding procedures to the test materials.

But opposite to the theory all test procedures work with different application of the tension speed. Thus, different test conditions are used to superpose thermal cycles by simulation or during welding and the tension for external deformation.

It still has to be clarified by which factors the test procedures are varying and how this is influencing the different test results.

Test Materials

The materials used for the comparison of test procedures were two austenitic stainless steels and one ferritic free-machining stainless steel, characterized by their Cr-equivalents and Ni-equivalents given in Table 1 and their chemical compositions given in Table 2.

Table 1. Tested stainless steels and their Cr-Ni-equivalents

AISI	Material	Mat. No	Cr-Equi WRC 92	Ni-Equi WRC 92	Cr-Equi 1948	Ni-Equi 1948
304	X5CrNi18-10	1.4301	19.13	13.48	19.00	10.78
316L	X2CrNiMo17-12-2	1.4404	20.32	13.90	19.35	11.50
430F	X14CrMoS17	1.4104	18.28	11.02	18.15	5.02

Table 2. Chemical composition of tested stainless steels

Mat. No	C [%]	Si [%]	Mn [%]	P [%]	S [%]	Cr [%]	Mo [%]	Ni [%]	Nb [%]	Ti [%]	N [%]
1.4301	0.05	0.37	0.96	0.034	0.048	18.2	0.29	8.68	0.018	0.006	0.09
1.4404	0.02	0.40	1.87	0.045	0.053	16.3	2.39	10.00	0.076	0.007	0.08
1.4104	0.12	0.42	2.64	0.015	>0.096	17.2	0.29	0.22	0.005	0.007	0.20

In Fig. 1 the test materials are characterized with their Cr-Ni-equivalents within the Schaeffler Diagram. This Schaeffler Diagram was also classified by Bystram [8] to show possible sensitivities and failure during welding of stainless alloys.

While both austenitic stainless steels would tend to hot cracking because of the small portion of δ-ferrite, the ferritic free-machining stainless steel could be sensitive to hardening brittleness because of its higher C-content. The contents of 0.6 % sulphur and 0.2 % nitrogen are interesting in view of hot cracking resistance.

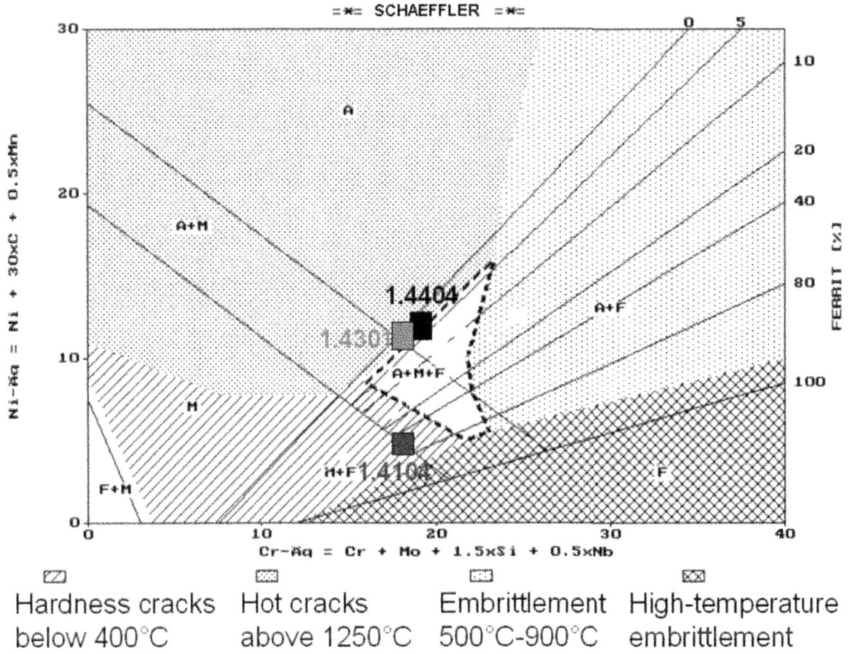

Fig. 1. Characterization of test materials within the Schaeffler-Diagram referring to Bystram [7]

Test Procedures and Results

Evaluation of Solidification Crack Initiation During Welding

Investigations of centerline solidification cracking in ship building industries have stimulated the calculation and measurement of cross displacements during tandem-submerged arc welding of larger plates. The measurements were taken for solidification crack-free welding design. They resulted in the assessment of the critical value for solidification-crack initiation, too [9].

A welding process-associated hot-cracking test method was created (Fig. 2) to measure the value and the speed of cross displacement which initiates the local solidification cracking. Such displacements are produced by plate movements due to assembly effects. The measuring device works with two cross gauges versus the welding direction.

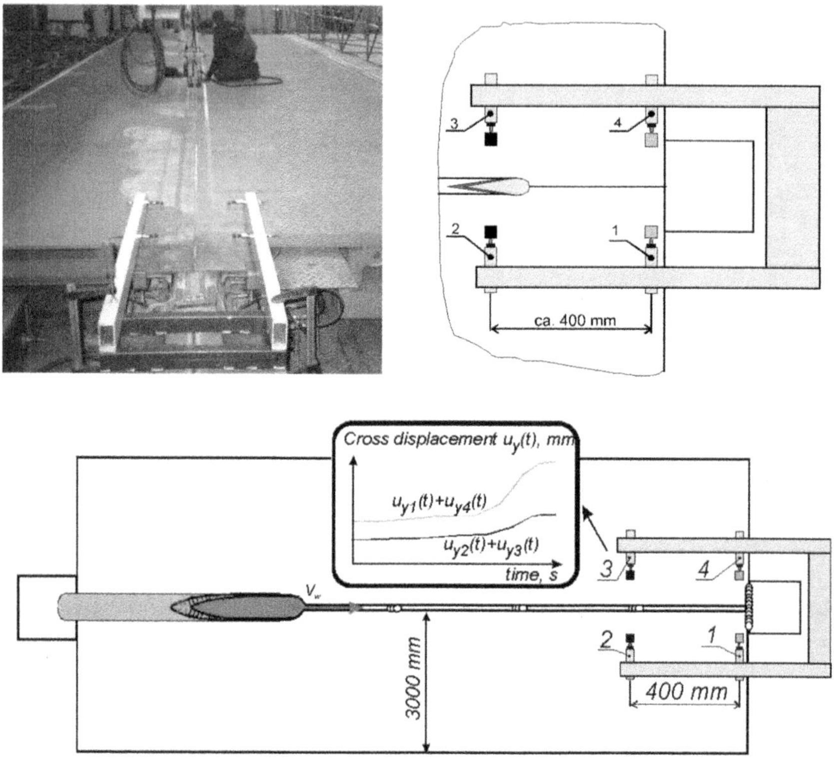

Fig. 2. Measuring device with two special cross gauges measuring the cross displacement u_y (t) versus time during one side welding of large components

The total displacements between the gauges across the weld seam were measured during the whole welding process over the seam length of about 10 to 18 m during welding fabrication in ship building industries. The speed of cross displacement v_q was calculated by its temporal change for this special welding process. The occurrence of hot cracking or the crack-free welding was examined by non-destructive weld inspection of a large number of fabricated large components.

The assessment of centerline solidification crack initiation during welding processing is based on the balance between the values of cross displacement and its changes versus time, distinguishing between welding assembly which initiates hot cracking and such which yields the movement of the gap without any hot cracking during welding [10].

The condition for hot crack initiation is given when the speed of cross displacement between the welding components is higher than the material-associated speed of cross displacement of the applied welding procedure.

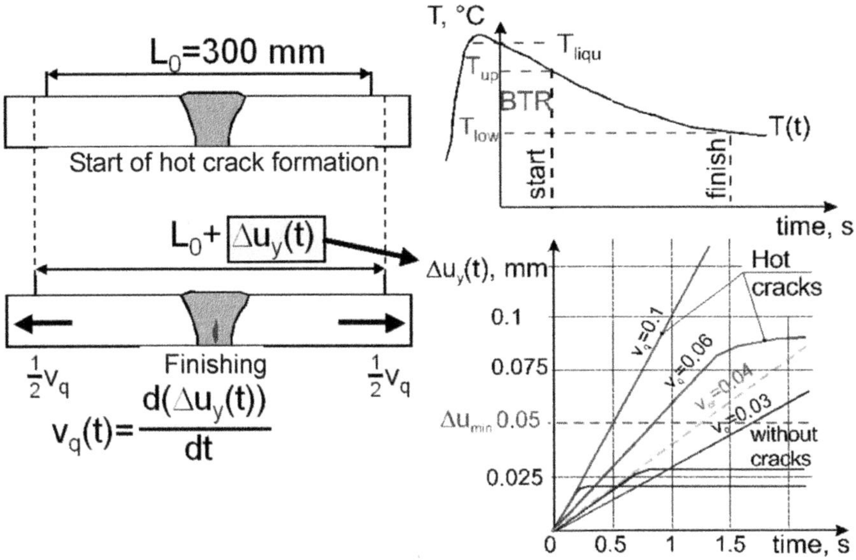

Fig. 3. Model and determination of the critical speed of cross displacement (tension speed) by the threshold curve within the diagram of cross displacement versus time during welding processing

The left hand side in Fig. 3 shows the model for the thermal tension vertical to the weld. The welding process on larger plates moves the components apart at the instant at which the observed point is passed by the welding torch. Above on the right, Fig. 3 shows the thermal cycle of the applied process.

The assessment diagram on the right shows the estimation of the critical speed of cross displacement (tension speed) during real welding. In this diagram the curves of measured speeds of cross displacement v_q are plotted as the cooling time of the weld metal versus the temporal change of the cross displacement $\Delta u_y(t)$ in [mm] and valuated according to the appearance of hot cracking.

A threshold curve separates the conditions for hot crack initiation (grey line) and hot cracking resistance (bold line), as a result of low amounts of cross displacement and different speeds of cross displacement.

For example, the critical tension speed for the applied tandem-submerged arc welding of 10 mm thick shipbuilding plates was found to be 0.04 mm/s, the critical amount of cross displacement was about 0.05 mm within the gap.

Hot Cracking Test Procedures Based on Analytical Simulation of Welding Procedures

The hot cracking test procedures which are based on analytical thermomechanical simulation use round tensile specimens with diameters of about 10 mm in contrast to the real weldment. The round specimens are arranged in accordance with the model of centerline solidification cracking in Fig. 3 and with the model of the Moscow MIS-test procedure in Fig. 4 created by Prokhorov from his theories. The resistant heated central zone of the round tensile specimen is in accordance with the tension direction vertical to the weld. The thermo-physical simulation with the Gleeble®-3500 System superposes the thermal cycle on the deformation using two different controlled test procedures.

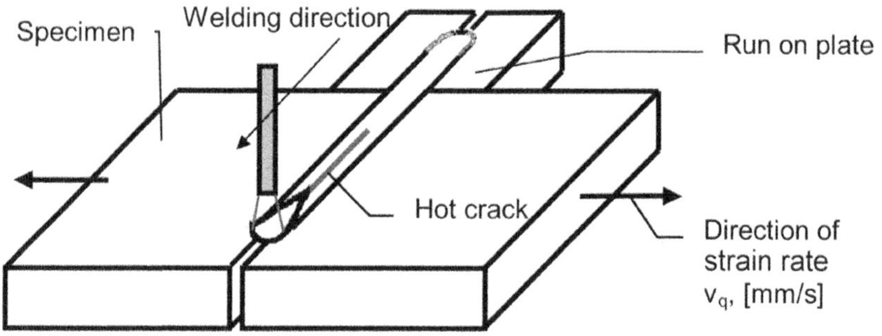

Fig. 4. Moscow hot cracking test procedure MIS, standardized by the Russian Federation

Determination of Hot Cracking Sensitivity by the Simulated Moscow Test Procedure

The simulation of the Moscow test procedure MIS needs several round test specimens which are heated up to the maximum temperature just above the liquation temperature. The same thermal cycle is applied to all tests. Each specimen is strengthened at another tension speed between 0.001 and 0.1 mm/s. The heating requires varied heating rates deviating from those of the real welding thermal cycle due to the specimen geometry. The thermomechanical superposition includes several steps (Fig. 5).

At first the heating rate is controlled by free expansion (Force = 0) up to 1000 °C, measuring the thermal extension. The slower heating rate up to the maximum temperature of about 1450 °C starts after setting the extension to zero.

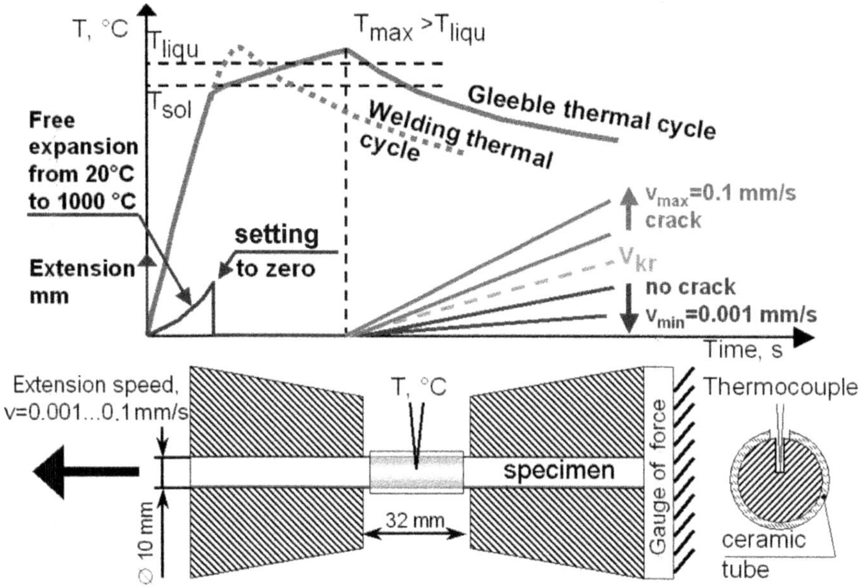

Fig. 5. Simulated Moscow test procedure for determination of critical tension speed for solidification cracking

The deformation at varied tension speeds of the other specimens starts after a short holding time at the maximum temperature under free cooling. The critical tension speed for solidification cracking results from the empirical approximation of crack-free yielding or hot cracking.

The effect of the simulated Moscow test procedure on the test materials is shown in Fig. 6–8 with the simulation records plotting only two varied tensions speeds near the critical one.

AISI 304 in Fig. 6 yields a tension speed of about 0.010 mm/s with increasing force without any hot cracking. The same material failures after loading with 0.012 mm/s tension speed.

The crack initiation is marked in the test record by the loss of force , determining a temperature for solidification crack initiation of about 1320 °C. Subsequently, the critical tension speed for crack initiation of the solidifying material is given by the threshold curve at v_{cr}= 0.011 mm/s. It also defines the lowest temperature of the BTR (brittleness temperature range).

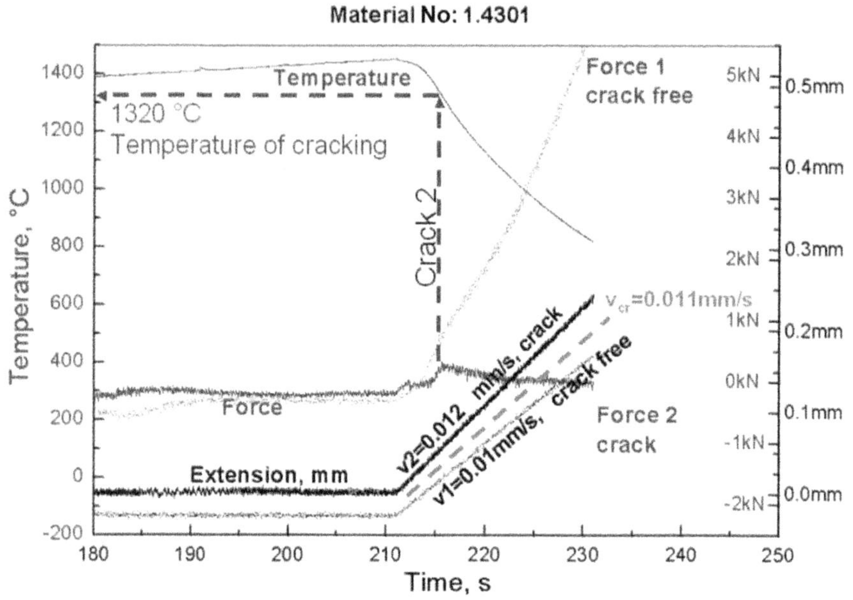

Fig. 6. Test record of the simulated Moscow test procedure of AISI 304

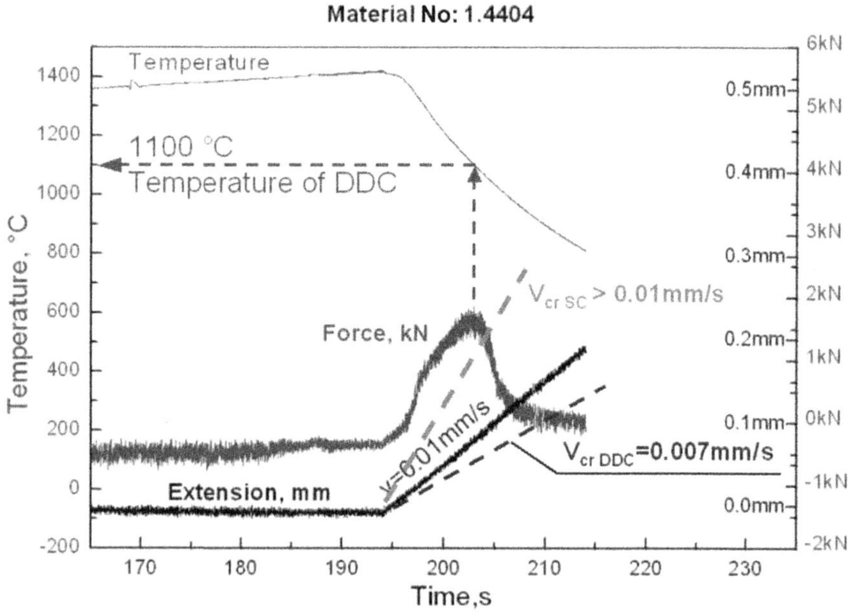

Fig. 7. Test record of the simulated Moscow test procedure of AISI 316 L

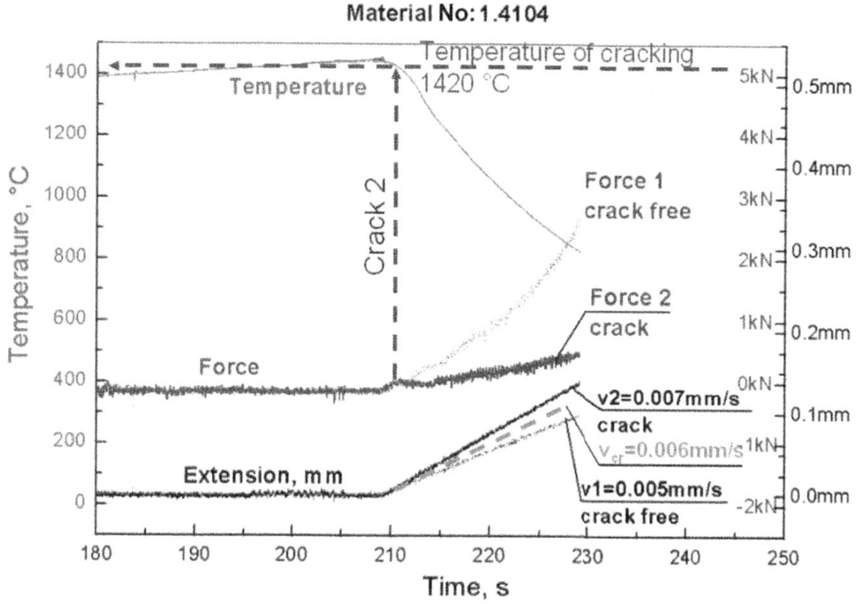

Fig. 8. Test record of the simulated Moscow test procedure of AISI 430 F

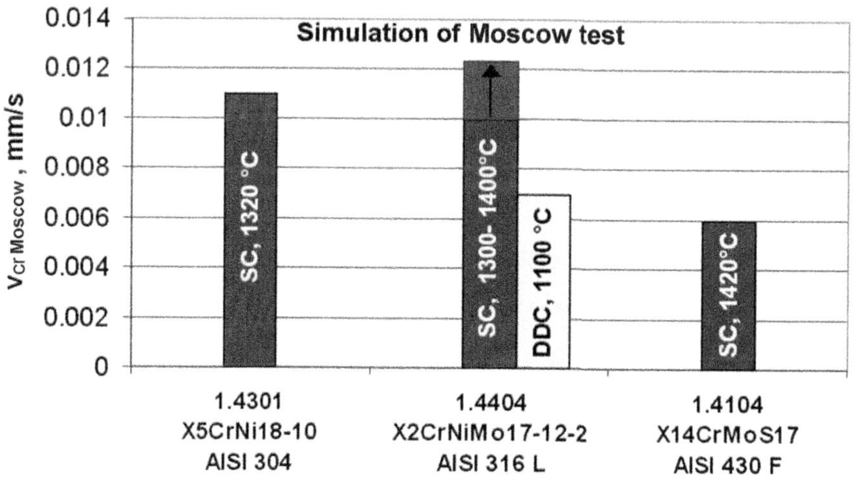

Fig. 9. Comparison of results obtained from the simulated Moscow test procedure

Various tension speeds induce two different crack mechanisms at the molten stainless steel AISI 316 L (Fig. 7). Solidification cracking was determined at a critical tension speed of about 0.015 mm/s.

Lower tension speeds in Fig. 7 result in failure at 1100 °C characterized by the type of ductility dip cracking (DDC). The critical tension speed for DDC was assessed to be 0.01 mm/s.

The hot cracking behavior of the ferritic stainless steel (Fig. 8) is not surprising considering the very low critical tension speed of about 0.006 mm/s, initiating solidification cracking already at 1420 °C. The graph in Fig. 9 compares the quantified hot cracking sensitivities determined in the simulated Moscow test for the three stainless steels.

Determination of Hot Cracking Sensitivity by the Standard Procedure of the Hot Tensile Test

The hot tensile test procedure is based on the analytical simulation of weld thermal cycle, distinguishing between controlled base metal heating up to the maximum temperature and cooling after exceeding the maximum temperature for the component section under examination.

The analyzed thermal cycle is interrupted at different chosen temperatures for each specimen by always the same high stroke rate (tension speed of about 50 mm/s) both during heating and during cooling (Fig. 10).

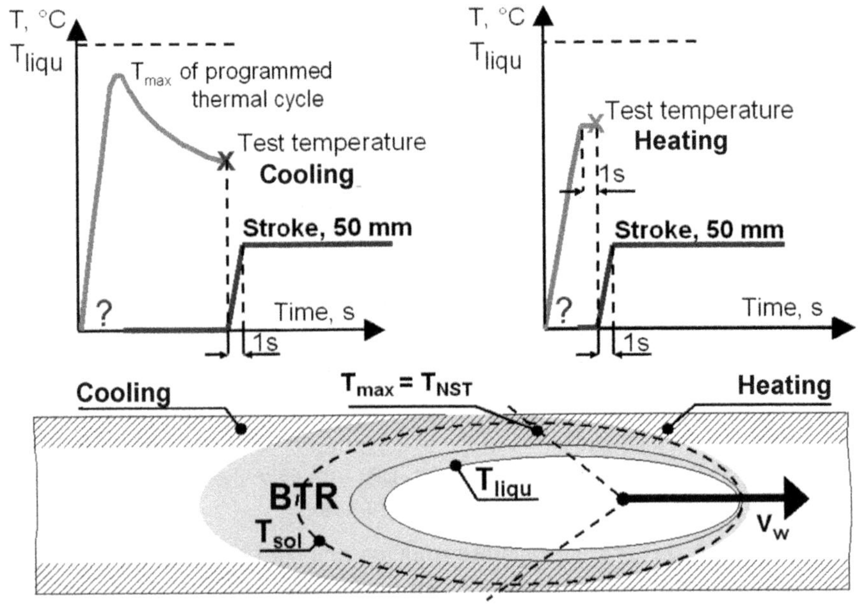

Fig. 10. Simulation programs for hot tensile tests during heating and during cooling (above),
analytical modeling of a thermal cycle for a welding process (below)

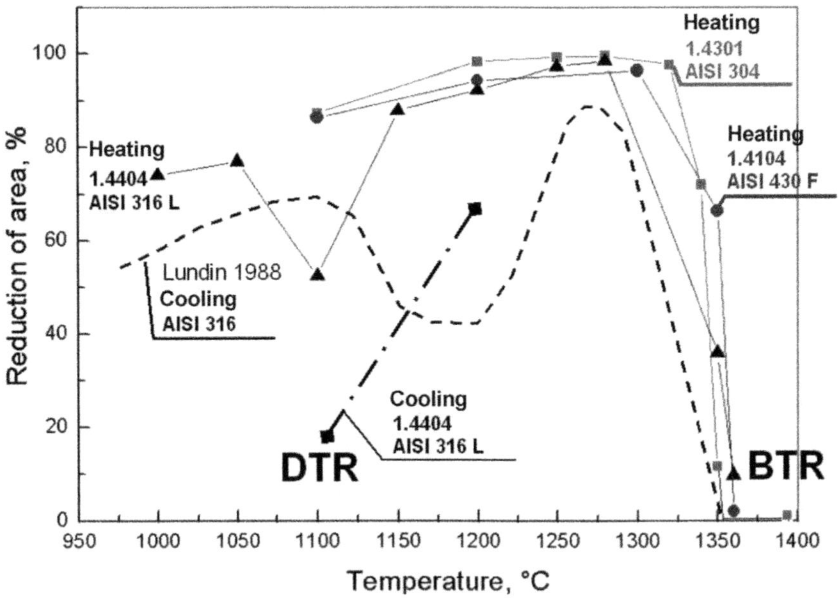

Fig. 11. Results of the hot tensile tests obtained during heating and cooling of the stainless steels

The test procedure during heating was carried out in correlation with the heating process of the simulated Moscow test by varied steps and onset of stroke after a holding time of about 1 s at test temperature.

The question mark in Fig. 10 notes that heating can be done by free expansion of with clamped specimen. This hot tensile test was carried out with clamped specimen.

The test results obtained during heating are plotted as the reduction of area versus the test temperature in Fig. 11. The nil ductility temperature (NDT) for all test metals was found to be near 1360 °C. The diagram is completed in principle for the cooling process by a curve of a modified AISI 316 from Lundin 1988 [11].

Added test results from cooling AISI 316 L down to two temperatures indicate a decreasing reduction of area within the ductility dip temperature range in comparison to the tests during heating. At 1100 °C, there is a 27 %-decrease in reduction of area during heating and cooling.

The ductility dip at this temperature correlates with grain growth and grain boundary effects (Fig. 12) during passing the maximum temperature of the thermal cycle and holding at a temperature exceeding 1100 °C. This results in significant grain boundary strengthening of AISI 316 L.

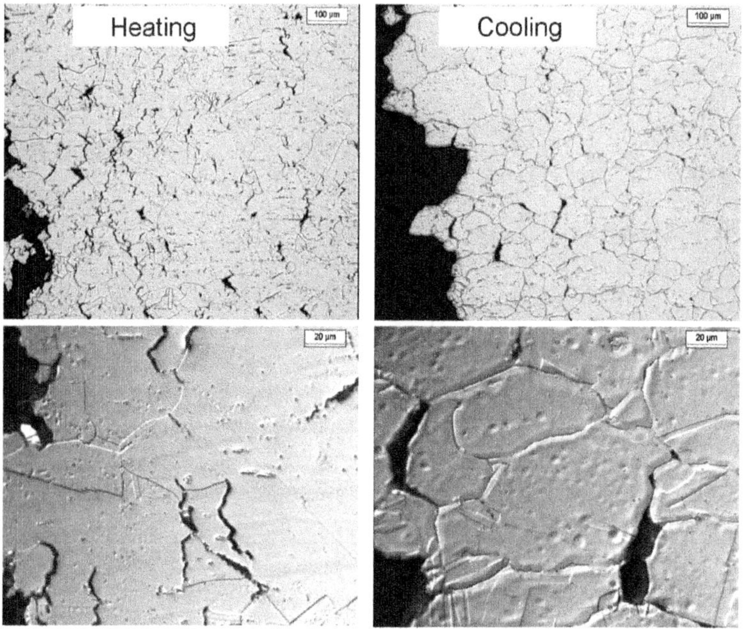

Fig. 12. Comparison of AISI 316 L microstructure sections for testing during heating and cooling, respectively

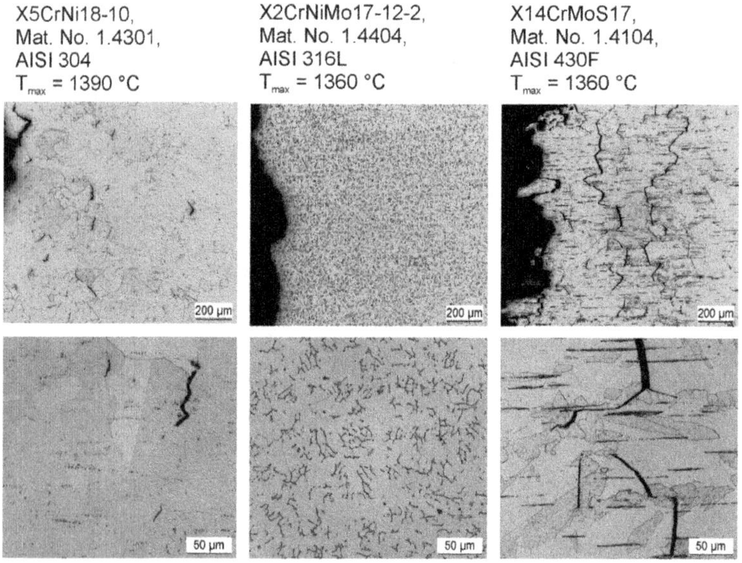

Fig. 13. Microstructure after testing during heating at maximum temperature

The applied maximum temperatures of the thermal cycles during heating result in characteristic microstructures of the pulled zones (Fig. 13).

While AISI 304 solidifies from 1390 °C as austenite with low embrittlement, AISI 316 L solidifies from 1360 °C as ferrite without any cracking. The remaining ferrite content at room temperature is about 6 %. The ferritic free-machining stainless steel embrittles at 1360 °C by partially liquated MnS particles.

PVR-Test (Controlled Deformation Crack Test, Longitudinal Tension Crack Test or Controlled Flat Tension Test)

The PVR test represents a test procedure for evaluating the effects of the welding procedure with regard to hot cracking [12]. Only one single test specimen is required to determine the hot cracking sensitivity of a base metal and/or a weld metal during welding.

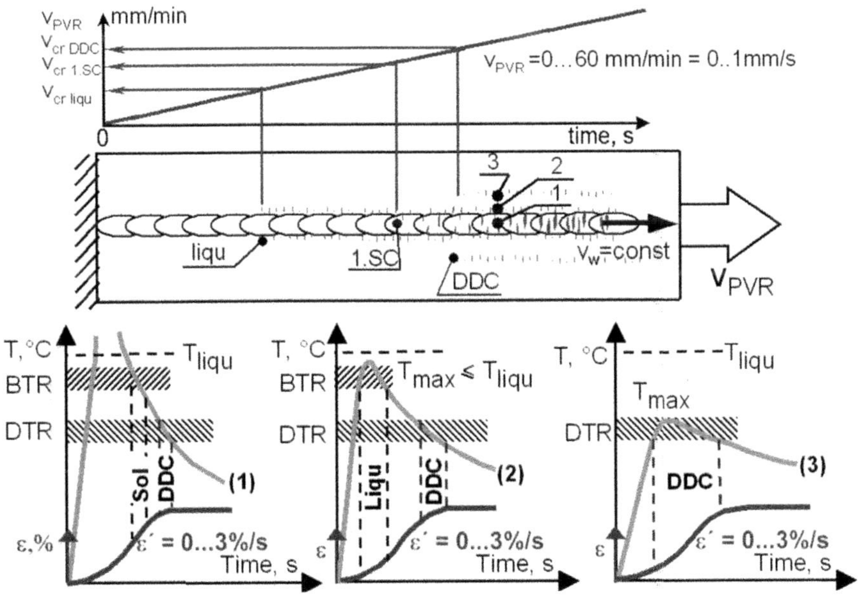

Fig. 14. Initiation of solidification cracking, liquation cracking and ductility dip cracking on one specimen (above) and effective superposition of real thermal cycles in the heated zones during the applied welding procedure with the deformation rate ε' induced by the linearly increasing tension speed during welding (below)

The ordinary PVR test procedure in Fig. 14 uses flat specimens with the dimensions of 40 x 10 x 300 mm (width x thickness x length), clamped into a special tension fixturing device, lengthened in a horizontal servo-hydraulic system of the test equipment. The welding process with a constant welding speed is superposed by a linearly increased tension speed v_{PVR} in welding direction.

The standard PVR test procedure is carried out at tension speeds linearly increasing from zero to 1 mm/s, using bead-on-plate TIG-welding with argon shielding gas and applying two types of heat input per unit length, i.e. about 7 kJ/cm (I_s = 180 A, U_s = 12 V, v_s = 19 cm/min) and 10 kJ/cm, respectively. The critical tension speed v_{cr} is the test criteria for the PVR test. v_{cr} corresponds to the length coordinate of the first hot crack detected visually on the specimen surface at a magnification of 25. For each of the hot crack types it is possible to determine solidification cracking, liquation cracking and ductility dip cracking (Fig. 14).

The possibility for to assessing all hot cracking types within one specimen depends on the superposition of the local heated zone beside the welding bead at the applied tension rate.

Different indices for the critical tension speed v_{cr} are used, depending on the size and number of visible cracks. For example: v_{cr1st} is the critical tension speed for the first microscopically visible hot cracking feature, v_{cr3} describes the first three hot cracks per 10 mm of weld bead and v_{cr9} determines the first nine hot cracks per 10 mm of weld bead. Modeling of the PVR test [13] has shown the correlation between test criterion, hot cracking theory by Prokhorov, and its applicability.

For a comparison with the tensile test results the PVR test is performed on round 10 mm diameter specimens. A TIG-melt run is simultaneously welded longitudinally at a superposed linearly increasing tension speed from zero to 1 mm/s in welding direction.

Fig. 15–17 combine the length position of cracks within the specimen both with their critical values to quantify the hot cracking sensitivity and with their microstructure for the classification of the hot cracking type. Primarily, all types of hot cracks disappear vertical to the welding and tension direction.

AISI 304 in Fig. 15 is sensitive to liquation cracking beside the fusion line because of the critical tension speed $v_{cr9liqu}$ of about 15 mm/min (0.25 mm/s). The resistance against solidification cracking is higher. The first micro-solidification cracks are initiated by the liquation cracks at $v_{cr\,1st\,SC}$ of about 36 mm/min (0.6 mm/s) directly at the fusion line. Macro-solidification cracks were not observed in the round specimens up to 60 mm/min (1 mm/s).

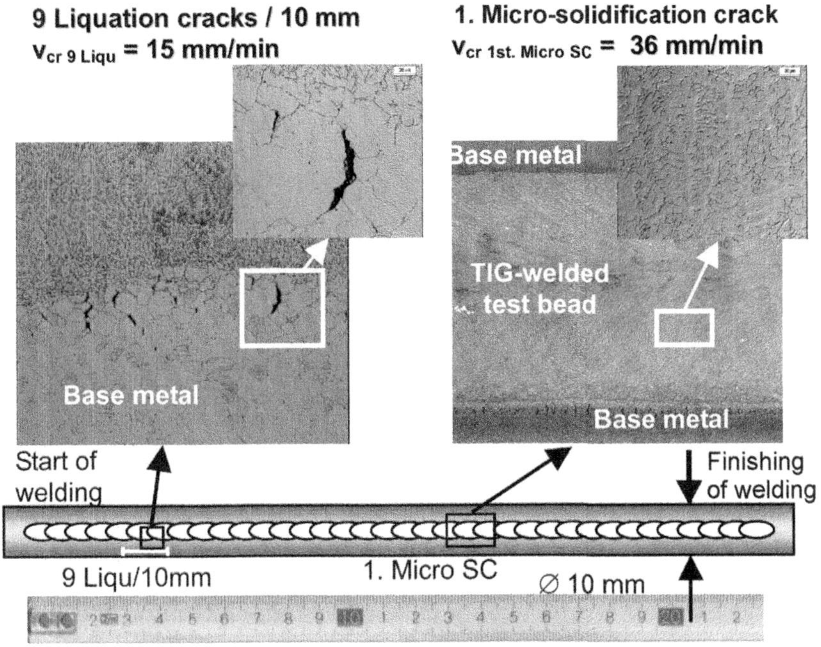

Fig. 15. Determination of v_{cr} of hot cracking in the PVR test of AISI 304

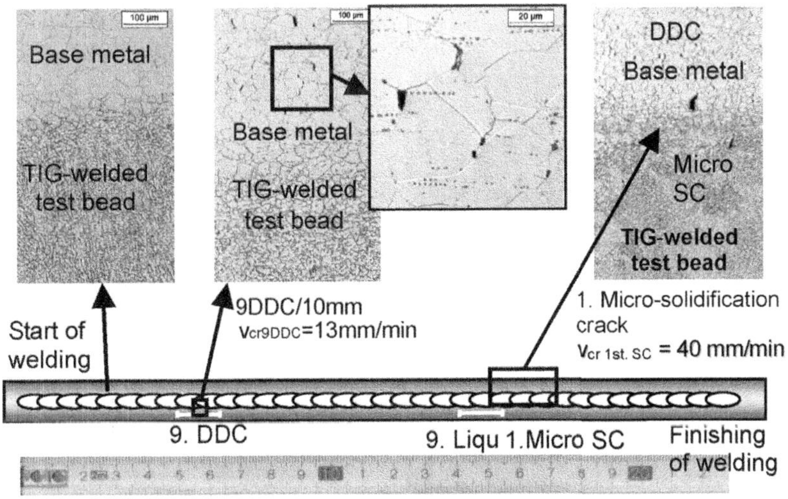

Fig. 16. Determination of v_{cr} of hot cracking types in the PVR test of AISI 316 L

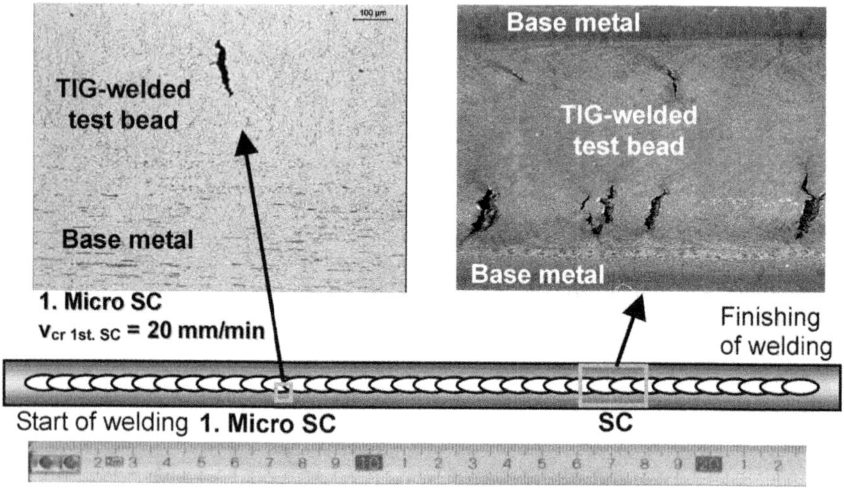

Fig. 17. Determination of v_{cr} of hot cracking types in the PVR test of AISI 430 F

AISI 316 L develops preferably DDC within the ductility dip temperature range in some distance from the fusion line already at $v_{cr\ DDC}$ of about 13 mm/min (0.22 mm/s) (Fig. 16). The micro-solidification cracks are detectable near the fusion line only above the high critical tension speeds of about 40 mm/min (0.66 mm/s). Macro-solidification cracks were not found in the specimens below 60 mm/min (1 mm/s).

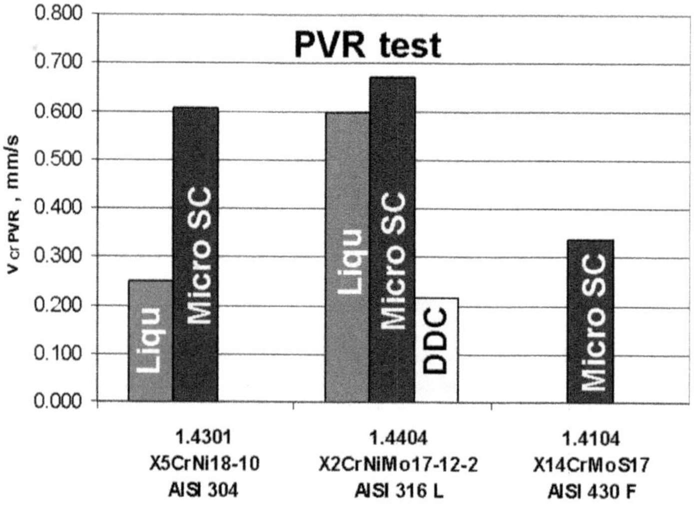

Fig. 18. Comparison of PVR-test results from tested stainless steel

The ferritic free-machining stainless steel AISI 430 F in Fig. 17 exhibits classical solidification cracking. The first micro-solidification crack occurs in comparison with the other materials at the critical tension speed of about 20 mm/min (0.33 mm/s) (Fig. 18) followed by macroscopic solidification cracks above v_{cr}.

Conclusion

The type of hot cracking test procedure influences the probability for quantifying the sensitivity to different hot cracking types by its specific superposition of thermal cycle and deformation rate.

Each test procedure programs the conditions for materials characterization in its own way. The material characterizing results depend on the success of superposition of the simulated welding thermal cycle or real welding with a specific deformation, independently of the question of whether the test was carried out using a tension speed or a deformation rate.
Each procedure has its preference:

- The simulated Moscow test procedure determines the sensitivity to solidification cracking and DDC as illustrated Fig. 5 globally. The critical tension speed for a specific hot crack type in the test material is estimated as a threshold curve from a number of tests carryied out at different tension speeds between 0 and 0.1mm/s;
- The classic hot tensile test procedure uses maximum temperatures below the liquidus temperature rupturing several specimens at a high tension speed of about 50 mm/s and different temperatures during heating and cooling as shown in Fig. 10 and 11. The test results are quantified by the deformation (reduction in area in %) and characterized by the microstructure and fracture behavior, classifying the material-specific crack mechanisms (liquation, solidification, grain boundary migration, segregation, precipitation, reheating, e.g.) by means of additional testing techniques;
- The PVR test determines the critical deformation rates for hot crack initiation quantifying the first solidification crack initiation in the weld metal and the different crack types within the HAZ during one single test welding procedure globally.

The application of different hot cracking test procedures results in equal ranking of material characteristics, independent of the deformation rate and of the specific test criteria. The hot cracking sensitivities, quantified with the help of different test criteria on three known stainless steels, agree with the tendency of hot cracking types detected in practice on the condition that suitable deformation rates and characteristic thermal cycles of the applied three test procedures are used.

References

1. Prokhorov NN (1962) The technological strength of metals while crystallisation during welding. Welding production Vol 9: No 4 April: 1–8
2. Matsuda F, Hashimoto T, Senda T (1969) Fundamental investigation on solidification structure in weld metal. Trans Nat Res Inst Metals 11: 43–58
3. PrEN ISO 17641-1, Nov 2003. Destructive tests on welds in metallic materials – Hot cracking tests for weldments – Arc welding processes – Part 1: General (ISO/FDIS 17641-1:1003)
4. PrEN ISO 17641-2, Nov. 2003. Destructive tests on welds in metallic materials – Hot cracking tests for weldments – Arc welding processes – Part 2 : Self-restraint tests (ISO/FDIS 17641-2:1003)
5. prCEN ISO/TR 17641-3 Destructive tests on welds in metallic materials – Hot cracking tests for weldments – Arc welding processes – Part 3 : Externally loaded tests (ISO/DTR 17641-3:1003)
6. The Japan Welding Society (1986) Hot cracking susceptibility evaluation of austenitic stainless steel weld materials. International Institute of Welding (Doc IIW IX-1395-86)
7. Wilken K (1999) Investigation to compare hot cracking tests – Externally loaded specimen. International Institute of Welding (Doc IIW IX-1945-99)
8. Bystram MCT (1956) Some aspects of stainless alloy metallurgy and their application to welding problems. British Welding Journal. Febr: 41–46
9. Herold H, Streitenberger M, Pchennikov A (2001) Prevention of centreline solidification cracking during one side welding. International Institute of Welding (Doc IIW IX-2000-01 (and II-C-220-01)
10. Herold H, Pchennikov A, Streitenberger M (2003) Current problems in hot cracking research described on the example of PVP test. Materials Science Forum 426–432: 4093–4098
11. Lundin CD, Lee CH, Menon R, Osorio V (1988) Weldability evaluations of modified 316 and 347 austenitic stainless steels: Part I – Preliminary Results. Welding Research Supplement 67 Febr.: 35-s – 46-s
12. Folkhard E (1984) Metallurgie der Schweißung nichtrostender Stähle. Springer Verlag, Wien New York, p 153
13. Herold H, Streitenberger M, Pchennikov A (2001) Hot Cracking Theory by Prokhorov and Modelling of the PVR-test. Welding in the World 45, ¾: 17–22

Testing for Susceptibility to Hot Cracking on Gleeble™ Physical Simulator

S.T. Mandziej

Advanced Materials Analysis, Enschede, The Netherlands

Abstract

Hot cracks appear when thermal shrinkage together with deformation caused by restraint cannot be accommodated by plastic deformation. This happens during welding to such alloys, which segregate on heating and cooling at near-solidus temperatures, in particular when low-melting and mechanically weak phases form and occur over a wide range of temperatures. To check for susceptibility to the liquation cracking caused by the low-melting, weak phases, hot tensile testing can be used in combination with a thermal cycle resembling that of real welding. This procedure, which can be executed on a Gleeble™ [1] thermal-mechanical simulator, comprises tensile testing of a number of cylindrical samples at the temperatures below solidus and determining their hot strength and ductility. For measuring of the brittle temperature range (BTR), the nil strength temperature (NST) is determined and used as the peak point, down from which the ductility recovery temperature (DRT) is searched for. The ductility is measured after the tensile test as a reduction in area at fracture. An alternative to the hot tensile test during the simulated welding cycle is the strain-induced crack opening (SICO) test, in which a rod-like sample mounted in "cold" copper jaws of the Gleeble is heated by electric current and then compressed till formation of a bulge in its uniformly heated central portion and appearance of cracks due to secondary tensile strain developed along the maximum perimeter of this bulge. Next to the studies of liquation cracking and ductility-dip cracking of the reheat-type, the Gleeble procedures can be also used to determine sensitivity to solidification cracking

---

[1] Gleeble is a registered trademark of Dynamic Systems Inc., Poestenkill, NY, USA.

and for this the samples are melted and solidified in a controlled manner before the hot tensile testing or the SICO testing. The developed routines of the Gleeble testing allow accurate determination of temperatures at which the cracks occur as well as measurement of critical strains to fracture and strain rates which are associated with the hot cracking. At an appropriate geometry of samples and optimum setting of the experiments, the hot cracks generated in the samples have sizes comparable to those that occur in heat-affected zones of real welds or in weld metals.

Introduction

To date, over 150 weldability tests exist, many of which are designed to assess the susceptibility of welds to hot cracking. In general, they can be put into two categories as representative (self-restraint) and simulative (externally loaded) test techniques [1]. The representative test technique usually tells only "cracking" or "no-cracking" of a material when an actual welding situation is represented, which cannot quantify the cracking susceptibility of the material under different welding condition. The simulative test can follow a thermal-mechanical history of a material during welding, while an external strain is usually applied to produce cracks or to record material characteristics allowing quantification of cracking susceptibility.

Hot cracking of welds appears when shrinkages of the solidifying and thus stiffening weld metal and of the adjacent heat affected zone cannot be compensated or accommodated by at first the back-filling (liquid flow) of the molten material and afterwards by the visco-plastic flow of the solidified hot material of the welded joint. The process is dynamic and to avoid hot cracking the welded material must be able to plastically deform at critical sites of the welded joint with the strain rate exceeding that of the shrinkage during the relatively fast heat cycle of the welding.

Studies in the 1950's carried out in the USA by W.F. Savage at Rensselaer Polytechnic Institute and at Duffers Associates Inc., led to the development of the Gleeble thermal-mechanical simulator which allowed the hot ductility of welded materials during thermal cycles of the welding processes to be determined [2]. This study showed that the region most susceptible to hot cracking is the heat affected zone of the parent metal, in which contaminants entrapped at grain boundaries form liquid or low strength solid films while the grains become stiff and strong. It was also found that if such weak films exist over a large temperature range after solidification, the welded materials show hot cracks in the HAZ.

To determine the range at which the weld HAZ is prone to hot cracking, a concept of nil strength temperature was introduced as the higher temperature of the brittle range, and appropriate attachments were designed to measure it. The lower temperature of the brittle range, so-called nil ductility temperature, was then taken as that at which 5 % reduction in area on hot tensile samples appeared [3].

The Gleeble testing procedure required a large number of samples to be hot tensile tested with strain rates representative of various welding methods (heat inputs), and this stimulated a study to develop another, simpler test, compatible with the deformation rates already known from the Gleeble testing. As a result, the Varestraint test was proposed in the early 1960's by W.F. Savage and C.D. Lundin [4], and applied to study the hot cracking susceptibility of welded alloys.

The Varestraint test comprises bending a test plate while the weld bead is being made on the long axis of the plate. The original Varestraint test had some limitations, e.g. difficulty in controlling the real amount of strain at the outer bent surface due to the position of a neutral bending axis, which varied depending on the strength and strain partitioning between the hot and cold parts of the sample during bending.

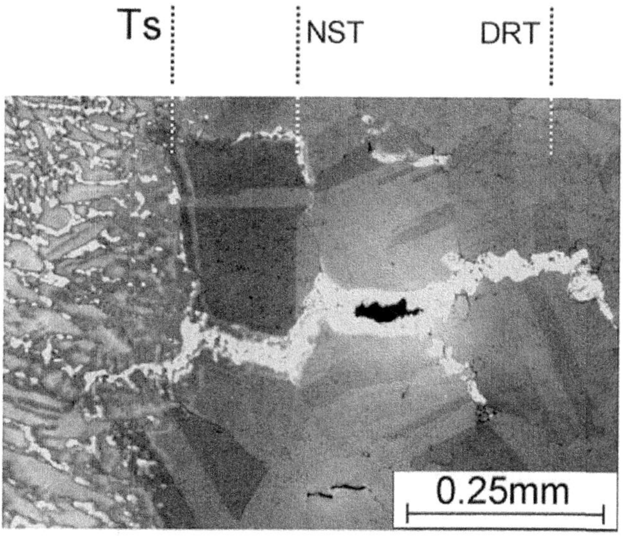

Fig. 1. Weld HAZ liquation crack appearing in a small distance from the fusion line i.e. nucleating at nil strength temperature (NST) lower than the solidus temperature Ts; marked ductility recovery temperature (DRT) near which the crack is closing

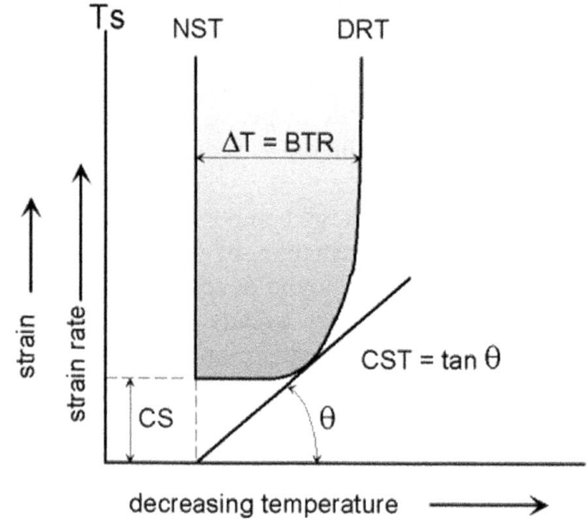

Fig. 2. Schematic presentation of the brittle temperature range (BTR) below solidus temperature (T_S) relating the hot crack susceptibility to critical strain (CS) and strain rate (CST)

The Gleeble testing method is dealing with the occurrence of the brittle temperature range (BTR) at high temperature during the solidification of the weld and the subsequent cooling. The BTR reflects the general appearance of the cracks occurring in heat-affected zones of welds (Fig. 1), and can be illustrated on the graph proposed by Prokhorov [5], as shown in Fig. 2. The graph shows that for any welded alloy there is some critical strain value, which can be accommodated at very low cooling rates by, for example, self-diffusion controlled visco-plastic flow of the hot metal. The crack in HAZ does not form exactly on the fusion line between weld metal and base metal; there is always certain space in which high diffusion rate and back-filling of the nucleating cracks or voids allow healing of the base metal. The hot cracks in HAZ occur in a small distance from the fusion line when the strain rate caused by shrinkage during cooling of the weld is sufficiently high for the solidifying material to "enter" the BTR. The wider the BTR of the alloy, the higher the hot crack susceptibility.

What is a Gleeble?

The thermal-mechanical simulator called Gleeble originated from the welding thermal cycle simulator built in 1948 at Rensselaer Polytechnic

Institute in Troy, NY, USA, which soon after (1951) was equipped with pneumatic deformation system and more recently with dynamic servo-hydraulic system (1979) and computer controls (1980). Its developer, Dr. W.F. Savage, wrote in one of his review articles [6]: "*In 1946 Nippes and I attempted to evaluate the notch toughness of various regions of the heat-affected zone of ship steel weldments by careful placement of the notches in Charpy V-Notch specimens machined from weldments. However, the results were biased by the sloping nature of the heat-affected zone, which caused the fracture to traverse more than one portion of the zone. Consequently, we decided to develop a thermal simulator and utilize temperature measurement data accumulated during the earlier cooling-rate studies at Rensselaer.*"

By its design, Gleeble has been dedicated to reproduce weld thermal cycles characteristic of the heat-affected zones with consequences related to thermal-mechanical effects caused by thermal gradients, restraints and shrinkage deformations. For this the up-to-date Gleeble uses "bulk" samples of 10 mm or more diameter and in appropriate parts of the thermal cycle can exert tensile or compressive deformations with strain rates adequate to those of the real welding process. The amounts of deformation and strain rates resulting from cooling rates and thermal gradients can be calculated using contemporary computer modeling techniques. The last should take into account apparent strengthening caused by thermal gradients and strain rates [7].

The thermal-mechanical situation in HAZ during welding comprises in the thin element of the base metal parallel to the fusion plane and longitudinal to the welding direction twice crosswise and lengthwise compressions on heating and on cooling and once on cooling a tension in the direction normal to the fusion plane, which is also the main direction of the heat flow.

For the electric arc welding it also involves the presence of electric current and related electro-thermal effects. Gleeble accounts for these by using AC electric resistance heating uniform through the cross section of the specimen.

This situation is schematically presented in Fig. 3. To simulate the HAZ in Gleeble, rod-like specimens mounted in water-cooled copper grips are heated at a rate resembling the weld thermal cycle and a short span between the grips assures that the heated zone is about twice the width of a real weld HAZ (Fig. 4).

When necessary, controlled tensile or compressive strains can be added in any point of the thermal cycle. The thermal cycle in the Gleeble is controlled by a thermocouple, percussion welded to the surface usually in the middle of the specimen.

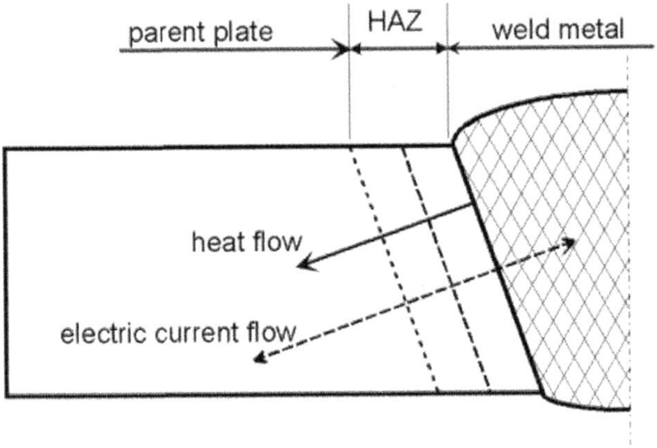

Fig. 3. Schematic presentation of the heat affected zone formed during electric arc welding with heat flow from the weld metal to parent plate causing thermal gradient

Fig. 4. Simulation of HAZ situation on a rod-like specimen mounted in water-cooled copper jaws of the Gleeble

The ability of Gleeble to deform heated specimens can be effectively used for hot tensile testing. The length of the uniformly heated zone in the middle of the Gleeble's specimen results from the balance between electric current heating the specimen and the heat flow towards the mounting grips.

Fig. 5. Stainless steel "hot" jaws of Gleeble for mounting of hot tensile specimens; note reduced contact portion between the jaws and the specimen

Depending on the span between the grips, as well as on conductivity of the grip material and on size of the contact area between the specimen and the grips, this length can be manipulated to a reasonable extent.

The photographs in Fig. 5 and in Fig. 6 show an example of the "hot" Gleeble jaws made of stainless steel, and the resulting substantial length of the uniformly heated zone on the specimen mounted in such "hot" jaws.

Fig. 6. Elongated uniformly heated zone on a plain carbon steel specimen mounted in the "hot" stainless steel jaws of the Gleeble

Hot Ductility and Hot Cracking

As the hot cracks form on-cooling when tensile strains caused by shrinkage and assisted by restraint cannot be compensated by ductility of an alloy, then method of studying susceptibility of an alloy to hot cracking should involve tensile testing at the conditions simulating these of the real welding (or casting) process. The use of thermal-mechanical simulator like Gleeble, able to reproduce on a specimen the welding thermal cycle and impose a strain in a controlled manner, allows achieving this goal.

Testing on a Gleeble for susceptibility to weld liquation cracking / HAZ hot cracking, means hot tensile testing of a number of specimens on-heating and then on-cooling, and determining their hot ductility measured as a reduction in area at the specimen's neck portion after the test. This procedure is schematically presented in Fig. 7 [1].

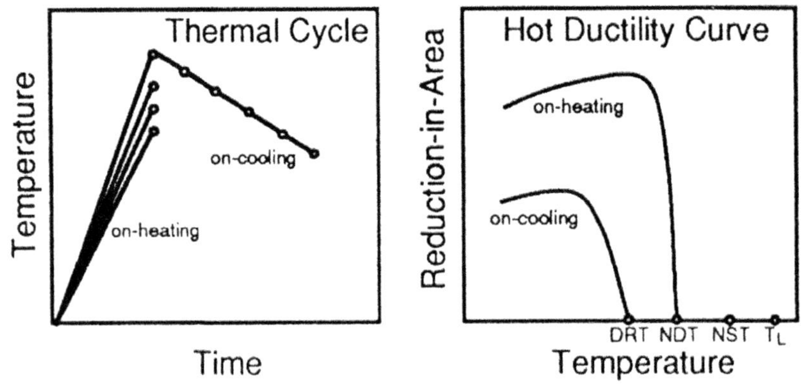

Fig. 7. Schematic of Gleeble's procedure for hot ductility testing, including hot tensile testing on-heating up to NST and then on-cooling after weld thermal cycle with NST as the peak temperature; note the NST being lower than the melting point T_L

The hot ductility of a weldable alloy increases gradually with increase of testing temperature from ambient towards melting point, however before reaching this point it drops abruptly from certain maximum to nil. Just above this nil ductility temperature (NDT) appears the nil strength temperature (NST) at which the alloy looses its strength due to formation of weak or liquid phases along grain boundaries. The real physically measurable melting temperature of such alloy $-T_L$, is higher than NST, as shown in Fig. 7.

On-cooling from the melt or from the NST, the ductility does not recover exactly at NDT but below it at so-called ductility recovery temperature – DRT. The temperature span from NST to DRT is considered to be the brittle temperature range – BTR.

The extent of BTR can be used as a rough criterion of the susceptibility to hot cracking, however more exact criterion is the measure how fast does the ductility recover on-cooling as compared with its decrease on-heating. As the reference point for this measurement the maximum of ductility from the on-heating ductility curve is taken and the representative areas below the on-heating and on-cooling curves are compared (Fig. 8) [8]. Arbitrarily considering 5 % of reduction-in-area on-cooling as the ductility recovery point and comparing the hot ductility curves on-heating and on-cooling, the nil ductility range (NDR = BTR), ductility recovery rate (DRR) and ratio of ductility recovery (RDR) can be determined to exactly characterize the susceptibility of an alloy to hot cracking.

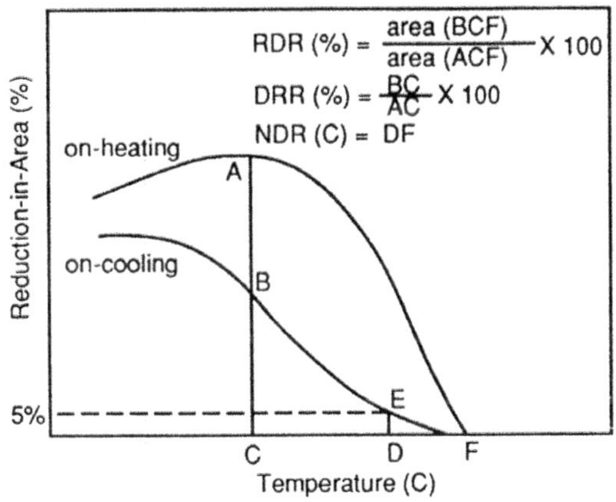

Fig. 8. Evaluation of hot ductility curves for hot cracking susceptibility [8];
– RDR = ration of ductility recovery,
– DRR = ductility recovery rate,
– NDR = nil ductility range (or BTR),
– Point F = nil ductility temperature (NDT) on-heating,
– Point D = ductility recovery temperature on-cooling

The reference point for the ductility measurements determining the susceptibility to liquation cracking is the nil strength temperature. This temperature has to be used as a peak of the welding thermal cycle, on-cooling after which the hot tensile tests should be run. To measure the NST on Gleeble an attachment is used, schematically presented in Fig. 9.

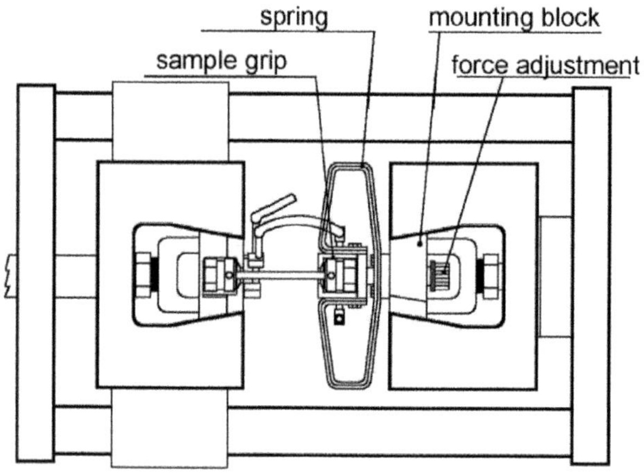

Fig. 9. Nil strength attachment of Gleeble using trapezoidal spring grip maintaining constant small tensile load on specimen during heating

This NST attachment keeps the specimen under a constant tensile load of about 50N while allowing specimen heating with an initial heating rate the same as of the welding thermal cycle to be simulated up to a temperature 50 °C below the solidus temperature of the material, then change to a heating rate of 2–5 °C/s until the NST is reached, as shown in Fig. 10.

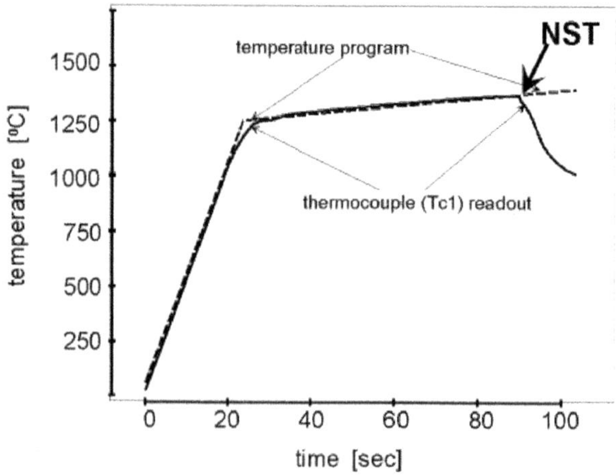

Fig. 10. Schematic time–temperature graph of Gleeble test for nil strength temperature measurement

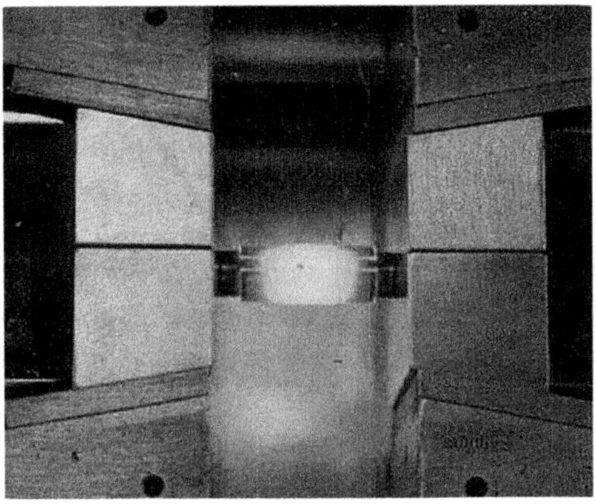

Fig. 11. Gleeble setup for controlled melting and solidification study

To avoid mixing of phenomena related to liquation cracking with those of solidification cracking, in the weld thermal cycles simulated to measure hot ductility for the liquation cracking the NST must not be exceeded.

To test for the susceptibility to solidification cracking, controlled melting and solidification of a rod-like specimen is carried out on Gleeble, and after the solidification the hot ductility is determined in the manner as described above. Here the ability of Gleeble to melt electrically conductive specimens is used. In this test the central portion of the specimen protected by a crucible / quartz sleeve is brought to a temperature above solidus and then this crucible contains the molten / semi-liquid metal, as shown in Fig. 11.

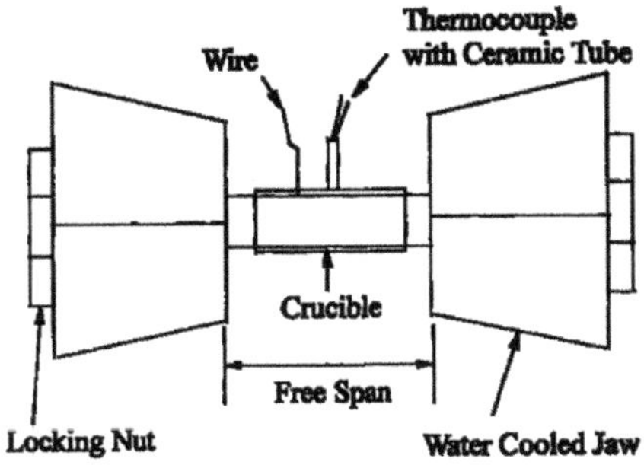

Fig. 12. Schematic of Fig. 11, with end nuts for hot tensile testing

The thermal gradient between the molten portion and mounting jaws prevent the metal from flowing out of the crucible while controlled thermal cycle allows conducting the solidification in a manner similar to that of real casting or welding. Schematic of the assembly used for this purpose in Gleeble is given in Fig. 12.

More details of the Gleeble testing for liquation and solidification cracking can be found in a Gleeble application note [9] available upon request from DSI in USA (info@gleeble.com). The Gleeble hot tensile testing procedures for liquation and solidification cracking are also mentioned in the technical report CEN ISO/TR 17641-3:2003, under the chapter "Hot Tensile Test" [10].

Fig. 13. SICO sample tested in Gleeble

An alternative procedure to study susceptibility to solidification cracking on Gleeble is the strain-induced crack opening test - SICO, developed at Dynamic Systems Inc., in USA, for studying hot deformability of alloys. In this test the central hot portion of the Gleeble specimen is compressed to form a bulge (Fig. 13), on outer perimeter of which cracks appear at the critical secondary tensile strain, as shown in Fig. 14.

The critical strain to fracture in a SICO test is defined as the hoop strain at onset of cracking in the bulge zone:

$$\varepsilon_c = \ln (D_f / D_o), \tag{1}$$

where D_o is the initial diameter of the specimen, while D_f is the final maximum diameter in the bulge zone.

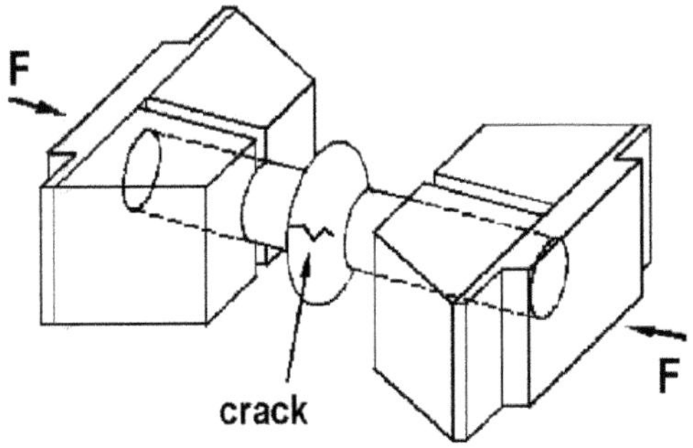

Fig. 14. Schematic of SICO test showing crack formed on bulge portion of sample

As during the controlled melting and solidification in the Gleeble the dendritic crystals grow mainly in the direction of the heat flow i.e. in the axial direction, the mid-span segregation may occur in the specimen causing deep and partly hidden cracking of the SICO specimen along the central plane perpendicular to the compression axis (Fig. 15). In such situation it is advised to check for the true critical diameter on the cross-section of SICO sample, like it is shown in Fig. 16.

As the hot tensile testing on Gleeble gives the adequate characteristics of an alloy regarding its hot cracking susceptibility, the SICO test appears to be more accurate for measuring of critical strains to fracture and related critical strain rates.

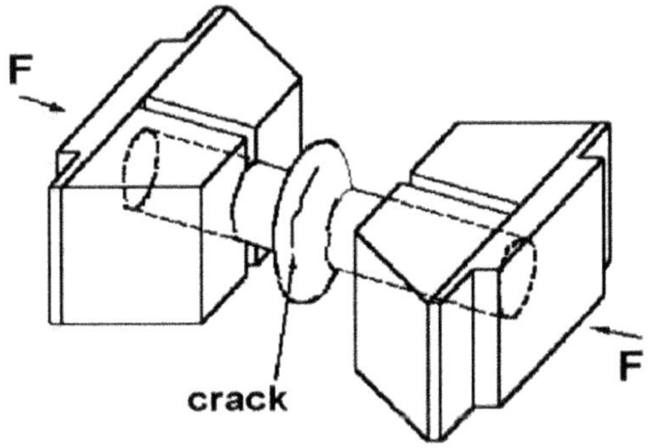

Fig. 15. Central plane crack in SICO often appearing after melting and solidification

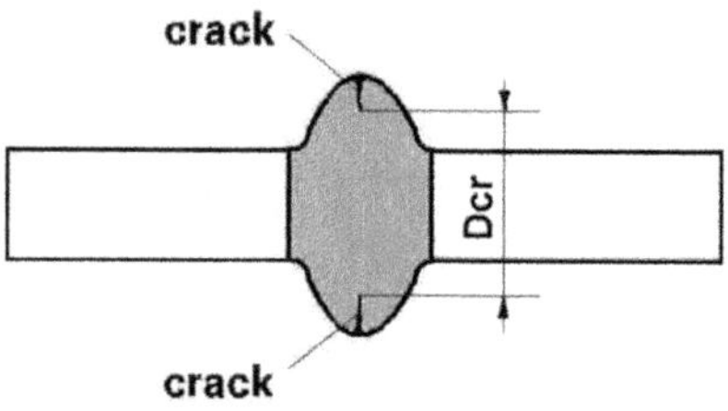

Fig. 16. Method of measuring critical diameter D_{cr} when central plane crack appears

Hot Cracking of Austenitic Steels and Alloys

Austenitic stainless steels, Ni-base alloys and other metal alloys having large thermal expansion coefficients are susceptible to hot cracking during welding. In most of these alloys the high content of alloying elements extends their solidification temperature range thus increasing segregation of impurities along grain boundaries. These alloys are susceptible to solidification cracking in weld metal and to liquation cracking in the HAZ, depending on the relative strength of the HAZ and weld metal in the hot range, and to avoid cracking both the HAZ and weld must be able to plastically accommodate the shrinkage strains.

The hot strength of austenitic stainless steels and Ni-base alloys below solidus is almost zero in a certain range of temperature till NST, and below the NST the metal becomes rigid and strong but its ductility is still be very low. During cooling through the low ductility range, the thermal shrinkage may cause hot cracks, and the susceptibility of steel to this cracking can be expressed as a difference between the NST and the ductility recovery temperature - DRT. Below the DRT, the steel is able to plastically accommodate the shrinkage thus avoiding cracks.

To determine the hot ductility for some austenitic alloys and their susceptibility to hot cracking, Gleeble hot tensile tests were carried out at DSI for a collaborative program of IIW – Commission II, of which some results are presented here. The full report of this testing [11] is available from DSI upon request at info@gleeble.com. Four austenitic steels and alloys of the composition given in Table 1 were examined. Test specimens were 6mm diameter by 100 mm long. They were heated to fracture in a nil-strength test fixture under a load of ~8kgf. The measured NST for each alloy, as well as the determined BTR and RDR, are listed in Table 2.

Table 1. Chemical composition of tested alloys, Fe is balance to 100 %

	chemical composition [wt%]										
Alloy	C	Si	Mn	P	S	Cr	Mo	Ni	Ti	Al	Others
800H	.068	.38	.70	.011	.002	20.40	_	30.55	.33	.28	
AC66	.072	.21	.50	_	_	27.35	_	31.45	_	.014	Nb=.83, Ce=.085
926	.010	.32	.82	.017	.003	20.85	6.38	24.80	_	_	Cu=.91, N=.196
825	.006	.32	.69	.015	.003	22.25	3.16	39.15	.77	.09	Cu=1.9

Table 2. Nil-strength temperature (NST), brittle temperature range (BTR) and ratio of ductility recovery (RDR) for each alloy

Alloy	800H	AC66	Alloy 926	Alloy 825
NST	1365°C	1331°C	1315°C	1343°C
BTR (°C)	115	94	40	71
RDR (%)	29.9	21.6	71.7	28.2

The results, given in Table 2, have shown a substantial difference in the behavior of Alloy 926 and 800H, and they also showed a good agreement with the hot cracking susceptibility data available from the modified Varestraint test and from practical observations [12].

Selected examples from these hot strength / hot ductility tests are given below on the graphs of Alloy 926 and Alloy 800H (Figs. 17 and 18), and one may see from them that the real hot ductility curves are not exactly so "smooth" like these in Figs. 7 and 8. The overlapping and competing effects of liquation and homogenization may result in a solid-state diffusion healing dependent on applied time / temperature / strain rate of the executed hot tensile test. Thus, for the applied strain rate representative to the shrinkage rate of the weld thermal cycle ductility dips were observed at 1240 °C and 1170 °C for alloys 800H and 926 respectively, while for slower strain rates these dips were absent.

This last observation calls for metallographic verification of the tests, in particular for comparing the microstructures of test specimens with these of real welds containing hot cracks and for identifying the micromechanisms of cracks formation at the conditions of the hot tensile tests.

What follows, is a part of metallographic verification of the specimens from the mentioned IIW testing programme, which revealed various microstructural features coinciding with the microstructures of the real welds as well as some discrepancies further used to modify the Gleeble testing conditions.

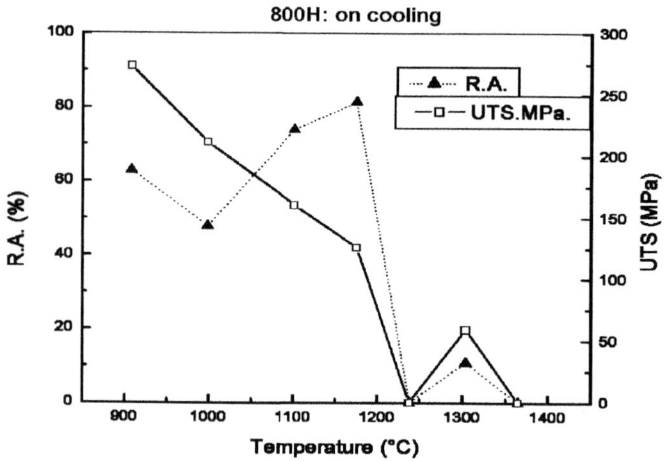

Fig. 17. Hot strength and hot ductility curves of alloy 800H

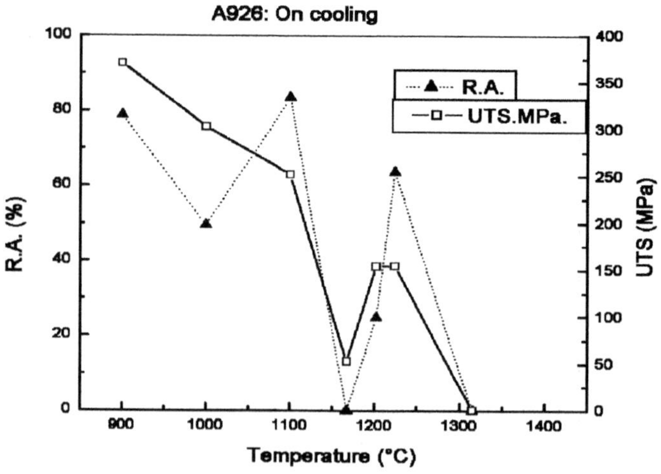

Fig. 18. Hot strength and hot ductility curves of alloy 926

In Alloy 926, very well resistant to hot liquation cracking, substantial grain growth appeared near to NST however no grain-boundary segregation / liquation could be noticed.

Samples tested at 1250 °C showed existence of long segregation bands filled with weak / low-melting eutectics (Fig. 19) however when tested at 1200 °C these bands almost entirely disappeared due to diffusion healing (Fig. 20).

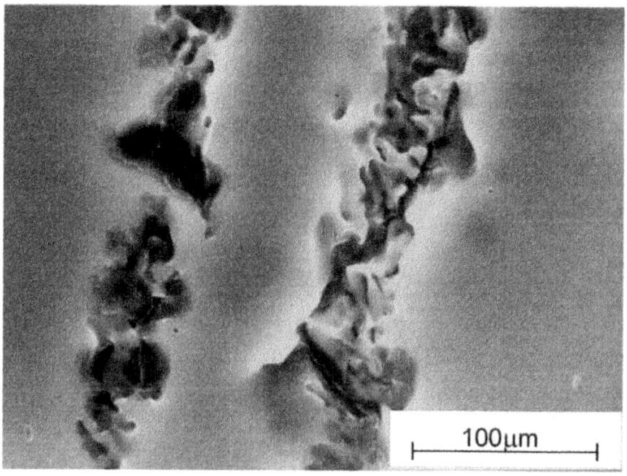

Fig. 19. Longitudinal cavities containing eutectic near to fracture zone of alloy 926 tensile specimen tested at 1250 °C

Fig. 20. Isolated longitudinal cracks near to fracture zone of alloy 926 tensile specimen tested at 1200 °C

In a more susceptible to hot cracking alloy 800H, a rapid grain growth occurred near to the NST (Fig. 21) which resulted in formation of thin films as well as voids along grain boundaries (Fig. 22).

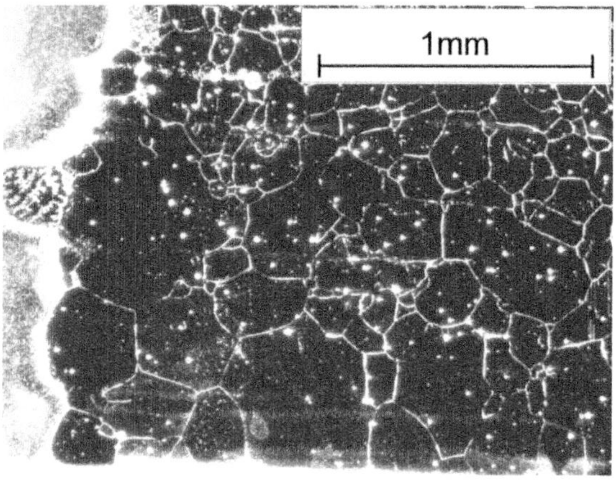

Fig. 21. Abnormal grain growth near to surface of specimen in alloy 800H at nil strength temperature (NST)

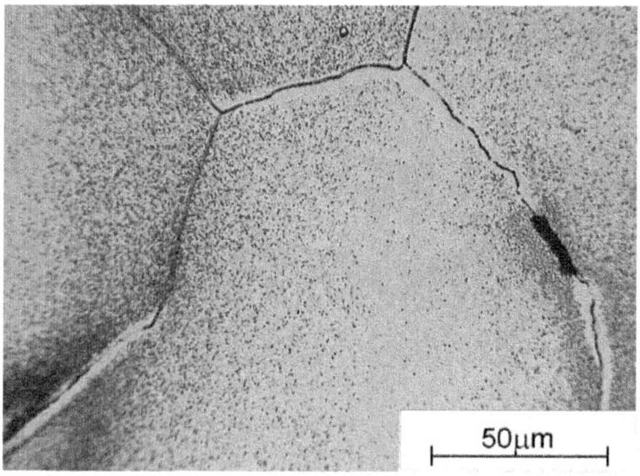

Fig. 22. Formation of grain boundary films and cavities in alloy 800H near to NST

Also chains of equiaxial voids appeared along segregation bands, however at the highest testing temperatures these voids did not easily coalesce to cause the failure – the weakest link of the microstructure appeared to be the grain boundaries (Fig. 23).

At temperatures lower than NST, i.e. at 1250 °C and below, elongated cracks appeared along the segregation bands and from them intergranular

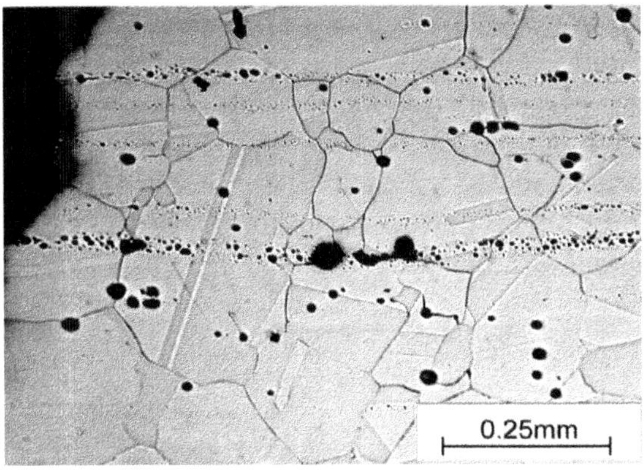

Fig. 23. Cavities in segregation bands close to fracture in alloy 800H tensile tested at NST

cracks extended in transverse directions (Fig. 24). The behaviour of alloy 800H indicated that various micromechanisms could operate at different testing temperatures and that the hot crack susceptibility is strain rate dependent.

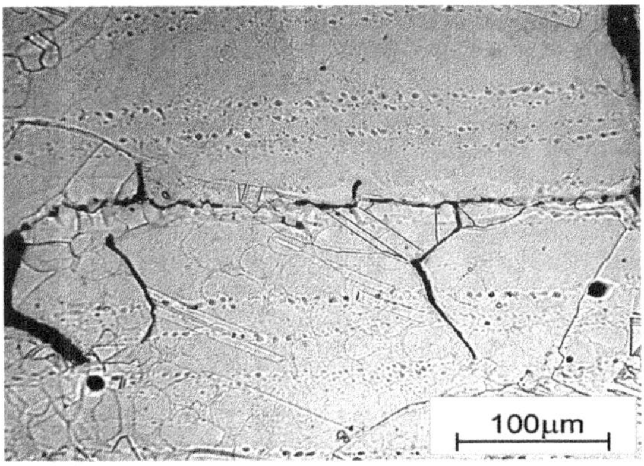

Fig. 24. Longitudinal cracks along segregation bands and transverse intergranular cracks in alloy 800H tested at 1200 °C

It was possible to find on the longitudinal sections of the hot tensile specimens the regions at which the highest density of intergranular transverse cracks appeared and using the formula:

$$\varepsilon = 2 \ln (D/d) \tag{2}$$

to determine the local critical strains as well as strain rates for the hot cracking at particular temperatures.

An example below (Fig. 25), shows the section through neck portion of hot tensile specimen, on which this relation is drawn and the following two pictures (Figs. 26 and 27) present cracks, which appeared in alloy 800H at temperature 1200 °C and strain rate 1.20 /sec while no cracks formed at the same temperature and strain rate 0.30 /sec.

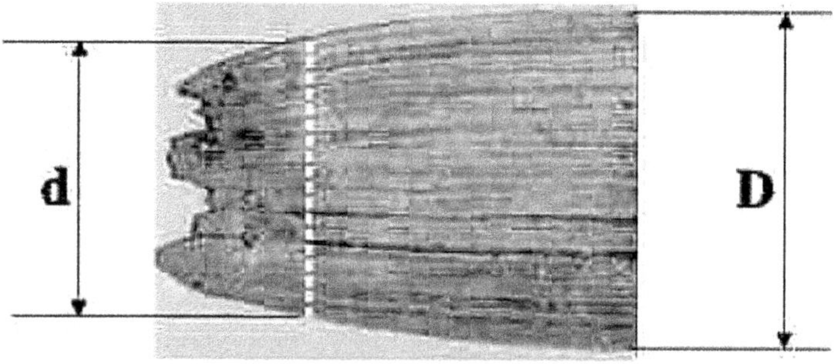

Fig. 25. Method of determining strain from the neck portion of tensile test specimen: $\varepsilon = 2 \ln (D/d)$

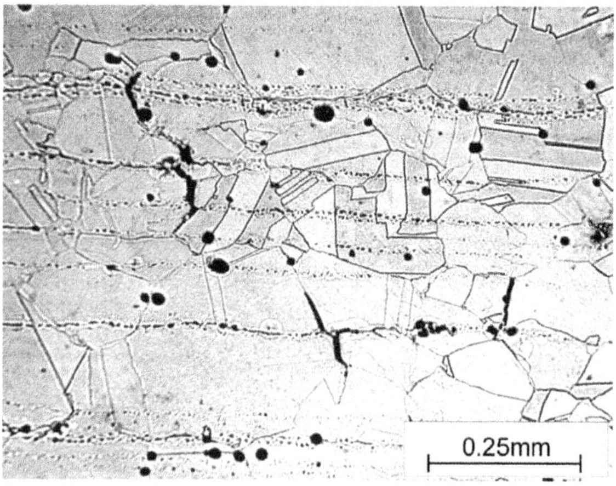

Fig. 26. Cracks revealed in tensile test specimen of alloy 800H tested at 1200 °C, at region deformed with strain rate ~ 1.20 /sec

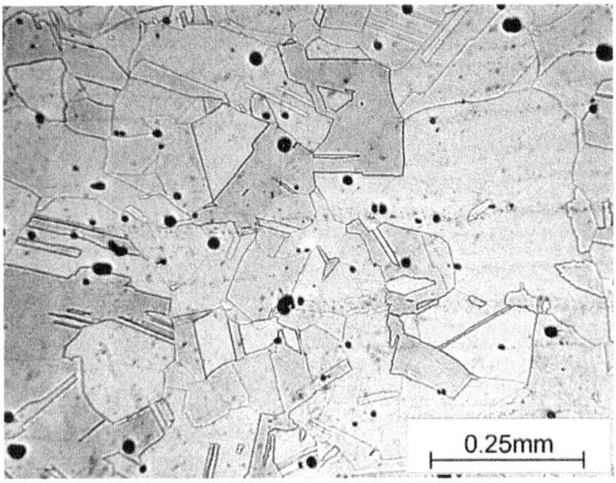

Fig. 27. No cracks visible in tensile test specimen of alloy 800H tested at 1200 °C, at region deformed with strain rate ~ 0.30 /sec

In another case, of alloy AC66, the loss of ductility due to liquation was quite severe and the recovery of ductility on-cooling delayed. Here, the liquation caused a permanent precipitation of a carbide eutectic, which evidently hampered deformability of the alloy.

The precipitates appeared in segregation bands (Fig. 28), assisting in the hot tensile test formation of transverse cracks nucleating from voids at these precipitates.

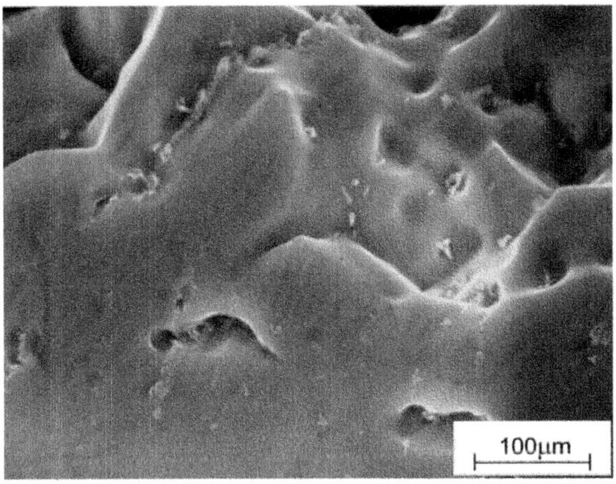

Fig. 28. Carbide eutectic near to fracture and on the fracture surface of hot tensile specimen of alloy AC66 tested at 1225 °C

In Fig. 29 carbide eutectic is shown, which separated during the test forming a deep transverse void. These precipitates in alloy AC66 visibly assisted the hot cracking for more than 100 °C down from the NST, mainly by the formation of voids around them. Fracture of the hot tensile specimens near NST occurred mainly by joining of these voids in the transverse direction thus indicating additional weakness of grain boundaries at this temperature (Fig. 30).

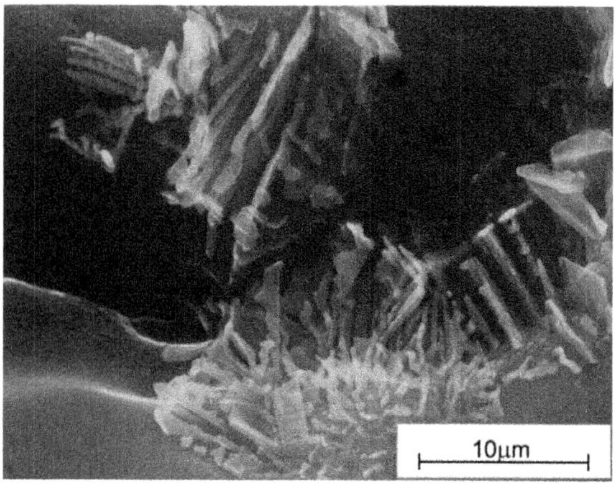

Fig. 29. An example of a cavity formed during the hot tensile test across the carbide eutectic in alloy AC66

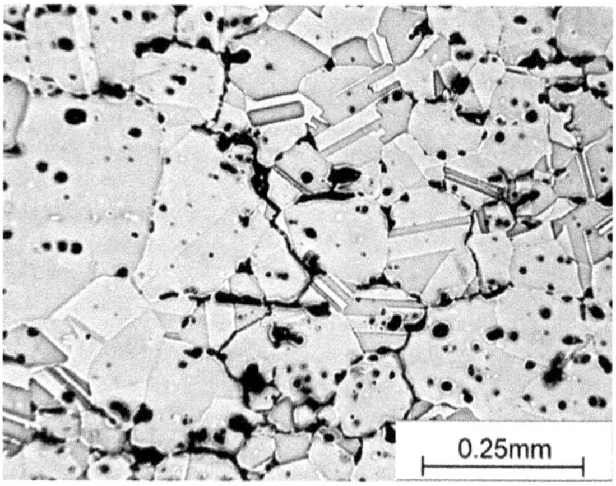

Fig. 30. Extended transverse crack near to the fracture portion and numerous voids in alloy AC66 tensile tested at NST

The eutectics appeared elongated in the rolling direction of the tested material (Fig. 31), suggesting its metallurgical low quality. Nevertheless, at lower temperatures a substantial recovery of ductility appeared, due to thermal-mechanical grain refinement (Fig. 32), most probably resulting from a dynamic recrystallization in the regions of strain localization and concentration between the elongated fields of the eutectics.

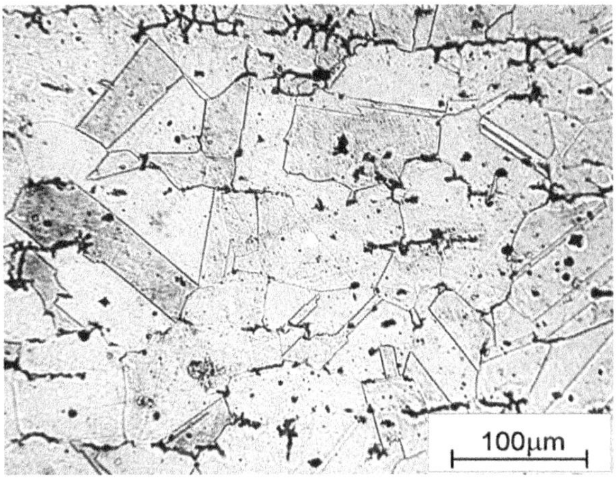

Fig. 31. Elongated carbide eutectics in alloy AC66 after heating up to NST and cooling down; total strain ~ 0.05

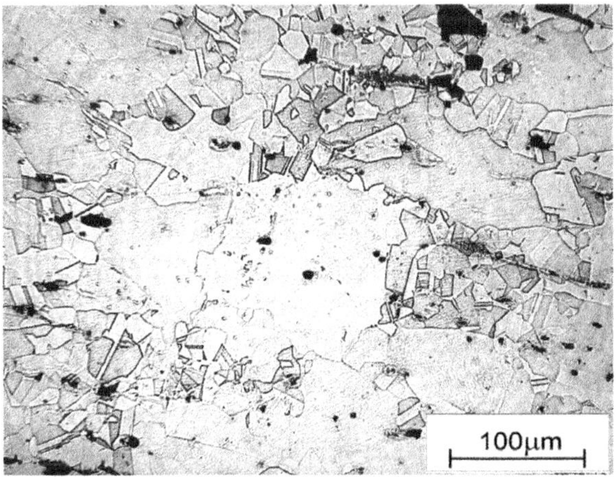

Fig. 32. Refined grains and coarse primary grains formed next to the precipitates of eutectics in alloy AC66 at the portion of tensile specimen deformed with strain rate ~ 0.50 /sec at 1175 °C

This last phenomenon "successfully" competed with the formation of voids near to the eutectics, giving a high value of reduction-in-area and simultaneously generating in the neck portion of the specimen a large amount of voids surrounded by fine recrystallized grains (Fig. 33).

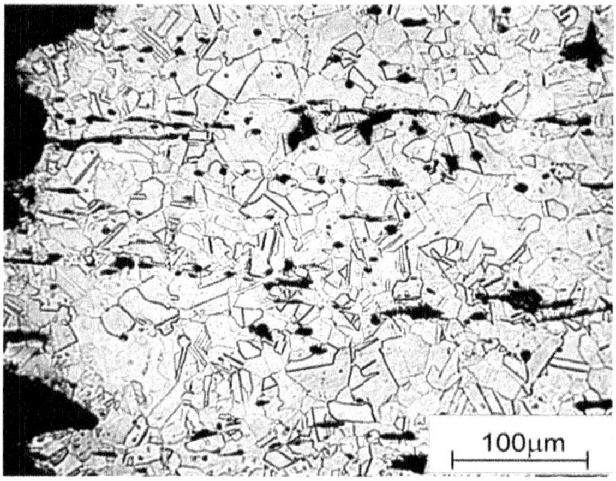

Fig. 33. Refined grains and numerous voids formed in the neck portion of hot tensile specimen of alloy AC66 deformed with strain rate ~ 2.50 /sec at 1175 °C

In conclusion it may be said that the metallurgical quality of an alloy is important as regards its susceptibility to hot cracking and that sometimes the "standard" test conditions may need to be adjusted in order to reach correct results.

Microfissuring in Multi-Bead Welds

Microfissures are reheat-type fine cracks of length about 1 mm or less when visible on transverse sections of welds and up to a few millimeters in the length direction of the weld. They are often of a ductility-dip origin and form in inter-bead heat affected zones of multi-layer, multi-bead welds, almost exclusively in the upper layers of these welds, i.e. when underlying portion of the weld is stiff enough to provide adequate restraint.

The microfissures are related to thermal-mechanical history of the weld manufacturing and assisted by the primary segregation of weld metal solidification and the secondary reheating liquation as well as by solid-state embrittling processes resulting in ductility dips.

They form in the multi-pass welds in zones of adjacent (underlying) weld beads, which constitute the heat-affected zone of a subsequent pass; an example is given in Fig. 34.

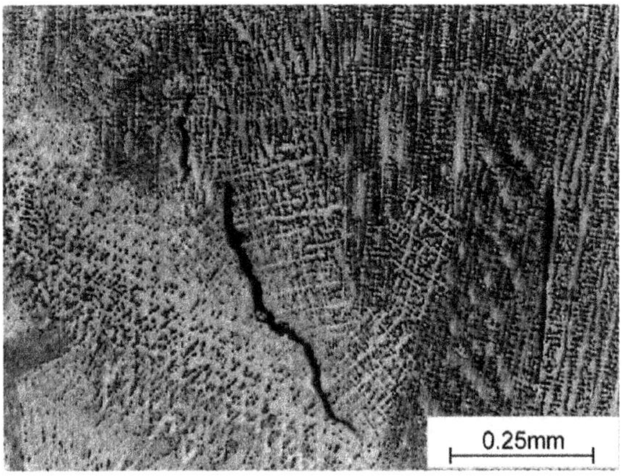

Fig. 34. An example of microfissure(s) appearing in inter-bead heat-affected zone of a multi-bead weld, visible on a cross-section of the weld

To explain the formation of microfissures the following model [13] can be used (Fig. 35) in which the inter-bead heat-affected zones are divided into two portions: the first marked (a) representative to higher temperature and coarse grain microstructure and the second marked (b) representative to fine-grained microstructure and incomplete recrystallization.

While welding with the bead sequence: (1) → (2) → (3) → (4) → (5) → (6) and so on, after laying down the bead #(6) the most potential site for microfissure formation will be reheated by the bead #(6) coarse-grained zone of the HAZ below the bead #(5) at the site where it overlaps with the HAZ of the bead #(6). The (a) and (b) zones of the interbead HAZs can be also treated as the upper one (a) in which the diffusion healing and the nil-ductility appear, and the lower one (b) in which the ductility has recovered substantially however strain hardening may occur. Thus at the bottom of the lower layer the thermal cycling may results in a slight strengthening due to the generation of dislocation and their incomplete annihilation / recovery, without any substantial annihilation of the primary crystalline lattice defects being mainly vacancies in the "as-frozen" weld metal.

In general, the microstructure of the as-solidified weld metal beads is far from thermodynamic equilibrium by containing large amounts of crystalline lattice defects, such as vacancies and dislocations as well as planar defects, which are formed to compensate shrinkage effects and which tend to annihilate during subsequent thermal cycling. Thus, when during the laying down of the bead #(6) such microstructure in the bead #(5) is heated up to a temperature near to solidus, a large number of defects annihilate and this process is affected at first by thermal expansion of the whole interbead HAZ, and then by shrinkage restrained in some particular directions. An intensive annihilation of vacancies and grain growth causing local reduction of volume is expected to occur during heating in the upper layer (a_6) of the HAZ by the overlying bead (6), i.e. close to the fusion line, due to restrained expansion.

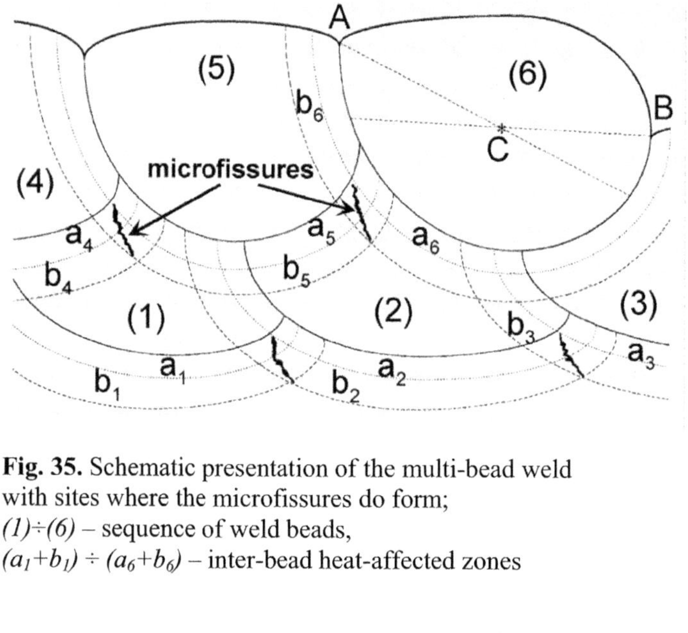

Fig. 35. Schematic presentation of the multi-bead weld with sites where the microfissures do form;
(1)÷(6) – sequence of weld beads,
$(a_1+b_1) \div (a_6+b_6)$ – inter-bead heat-affected zones

Then on cooling, this zone shrinks and may be particularly prone to nucleation of cracks if its ductility is low. However, the simultaneous shrinkage of the already solidified weld metal in the bead #(6) acts to compensate this effect.

The expansion and its resulting compressive strain may have a positive effect that is the acceleration of homogenization and partial healing in the upper layer (a_6) of the HAZ, and these may reduce the sensitivity to liquation cracking. Additionally, the shrinkage after solidification of the weld metal bead #(6) may close partly the liquation cracks in the upper layer (a_6) of the HAZ.

When sensitization occurs, the microfissures nucleate in the upper or in the lower layer of the interbead HAZ, depending on which effect prevails. The critical sites for their formation are intersections with the coarse grained part of the HAZ (a_5) of the previous bead #(5) where laying down of this bead caused the liquation. In such site which is relatively remote from the centre 'C' of the weld bead #(6) as is visualized in Fig. 35, the shrinkage of the weld bead #(6) after its solidification and "anchoring" between points 'A' and 'B', can result in localization of a substantial tensile strain, thus opening the microfissure in the a_5-b_6 area. The location of microfissures at the intersection with the HAZ of previous beads and frequent propagation along boundaries of primary columnar grains into the underlying bead, indicate that the segregation and liquation are important factors for their formation.

To study the susceptibility of the multi-bead weld joints to microfissuring as well as the influence of microfissures on hot strength and ductility of the weld metals, SICO testing can be used on Gleeble. Samples for such testing are cylindrical bars of 10 mm diameter by 90 mm long, and are clamped from both sides in the Gleeble's cold jaws. The simulated weld heating cycle is then applied and the sample can be compressed immediately at the peak temperature or at any temperature on cooling from the peak.

To check for the presence of microfissures and their influence on hot ductility SICO tests were run on samples taken from the upper half and lower half of an austenitic stainless steel multi-bead weld, and the cylindrical bar samples were machined transversely to the weld length, like schematically presented in Fig. 36. Then in the Gleeble the samples were heated at a heating rate of 150 °C/sec to a test temperature, and compressed at a speed of 50 mm/sec to different reductions. The mean strain rate was approximately 3 /sec at a free span of 30 mm.

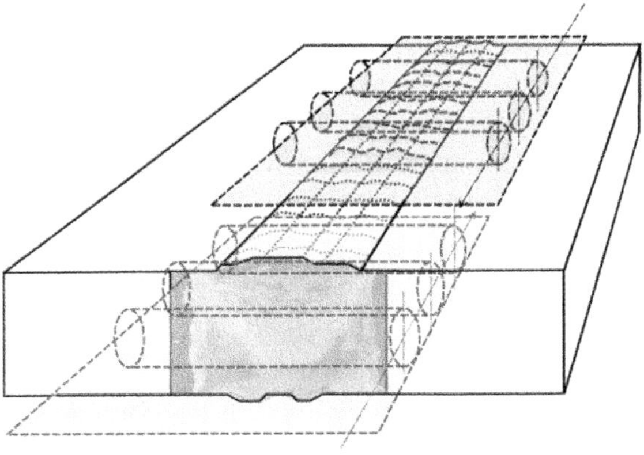

Fig. 36. Schematic presentation of SICO specimens' extraction from lower and upper portions of multi-bead weld

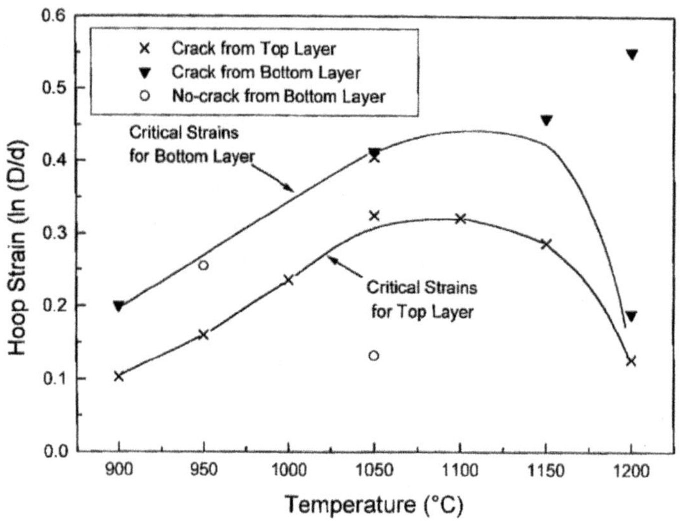

Fig. 37. Results of SICO testing for microfissure susceptibility of multi-bead austenitic stainless steel weld metal [13]

For the SICO test specimens cut from the top layer of the multi-pass weld metal, critical strains were lower at each temperature at the same testing condition than that from the bottom layer with rare microfissures, as shown in Fig. 37. Moreover, it was found that all the cracks appeared at the specimen surface directed to the top of the weld, which indicates the presence of microfissures in the top layers of the multi-pass weld joint. The microfissures simply act as nuclei or microcracks for crack opening during compression under secondary tensile stresses. This conforms with the applied physical model of microfissuring. It has been concluded that the strains at which the weld metal cracked at the temperature and strain rate of the SICO test, or the microfissures extended to the surface of the SICO sample and appeared visible, can be treated as characterizing the susceptibility of weld metal to microfissuring.

Summary and Conclusions

The examples of Gleeble thermal-mechanical simulation procedures presented in this paper, show the results of simulation that can be achieved if the fundamental physical phenomena occurring during welding is well understood and obeyed. They highlight an important role of thermal gradients, which occur in all known industrial thermal processes and resulting from these gradients, thermal-mechanical strains and strain accommodation phenomena which in turn affect strain hardening and recrystallization, phase transformations and/or precipitation processes. The combinations of different factors, and their importance, may vary substantially, and the correct physical simulation of welding must take into account a number of interacting phenomena appearing in the real application process, such as:

1. The balance between the heat input and electric current flow during heating, and controlled by thermal gradient heat flux, cooling rate and micro-deformation rate during the short thermal cycle of welding, in the case of heat-affected zone during arc welding.
2. The changes of hot ductility and of hot strength of welded material on heating and on cooling and its susceptibility to form liquid phase along grain boundaries at elevated temperatures below solidus, in the case of hot cracking.
3. The accommodation of strains occurring due to multiple welding thermal cycles and due to annealing microstructural restraints in multi-bead austenitic weld metals, in the case of micro-fissuring.

References

1. Lin W, Lippold JC, Baeslack III WA (1993) An Evaluation of Heat-Affected Zone Liquation Cracking Susceptibility. Part I: Development of a Method for Quantification. Welding Journal 72: 135–153
2. Savage WF (1955) An Investigation of the Hot Ductility of High-Temperature Alloys. Welding Journal 34: 183–196
3. Nippes EF, Savage WF, Grotke G (1957) Further Studies of the Hot-Ductility of High-Temperature Alloys. Welding Research Council, Bulletin 33
4. Savage WF, Lundin CD (1965) The Varestraint Test. Welding Journal 44: 433–442
5. Prokhorov NN (1968) Theorie und Verfahren zum Bestimmen der Technologischen Festigkeit von Metallen beim Schweissen. Schweisstechnik 1: 8–11
6. Savage WF (1987) A Historical View of Weldability. In: Welding Metallurgy of Structural Steels, TMS, Warrendale PA, USA, pp 3–19
7. Ferguson HS (1992) Fundamentals of Physical Simulation. In: Proceedings of the International Symposium on Physical Simulation, TU Delft, The Netherlands, pp 1–21
8. Lundin CD, Qiao CYP, Gill TPS, Goodwin GM (1993) Hot ductility and hot cracking behavior of Modified 316 Stainless Steels designed for high temperature service. Welding Journal 72: 189–200
9. Chen WC (1995) A Standard Procedure for Hot Cracking Test. Gleeble Application Note, Dynamic Systems Inc, Poestenkill, NY, USA, (IIW Doc II-C-042A/95)

10. CEN ISO/TR 17641-3:2003 (2003) Destructive tests on welds in metallic materials – Hot cracking tests for weldments – Arc welding processes – Part 3: Externally loaded tests
11. Chen WC (1995) Hot Tensile Ductility Test for HAZ Liquation Cracking Susceptibility Study. Report #950426, Dynamic Systems Inc, Poestenkill, NY, USA, (IIW-Doc II-C-042/95)
12. Wilken K (1994) Investigation to Compare Hot Cracking Tests – external loaded specimens. BAM, Berlin, Germany, IIW Doc IX-H-305/94
13. Mandziej ST (1997) Physical Simulation of Welding. In: Welding and Joining Science and Technology, ASM International Europe, Brussels, pp 253–268

Scientific Bases of the International Standardization Project "Hot Cracking Tests for Welds"

B. Yakhushin, D. Semin

Baumann University, Moscow, Russia

Abstract

Microcracks which are hardly detectable by non-destructive test methods represent a potential risk for the reliability of components or structures. More important than improving the postweld failure detection procedures is the improvement of the material-inherent resistance against hot cracking and the optimization of the welding process. This requires respective consistent test procedures which are currently being internationally standardized. The present contribution provides some remarks which might be considered for this standardization project.

Introduction

For the assessment of hot cracking resistance, national standards for respective test procedures have been released in many countries. However, their operation principles vary greatly and the applied hot cracking criteria are conditional and *only to a limited degree* quantitatively comparable. In this connection it is worthwhile to notice that a comparison between hot cracking test results of four different alloys investigated in the USA, France, Austria, Germany, Slovakia and Russia is described in the final report of K. Wilken [1].

Non-standardized test devices and the lack of statistical and scientific data may even limit the free trade of base material and filler material within the framework of the World Trading Society.

It is therefore the objective of the recent investigations to gather the respective "hot cracking test experience" from different countries and to work out a scientific concept regarding the ongoing multinational standardization discussion.

In principle, a general standard should allow comparative and quantitative assessment of the hot cracking resistance of base materials, filler materials and their common behavior under real welding conditions [2]. The hot cracking theory has to be understood as a basis for such standardization and a method to really measure the hot cracking resistance has to be derived and welding technology has to be assessed.

The Hot Cracking Criterion and Theory

According to the frequently cited theories, hot cracks (HC) are caused by a special kind of high-temperature brittleness originating from the non-equilibrium states of a material during welding. Like any brittleness-related material defect, a hot crack can thus be regarded as the consequence of insufficient material plasticity. The following respective criteria must occur simultaneously to initiate hot cracking:

1. Existence of a brittleness temperature range (BTR) at temperatures close to or within the melting range
2. The minimal plasticity δ_{min} is exceeded within this BTR ($\delta_{min} > 0$)
3. The critical deformation rate $d\varepsilon/dt$ is exceeded within the BTR, whereby the deformation rate can also be related to the cooling rate by $d\varepsilon/dT$.

The function of the temperature related brittleness is schematically plotted in Fig. 1 following the theory of Prokhorov. It can be seen that the straight line E1, representing the strain increase during cooling, intersects the brittleness function, which means that the maximum ductility is exceeded and hot cracking is initiated.

Every alloy has a particular range of plasticity within its BTR. The plasticity can partly or completely be reduced by the thermomechanical influence of the welding process. The reason is that the thermomechanically induced material displacements lead to monotonously increasing tensile deformation behind the weld pool caused by weld metal shrinkage during cooling. The material-specific and quantitative criterion of any alloy is thus the critical deformation rate (B_{cr}). Exceeding the B_{cr} limit causes a complete loss of plasticity as well as hot crack initiation.

The value of B_{cr} combines the effects of the three basic criteria mentioned above. The smaller the BTR, the higher the minimal plasticity and the higher the critical deformation rate, the higher is the resistance of an alloy against hot cracking during welding measurable by a higher B_{cr}.

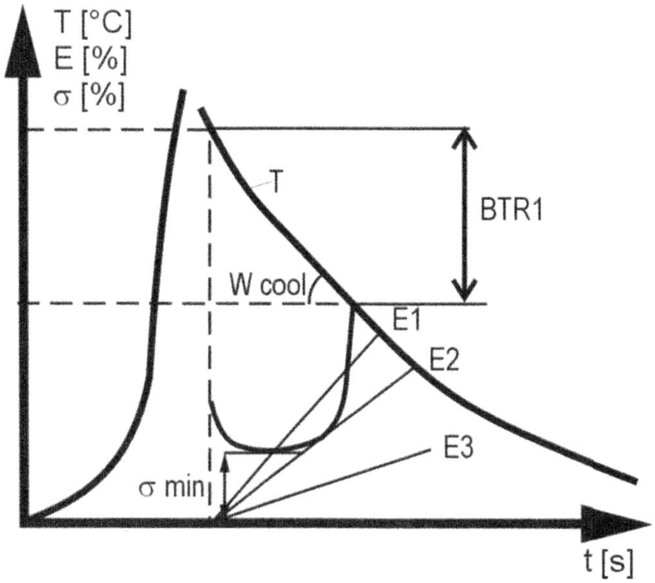

Fig. 1. Illustration of Prokhorov´s theory of hot cracking formation

Method for Measuring the Hot Cracking Resistance

During the past decades, different machine and technological test methods have been developed in which a specimen is deformed during welding at constant or increasing strain rates. Internal and external material deformations are thus superimposed. Such externally loaded hot cracking tests allow the determination of each of the three criteria separately.

But the measurement of the mechanical high-temperature properties and the calculation of the criterion $B_{cr} = \delta$ min/BTR - Δe/BTR is complex. Therefore, such methods which might be applicable in research cannot be recommended as a general standard.

There are especially some technical difficulties concerning the measurement of the minimal high-temperature plasticity. This means that the determination of criteria-related parameters under welding conditions is inexact.

A second variant by which defined specimen deformation is generated under isothermal conditions is more precise, but does not correspond to real welding processes [2].

A third variant of plasticity testing consists in creating high-speed deformation during welding, thus reducing the blurring influence of the moving heat source [3]. However, the plasticity of a material under high-speed deformation in solid-liquid state might be decreased to zero, which does not correspond to the theory as stated above. The contradiction between theory and experiment becomes especially clear under the consideration that hot cracks would be engendered in any alloys if zero plasticity exists, which has not been proven in practice, yet.

Integrated quantitative assessment seems to be more promising, however. The respective criterion is represented by the sum of all influencing factors and can be determined by a machine method if the above stated theory visualized by Fig. 1 is taken as a basis.

For this purpose, a specimen is deformed during welding with different strain rates, so that the weld metal and the heat affected zone (HAZ) are subjected to different mechanical loadings.

This means that from the upper limit of the BTR, which corresponds to a temperature at which the crystals get in touch with each other, up to the lower limit, which is between 100 °C and 200 °C below the solidus temperature, a material is subjected to different strain rates which gives clear and real information of its hot cracking resistance. This test variant preserves the kinetic adequacy of the real welding process and must therefore be considered as the superior method.

In this connection it has to be remarked that one fundamental dilemma of hot cracking testing still remains undissolved, which is the detection of hot cracks, the measurement of their geometries and the determination of the number of cracks.

Furthermore it has to be noticed that there might be a further classification for hot cracks, because cracks emerging within BTR1 are differently oriented compared to those emerging within BTR2 [4].

Additionally to weld metal testing, the machine test methods allow also the measurement of the critical deformation rate of the heat affected zone (HAZ) in the base material. The determination of the critical deformation rate of the HAZ is possible by subjecting a specimen to a simulated weld thermal cycle up to the temperature of "nil strength" (NST) and by applying variable strain rates during cooling.

The respective critical deformation rate is determined according to the following equation:

$$B_{cr} = v_{cr} / v_{btr} \text{ [mm/°C]}, \tag{1}$$

whereby v_{cr} represents the critical deformation rate (lowest rate at which hot cracks occur) and v_{btr} represents the cooling rate within BTR [4].

The Assessment of Welding Technology

The application of the B_{cr} criterion allows the assessment of the influences of different welding parameters on the hot crack formation.

The diagram of weldability shows schematically the functional dependency of B_{cr} on welding parameters (Fig. 2).

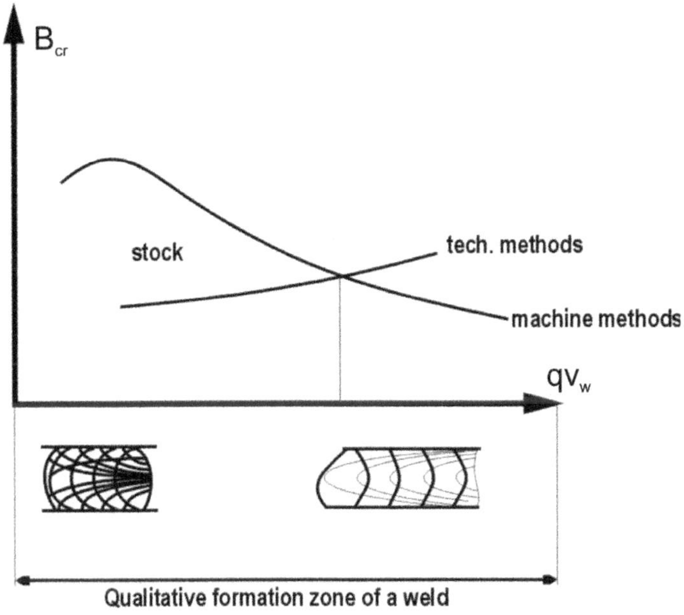

Fig. 2. Diagram of weldability

The general criterion of the welding condition $q \cdot v_w$ was applied for determining the crystallization formation and this function turned out to correlate with the critical deformation rate of an alloy, which has been proved for a wide range of welding conditions.

It was found, that the greatest influence on hot crack formation is exerted by the crystallization pattern which is described in detail elsewhere in this book [4].

The application of standard machine and technological methods allows the assessment of the hot cracking probability. For this purpose, testing of the corresponding standards specimens is performed using different welding conditions in order to detect critical welding parameters causing hot cracks.

Conclusions

The technological methods developed so far might be suitable for the assessment of technological parameters (especially welding parameters) causing a susceptibility to hot crack formation. But, by application of machine methods, in particular in terms of different, i.e. increasing, strain rates, an assessment of a material-inherent hot cracking susceptibility might be achieved providing the following advantages:

- The hot cracking resistance can be defined separately for deformations along and across the weld direction. The anisotropy of the crystallization pattern is therewith taken into account.
- The hot cracking resistance for homogeneous welds is defined both in a solid-liquid stage and in a temperature range below the solidus temperature. Especially hot cracking below the solidus temperature depends on specific alloying elements [4].
- The hot cracking resistance can be determined both for the weld metal and the HAZ, allowing the selection of an optimal base material / filler material combination

For these reasons, a general standard should specify a hot cracking test device capable of introducing universal external loads.

References

1. Wilken K (2000) Investigation to Compare Hot Cracking Tests – Externally Loaded Specimen. IIS/IIW – Subcommission IX-H/II-C, Final Report 2000, Doc.: IX-1945-99
2. GOST 26389 - 84 Welded joints: Test methods of metals on hot cracking resestivity during welding. By Nikolaev G, Yakushin B et al., M.: USSR Standards 1984- 23
3. Prokhorov N, Yakushin BF, Prokhorov NN (1968) Theorie und Verfahren zum Bestimmen der technologischen Festigkeit von Metallen während des Kristallisationsprozesses beim Schweißen. Schweißtechnik vol.19, N1: 8–11
3. Stan T, Mandziej. Physical Simulation of Welding. Advanced Materials Analysis, P.O. Box 3751, 7500 DT Enschede, Netherlands
4. Yakhushin B (2004) Morphology of Hot Cracks in Single-Phase Weld Metal. In this book

Discussion and Evaluation of Some Extraordinary Cases of Hot Cracking

K. Wilken

Germany

Abstract

Reasons for hot cracking formation are frequently present. Metallurgical aspects represent only one conceivable contingency. Some years after the active period there is time to clear up the writing desk. It is amazing and interesting how many unanswered and unclear cases exist. Some of them are worth to be reflected again. Four of them will be presented in this contribution:

1. What sort of influence on hot cracking is created by the welder himself?
2. Hot cracking occurs in a material which normally is hot crack-proof. What is the reason?
3. Is it possilble to weld an extremly hot crack-sensitive material nearly without hot cracks?
4. What is the reason for hot cracking in the heat affected zone of the base material?

Case 1 – What sort of influence on hot cracking is created by the welder himself?

Some years ago, a test program was carried out by IIW SC IX H and SC II C for comparative investigations and scatter studies with the new Longitudinal Bend Test (LBT) [1], now standardized in CEN ISO 17641 Part 2. Test coupons were welded in seven different countries. A straight and precise welding instruction had been established before. Each participant had to follow this instruction. The welding consumables to be used and additional instructions were handed out to each participant. Everybody had to cut out his own test specimen and to check the microfissure sensitivity indicator (MSI).

The results revealed a lot of differences [2], as shown in Table 1.

Table 1. Longitudinal Bend Test (LBT). Comparison of MSI results

Test coupon welded by	Number of runs top layer	MSI 10^{-3} mm^{-1}
Soudometal	3	0
Avesta	6	1.3
Esab	5	2.0
Böhler	4	14.3
BAM	4	16.6

MSI microfissure sensitivity indicator.

For the moment, the SC´s were helpless and it seemed that LBT should be condemned. But before doing that, it was decided to prepare cross sections of all test coupons.

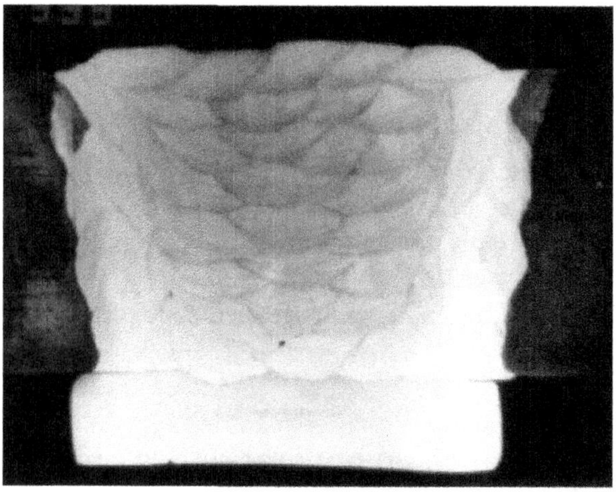

Fig. 1. Correct weld cross section following the instructions

Some of the welders followed the instructions, others did not. They used their ability to weld successfully without hot cracks. One conclusion on this is: The influence of the welder and the influence of the welding process are greatly significant [3]. All other factors commonly named „influencing factors“ are less important. A flat weld pool produced by electrode weaving and clever handling of heat during the welding process prevented hot cracking.

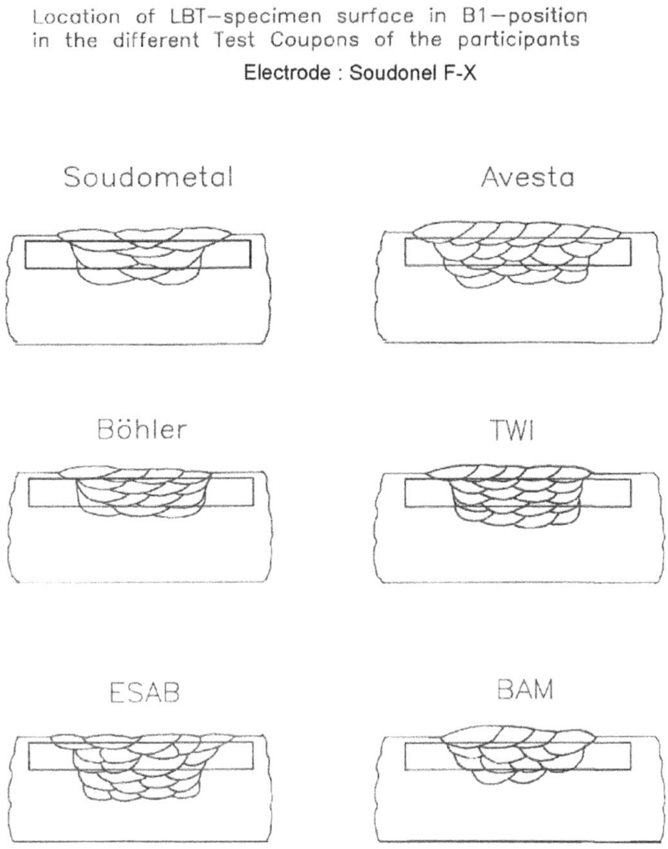

Fig. 2. Different number of top layer runs

Case 2 – Hot cracking occurs in a material which normally is hot crack-proof. What is the reason?

Hot cracking occurred in the heat affected zone inside a welded pipe. It was detected on occasion of a routine inspection in a nuclear power plant. The austenitic parent metal X 6 CrNiTi 18 10, material number 1.4541 normally is weldable without problems in the heat affected zone. The chemical analysis of the material was customary.

The calculated ferrite number FN was 4. The result of an MVT-specimen taken from the cross section of the pipe was: The material is not sensitive to hot cracking.

Fig. 3. Cross section of a pipe; segregation zone at the inner side

But in a detailed metallographic examination, a high amount of segregation was found on the inner side of the seamless manufactured pipe.

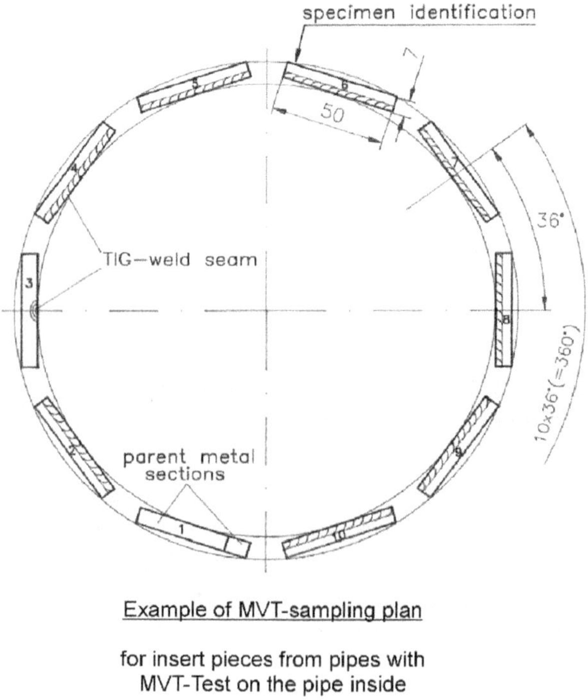

Fig. 4. Sampling plan

Specimens from the inner side of the pipe were taken and were prepared in order to be able to carry out the MVT-Test along a line of the inner pipe surface (Fig. 4). Together with this pieces a special MVT-specimen was prepared (Fig.5).

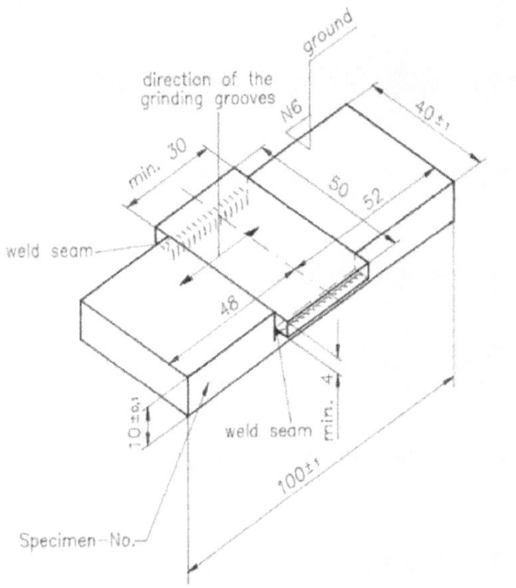

Fig. 5. Special MVT-specimen

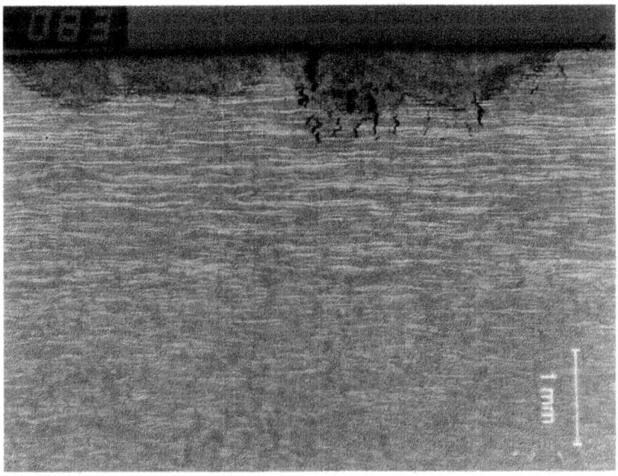

Fig. 6. Cross section of an MVT-specimen;
hot cracking in the segregated inner side of the pipe in the heat affected zone

The following result was obtained from the MVT-Test: The material is susceptible to hot cracking in the heat affected zone. Metallographic examination showed that primary austenitic solidification occured. And in this area hot cracks started [4]. This is shown in Fig. 6 to Fig. 8.

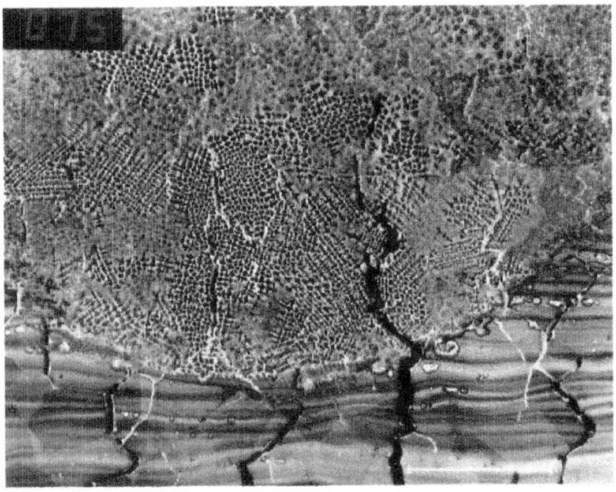

Fig. 7. Detail from Fig. 6; hot cracking in the primary austenitic zone

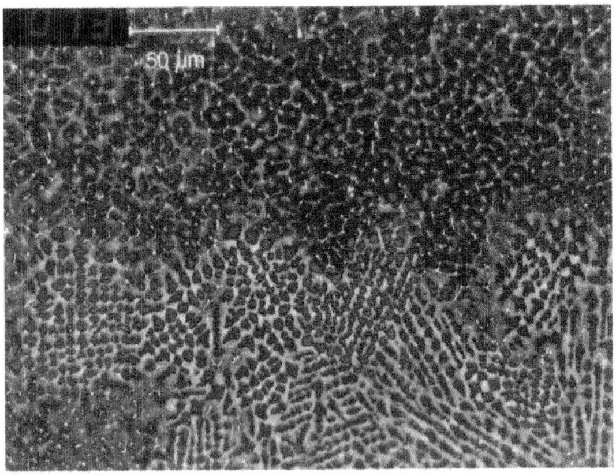

Fig. 8. Detail from Fig. 6; primary ferritic solidification above, primary austenitic solidification below

This damage at that time caused a lot of trouble in Germany. As a tentative conclusion one cannot trust in normal material data. Sometimes it is possible that other proportions exist in local areas. In this case the hot cracks inside the pipes were the starting points of crack growth through the wall thickness of the pipes.

Case 3 – Is it possible to weld an extremely hot crack-sensitive material nearly without hot cracking?

Railway tracks are subjected to abrasion, especially on bracking tracks. Therefore, surfacing by welding on top of the rails is necessary. The welding is possible in some breaks of the traffic or at night. Early replacement of a rail can thus be ensured. But the welding consumables used are extremely hot crack-sensitive.

Such a huge number of cracks is unacceptable. They are starting points of cracking across the rail. How is repair welding realized on the rail top without or nearly without hot cracks?

Hot cracks only occur in the Brittleness Temperature Range (BRT). It is necessary to have strains higher than the critical strain to cause hot cracks. That means: If strains are avoided, no hot cracking will occur.

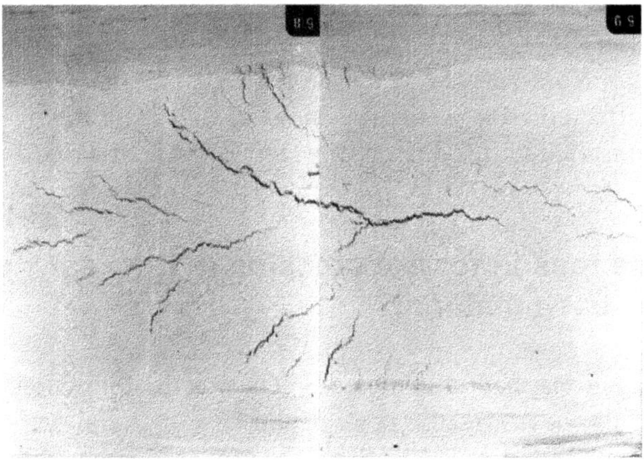

Fig. 9. Welding of rail material; MVT-specimen showing an extremely high amount of hot cracking

In general, welding of rail material is a complex process, because of the high carbon content of up to 0.7 %. Depending on the rail material, preheating between 350 °C and 500 °C and postheating between 350 °C and 500 °C for 5 min to 10 min are necessary.

But, a rail is a simple workpiece. So a simple bending device was built to initiate compression stresses in the top of the rail during welding. The preheating and postheating procedure was regulated in such a way that the rail foot was under tension and the rail top was under compression. Using this method was very successful.

a) b)

Fig. 10. Experimental welding of hot crack-sensitive rail material without hot cracking

The microcracking factor (similar to MSI, case 1) dropped down from the maximum of $7.56 \cdot 10^{-2}$ to 0. But one specimen showed a small amount of cracking in the heat affected zone of 0.2×10^{-2} mm.

Now the question was: What was the reason for not being completely successful? In the heat affected zone of the rail material, hot cracking occurs depending on the consumables used. However, this is the next case.

Case 4 – What is the reason for hot cracking in the heat affected zone of the base material?

Fig. 11 is about 30 years old. At that time, my comment was the following: The grain boundaries of the heat affected zone of the base material were remelted because of the heat of the weld metal. At the same time, the additional thermal strains initiated hot cracking. After that, the molten austenitic weld metal filled the hot cracks which is also called eutectic healing.

Now two other examples:

Fig. 12 shows the nickel base alloy 718, NiCr19NbMo, material number 2.4668. It seems that the molten material of the weld pool flowed into the opened crack in the heat affected zone of the parent metal. There was not enough liquid weld metal for complete healing, so that a liquation crack remained.

Similar to that figure is Fig. 13 which shows the copper nickel alloy CuNi10Fe1Mn. The molten material of the weld pool filled up the crack in the heat affected zone of the base material.

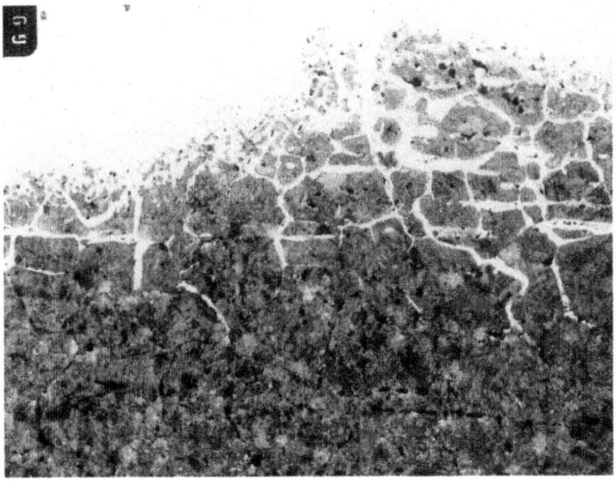

Fig. 11. Heat affected zone hot cracking in a rail steel, 18/8/6 CrNiMn Type electrode

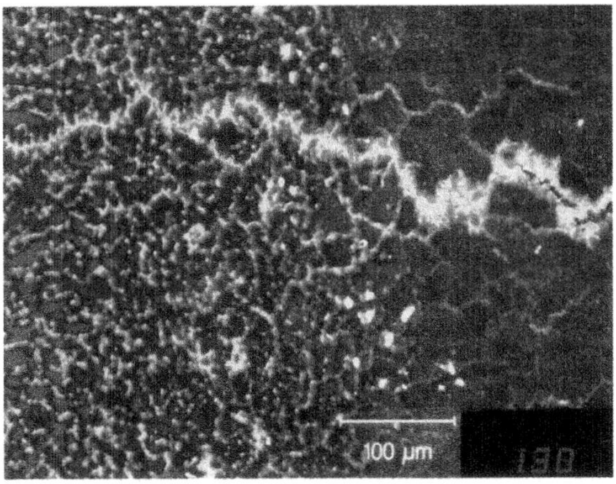

Fig. 12. Heat affected zone cracking in nickel base alloy 718

But there was a lack of material. The growing dendrites of both sides of the crack could not come together and a liquation crack remained. But are these explanations really true?

The next example may give the answer.

In a chemical plant, cracks occurred in the heat affected zone of welds between an austenitic steel (material number 1.4541) and a zinc coated mild steel. They were identified as liquid metal embrittlement (in German: „Lotbrüchigkeit").

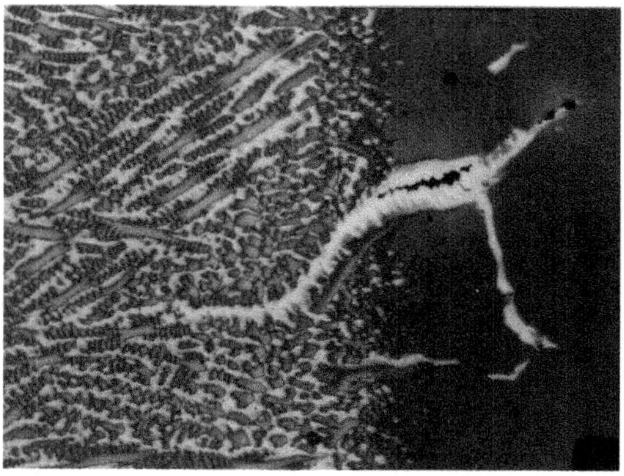

Fig. 13. Heat affected zone cracking in CuNi10Fe1Mn

It was intended to investigate this phenomenon using hot cracking tests i.e. the MVT-Test. Specimens of mild steel (S 355) and stainless steel were used which are normally unsensitive to hot cracking. These specimens were galvanized with copper (thickness: 2 µm) and zinc (thickness: 5 µm), respectively. Then the specimens were tested in the MTV-machine by producing a melt run and 2 % strain. Extreme cracking was found in the heat affected zone of both specimen types (Fig. 14).

On the specimen surface, the molten copper or zinc attacked the grain boundaries of the parent metal. This happened only in the bending zone of the specimens. In areas without bending, no cracking was observed.

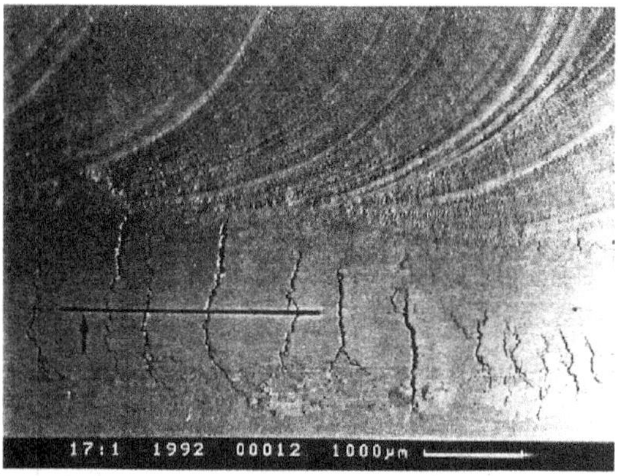

Fig. 14. Surface of an MVT-specimen, parent metal S355, coated with copper, hot cracking in the heat affected zone
(position in the cross section marked by the line)

The specimens exhibit extreme heat affected zone cracking. Metallographic examination revealed that copper is present on the surface and at the end of the cracks. Zinc is more difficult to be seen on the cracks, because most of the zinc was eliminated by the requisite etching of the specimen.

Fig. 15. Cross section (position see Fig. 14) of a copper coated specimen, copper situated at the end of the cracks

The testing time for an MVT-Test is about 0.3 s. During this short time the process of cracking is completely finished. The behavior of molten copper or zinc was extremely aggressive. The molten metal attacked the grain boundaries of the normally hot crack-unsensitive parent material S355 and X 10 Cr NiTi 18 9 (Fig. 16).

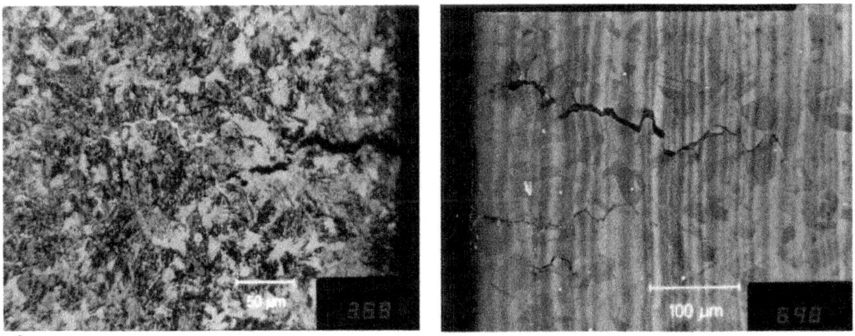

a) b)

Fig. 16. Cross section of **(a)** a zinc-coated S355 MVT-specimen and **(b)** stainless steel (X10 CrNiTi 19 9);
zinc situated on the cracks and at the crack ends

Would it be possible that this process is similar to stress corrosion cracking where a medium, i.e. an acid, attacks the grain boundaries and initiates a crack under tension condition? Could it be that these are the conditions for heat affected zone hot cracking?

This could be a future question for metal experts.

References

1. Wilken K (1995) Review of experiences with the Longitudinal Bend Test (LBT) for microfissuring susceptibility of deposited weld metal. IIW/IIS – 1208-93 (Ex doc IX-1722-93). Welding in the World, vol 35, no 5: 307–326
2. Wilken K (1994) Scatter of LBT-results. IIW Doc IX-H-228-94, II-C 009-94
3. Wilken K, Heuser H, Hoffmann Th (1997) Heißrißverhalten von Chrom-Nickel-Stählen und Nickelbasiswerkstoffen. DVS-Berichte, Band 183: 29–35
4. Wilken K, Bauer S (1994) Heißrißprüfung mit dem MVT-Test an Rohrabschnitten des Grundwerkstoffes X6 CrNiTi 18 10 (Mat.-No. 1.4541) zur Ermittlung des Chargeneinflusses. Unpublished report, BAM-6.43
5. Wilken K (1979) Untersuchungen zur Heißrißbildung neuartiger Schienenstähle. Unpublished report BAM-6.43
6. Ehmig R, Wilken K (1994) Nutzungsmöglichkeiten von Farbniederschlagsätzungen zur Erkennbarkeit und Bewertung von Heißrissen in Schweißverbindungen. In: Sonderband „Praktische Metallografie", Band 25, Carl Hanser Verlag, pp 33–37
7. Pohle C,Wilken K, Schuster J (1996) Prüfen der Lotbruchanfälligkeit beim Schweißen von Baustählen und austenitischen Chrom-Nickel-Stählen mit verzinkter Oberfläche. Schweißen & Schneiden, Band 48, Heft 9: 684–689

GPSR Compliance

The European Union's (EU) General Product Safety Regulation (GPSR) is a set of rules that requires consumer products to be safe and our obligations to ensure this.

If you have any concerns about our products, you can contact us on ProductSafety@springernature.com

In case Publisher is established outside the EU, the EU authorized representative is:

Springer Nature Customer Service Center GmbH
Europaplatz 3
69115 Heidelberg, Germany

Batch number: 10755244

Printed by Printforce, the Netherlands